Plant Science – A Technological Perspective

Plant Science –
A Technological Perspective

Edited by **Clive Koelling**

R CALLISTO
REFERENCE

New York

Published by Callisto Reference,
106 Park Avenue, Suite 200,
New York, NY 10016, USA
www.callistoreference.com

Plant Science – A Technological Perspective
Edited by Clive Koelling

International Standard Book Number: 978-1-63239-513-9 (Hardback)

Contents

Preface

Plant science is a diverse science dealing with the innumerous species of plants existing on the face of the planet. This book consists of contributions from researchers across the globe. It elucidates the relationship between environment and plants, specifically highlighting species-environment relationship and response of plants to distinct environmental stress conditions. It also presents both negative as well as positive impacts of microbes on plants. Furthermore, the book highlights present biotechnological research for development of novel technology in order to produce biologicals and enhance plant immunity in the current environmental conditions. This book will appeal to a broad spectrum of readers including scientists, researchers, professionals, students interested in studying in-depth about plant science and its various aspects.

The information shared in this book is based on empirical researches made by veterans in this field of study. The elaborative information provided in this book will help the readers further their scope of knowledge leading to advancements in this field.

Finally, I would like to thank my fellow researchers who gave constructive feedback and my family members who supported me at every step of my research.

Editor

Plant and Environment

Plant Phosphate Nutrition and Environmental Challenges

Mohammad Ali Malboobi, Ali Samaeian, Mohammad
Sadegh Sabet and Tahmineh Lohrasebi

Additional information is available at the end of the chapter

1. Introduction

Since the ancient times, food production was exercised in farms where animals and plants were grown together (Figure 1). However, as a part of Green Revolution, the use of synthetic fertilizers turned out to be an integrated part of industrial agriculture which has progressively encouraged disintegration between cropping and animal husbandry. As a result, the consumptions of fertilizers have remarkably increased since half a century ago [1]. Indeed, disruptions in the cycles of nutrients have brought about environmental challenge that caused irreversible damages to natural ecosystems while reasonably justified by the real needs for food security for growing human population. As one of the main challenges in the world of agriculture, provision of phosphorus (P) for plant nutrition requires a closer look from several points of views.

Figure 1. Painted grain and livestock growing together in ancient Egypt.

In this chapter, we firstly explain the importance of P in living organisms and the evolved adaptive mechanisms, particularly from the molecular and genomic aspects. Subsequently, the cycle of exchanging P between physical and biological worlds will be described to show the extent of disturbance by current agricultural practices. Then, possible solutions to the experienced problems in industrialized agriculture will be discussed. The needs for introducing less-energy demanding production and consumption methods for P provision and the use of new generation of fertilizers, particularly organic and biological ones in combination with chemical P fertilizers will be described in details to address integrative measures for sustainable agriculture.

2. Pi importance

P, in the form of phosphate ion (Pi), is the most vital element for all living organisms playing major roles in the structures of essential biomolecules such as nucleic acids, phospholipids and phosphosugars, in almost all metabolic reactions including photosynthesis and respiration, in energy delivering molecules such as ATP, ADP or NADPH and in transduction of signals within the cells. To ensure functional metabolic reactions, Pi homeostasis must be kept between 5 to 20 mM in the cytoplasm. Plants absorb P only in its soluble inorganic form of Pi, $H_2PO_4^-$ or HPO_4^{2-}, which occur in the soil between 0.1 to 1 μM [2-4]. Therefore, it is one of the most needed nutrients for plant growth and development and considered as a major limiting factor in crop yield.

3. Soil Pi and plants uptake

Most soils contain a significant amount of P compounds, ranging from 200 to 3000 mg/kg, averaged at 1200 mg/kg [5]. P compounds in soil comprise a wide variety of organic and inorganic forms [6]. However, only a small proportion (generally less than 1%) is immediately available to plants as free Pi. The majority of inorganic compounds are predominantly associated with calcium (Ca) in alkaline soils or with iron (Fe) and aluminum (Al) in acidic soils [6]. Organic Pi accounts for 30 to 80 percent of soil P, among them monoester P occurs predominantly as cation derivatives of inositol hexakisphosphates (mainly as phytate), whereas sugar phosphates and diester phosphates (e.g. nucleic acids and phospholipids) constitute only a small proportions (~5%)[7, 8]. Factors that contribute to the accumulation and turnover of different forms of in/organic P in soil are complex and controlled by various competing processes that have been the subject of several reviews [6, 9-13].

High concentrations of Pi are generally found in the surface layer of soil profiles or in nutrient-rich patches. Plants usually produce more roots in the surface soil than the subsoil. For instance, an analysis of traits associated with the rate of Pi uptake in wheat showed that root length density in the surface soil was the most important trait for Pi acquisition [14]. Because of that, drying the surface soil can cause ceased Pi uptake or 'nutritional drought' [15].

Low solubility of Pi in water (0.5 mg/lit), slow diffusion rates of Pi in soil (10^{-12} to 10^{-15} m/s) and limited capacity for replenishment of soil Pi-solution are major factors that contribute to its deficiency in plants [4, 16-18]. It is also influenced by biological processes such as the hydrolysis of ester bonds in organic Pi compounds by phosphatase enzymes.

The uptake of Pi from soil by plants depends on both the rate of diffusion of Pi towards roots and the growth of the root system to access unexploited soil [19]. Roots rapidly deplete Pi in the soil solution so that its concentration at the root surface is estimated around 0.05–0.2 mM [19]. Although this establishes a Pi-diffusion gradient from the rhizosphere to bulk soil [19-21], low Pi diffusion rate effectively limits its uptake [22]. It is believed that proper application of the fertilizers not only provides Pi, but also promotes root growth into unexploited soil [18].

4. Pi uptake and reallocation in plants

Under experimental conditions, both high and low affinity Pi uptake mechanisms have been recognized in plants [23,24,25]. Nevertheless, it is generally accepted that if Pi level is within the micromolar range (1–10 μm), which corresponds to Pi concentrations in most cultivated soils, the high-affinity transporters handles Pi uptake. The K_m for high-affinity transporters varies from 1.8 to 9.9 μM [25]. It is an energy mediated co-transport process, driven by protons generated by a plasma membrane H^+-ATPase [23, 26]. Additional evidence for the involvement of protons in Pi uptake comes from the use of inhibitors that disrupts proton gradient across membranes causing the suppression of Pi uptake [25,27].

Both experimental data and genome sequence analyses indicate that plants possess families of Pi transporter genes [24, 28-31]. Current data suggest that members of the PHT1 Pi transporter family mediate transfer of Pi into cells, whereas members of the PHT2, PHT3, PHT4, and pPT families are involved in Pi transfer across internal cellular and organelle membranes [32-35].

Members of the PHT1 Pi transporter gene family have been identified in a wide range of plant species including Arabidopsis, rice, medicago, tomato and soybean [28, 36-41]. Analysis of Arabidopsis whole genome sequences revealed a set of nine PHT1 transporters. Eight of them expressed in roots from which four are expressed in the epidermal cells. In contrast, there is less redundancy in the aerial tissues [42]. In rice, at least 10 of the PHT1 transporters are expressed in roots [40].

Overlapping expression patterns have also been reported for the PHT1 Pi transporters in other plant species [38,41,43-46]. The function of some PHT1 transporters have been analyzed either by expression in yeast Pi transport mutants or in plant cells [27,30,36,43,47]. In Arabidopsis two Pi transporters, PHt1;1 and PHt1;4, mediate 75% of the Pi uptake capacity of the roots system in a wide range of environmental conditions [48].

After uptake into the roots, Pi moves symplastically from root surface to xylem at a rate of about 2 mm/h and to the other organs afterwards [2]. Entering into the xylem for long-distance translocation to the shoot is facilitated by another set of transporter-like proteins [49-50]. Most of the absorbed Pi by the roots is transported through xylem to growing leaves of Pi-fed plants. In Pi-starved plants. Stored Pi in older leaves is retranslocated to both younger leaves and growing roots, from where Pi can again be recycled to the shoot [51]. Consequently, the uptake and allocation of Pi in plants requires multiple transport systems that must function in concert to maintain homeostatic level of it throughout the plant tissues [52].

5. Plant adaptive strategies toward low Pi

5.1. Morphological changes in root architecture

Factors that affect the initiation and activity of the meristems have a large effect on the three dimensional patterns of roots in space, the so-called Root System Architecture (RSA) [53] which is greatly influenced by surroundings soil and particularly the availability and distribution of nutrients, including Pi [54]. Several studies demonstrated that *Arabidopsis thaliana* exposed to low available Pi have reduced primary root growth and at the same time increased lateral root formation and growth and, also, root hair production and elongation [55]. Under sever Pi starvation, root hairs disappeared entirely in tomato [56]. Modification of RSA enable plant roots to explore the upper parts of the soil, a strategy described as 'topsoil foraging' [57]. Symbiotic associations with fungi (Vesicular-Arbuscular Mycorrhizae; see below) and formation of cluster roots are adaptive responses to increase Pi uptake in many plants which allow competent exploration of soils for fixed Pi [58-63].

Detailed analysis demonstrated some differences in RSA responses among ecotypes [64]. Among 73 Arabidopsis ecotypes, half showed reduced primary root growth on low Pi suggesting that root growth inhibition is determined genetically rather than being controlled metabolically only [65].

The first visible event upon Pi starvation is reduction of primary root cell elongation followed by a reduction of cell division as traced by rapid repressonof the cell cycle marker CYCB1;1. This is accompanied by a loss of quiescent center identity as detected by the QC46 marker [66].

Transcriptomics approach as well as mutations analyses have revealed an inventory of genes which are repressed or induced during Pi starvation. For example, PRD (Pi root development) gene is rapidly repressed in roots under low Pi conditions [67-70]. In this context, PRD repression mediated primary root growth arrest [66].

5.2. Metabolic adaptations to Pi deficiency conditions

As mentioned above, Pi is an essential macronutrient that plays a central role in virtually all metabolic processes in plants. This was clearly illustrated for Pi-induced inhibition of a major regulatory enzyme of starch biosynthesis, ADPGlc pyrophosphorylase. Similarly, a vacuolar acid phosphatase (APase) displayed strong activity inhibition by sufficient Pi [71-73]. Conversely, the depletion of vacuolar Pi pools by extended Pi deprivation effectively relieved the inhibition of some APase expressions [74].

A common feature of the plant response to long-term Pi starvation conditions is the development of dark-green or purple shoots due to anthocyanin accumulation. It is brought about by Pi-induced biosynthetic enzymes in each step of the pathways leading to the synthesis of cyanidin, pelargonidin, flavonoids and anthocyanin [74-76].

A reduction in the phospholipid content of Pi-starved plant membranes coincided with increased sulfolipid sulfoquinovosyldiacylglycerol and galactolipid digalactosyldiacylglycerol membrane lipids. SQD1 and SQD2 are Pi starvation inducible enzymes required for sulfolipid biosynthesis in Arabidopsis [74,77]. Consistently, an Arabidopsis *sqd2* T-DNA insertional mutant showed reduced growth under Pi starvation conditions [78]. Galactolipid digalactosyldiacylglycerol accumulation was reduced in the roots of Pi-starved *pldz1* single and *pldz1/pldz2* double mutants [79]. PLDz generates phosphatidic acid that can be dephosphorylated by an APase to release Pi and diacylglycerol serving as a second messenger. The latter activates a protein kinase-mediated protein phosphorylation cascade which controls root growth. In contrast to PLDz function in roots, a non-specific phospholipase C5 is responsible for phospholipid degradation in leaves during Pi starvation [80].

As a consequence of harsh Pi stress, large (up to 80%) reductions in intracellular levels of ATP, ADP, and related nucleoside phosphates also occur [81-82]. A noticeable feature of plant metabolism alterations is that some step in metabolic pathways could be bypassed to reduce dependence on Pi or ATP. This was confirmed by silencing of genes encoding enzymes traditionally considered to be essential. The growth and development of resulting transgenic plants were more or less normal [82] while it was expected to have inhibition of C flux. Protein phosphorylation and glycosylation were found responsible for controlling the activity and/or subcellular targeting of some enzymes involved in bypassed metabolic reactions in response to Pi deprivation [83-85].

5.3. Acid Phosphatases and Pi recycle

A key plant response to Pi deprivation is the up-regulation of a large number of intracellular and secreted APase enzymes that hydrolyze Pi from a broad range of Pi compounds. Secreted APases are believed to function in scavenging nutritional Pi from many exogenous organic Pi substrates, including phytate, RNA, DNA, ATP, 3-phosphoglycerate, and various hexose phosphates that typically constitutes 20-85% of P compounds in soil [17,73-75,86-88]. Similarly, intracellular APases scavenge and remobilize Pi from expendable intracellular Pi monoesters and anhydrides. This is accompanied by marked reductions in levels of the cytoplasmic Pi-containing metabolites during extended Pi deprivation [75,81].

It is noteworthy that APase activity in rhizosphere or soil solution may also originate from fungi such as Aspergillus [89] and mycorriza [90] or from bacteria [89,90]. Microorganisms may produce both acid and alkaline phosphatase [89] while plants secrete APases only [89,92].

5.4. Organic acid biosynthesis and secretion

An adaptive strategy for Pi acquisition is the excretion of proton and organic acids from roots which results in acidification of rhizosphere. The importance of this mechanism was unknown until plasma membrane H^+-pumping ATPases were shown to be involved in plant adaptation to Pi starvation [93]. Acidification was also correlated with the up-regulation of novel membrane channels needed to transport anions such as citrate and malate from root cells into the rhizosphere [94]. Organic acid excretion results in the chelating of metal cations that immobilize Pi (e.g. Ca^{2+}, Al^{3+}, $Fe^{2/3+}$), thus, increasing free Pi concentrations in soil upto 1000-fold. Acidification of rhizosphere also enhances the hydrolysis of organic Pi by secreted APases. As well, organic acids could function as carbon source for symbiotic rhizobacteria that facilitate root Pi acquisition [17,74,75,86].

The amounts of exuded carbon as organic acids can be enormous, ranging from 10% to greater than 25% of the total plant dry weight [75]. Enhanced synthesis of organic acids in Pi-starved plants has been correlated with up-regulation of phosphoenolpyruvate carboxylase and its activation by reversible phosphorylation as well as malate dehydrogenase and citrate synthase and elevated rates of dark CO_2 fixation [75,83].

6. Genomic analysis of Pi adaptation mechanisms

As sessile organisms, plants stand on their own potential to retrieve Pi from their surrounding soil and utilize it as efficient as possible. Perhaps, this is why they carry numerous loci encoding APases and Pi transporters. Some representative genomes are shown in Table 1.

Organism	Genome size (Mb)	APases	Pi transporters
Oryza sativa	450	40	14
Arabidopsis thaliana	125	58	15
Glycine max	975	128	35
Populus trichocarpa	10	51	22

Table 1. Genome size and the number of APase and Pi transporter-encoding genes in four plant genomes with annotated sequenced.

Genome-wide profiling methods has been employed to compare the transcriptional profiles of Pi-starved and Pi-fed plants [67-69,95-99]. These studies have shown that the changes in gene expression can be detected within hours after exposure to Pi starvation [67,68,95]. Wu et al. [67] found that within 72 h from the onset of Pi starvation, the expression of 1800 of 6172 surveyed genes were changed over two-fold, which include more than 100 transcription factors and cell signaling proteins. Furthermore, differential expression patterns in leaves and roots demonstrated distinct responses to Pi starvation in those organs. A similar conclusion was obtained from the microarray analysis of the Pi-starved rice seedlings [100]. Using an Arabidopsis whole genome Affymetrix chip which includes 22,810 genes, Misson et al. [68] found that the expression of 612 genes were induced while 254 genes were suppressed under Pi-limiting conditions. In addition to the Pi transporters, RNase and APase genes, the induced genes include those that function in sulfate and iron transport and homeostasis, Pi salvaging from organic compounds, phospholipids degradation and galacto- and sulfolipid synthesis, anthocyanin synthesis, phytohormone responses, signal transduction, transcriptional regulation, protein degradation, cell wall metabolism and so on. The suppressed genes are involved in lipid synthesis, reactive oxygen controlling and protein synthesis. In another research, Morcuende et al. [69] showed P deprivation led to transcriptional alterations in over 1000 genes involved in Pi uptake, the mobilization of organic Pi, the conversion of phosphorylated glycolytic intermediates to carbohydrates and organic acids, the replacement of phospholipids with galactolipids and the repression of gene implicated in nucleotide/nucleic acid synthesis which were reversed within 3 h after Pi re-supply. In addition, analysis of metabolites confirmed that P deprivation leads to a shift towards the accumulations of carbohydrates, organic acids and amino acids. Pi deprived plants also showed large changes in the expression of many genes involved in secondary metabolism and photosynthesis. Hammond et al. [101]used an oligonucleotide potato microarray to investigate the transcriptional profile of potato leaves under Pi deficiency and compare their data with previously described transcriptional profiles for the leaves of Arabidopsis and rice. They identified novel components to these profiles, including the increased expression of potato patatin genes- with potential phospholipase A2 activity- in the leaves of Pi deficient potatoes. A set of 200 genes were identified that show differential expression patterns between fertilized and unfertilized potato plants.

Müller et al. [102] investigated the effect of interaction of Pi and sucrose signals on the gene expression pattern in Arabidopsis. They found several genes that were previously identified to be either sugar-responsive or Pi-responsive genes. In addition, 150 genes were synergistically or antagonistically regulated by the two signals.

In a comprehensive analysis, Lin et al. [103] conducted time course microarray experiments and co-expression-based clustering of Pi-responsive genes by pair wise comparison of genes against a customized data base. Three major clusters enriched in genes functioning in transcriptional regulation, root hair formation and developmental adaptations were distinguished in this analysis. The genome-wide transcriptional approach may be used to infer inclusive scenarios for involved mechanisms in the signaling and adaptation of plants to Pi deficiency.

7. Pi sensing and gene expression

Metabolic adjustments to Pi limitation are largely cellular responses to sensed by internal Pi status which trigger a systemically integrated regulating mechanisms involving microRNAs, non-coding RNAs and PHO2 downstream of PHR as revealed in recent studies. PHR1, a MYB transcription factor, binds to the promoters of most of the Pi-responsive genes that are positively or negatively affected by Pi starvation [104]. Despite its central role in controlling the expression of numerous Pi-responsive genes, the phr1 mutant showed no major phenotypic defects except for a slight difference in the root-to-shoot ratio and root hair induction [105]. Indeed, PHR1 works with another MYB factor, PHL1, to control most transcriptional activations and repressions in responses to Pi deficiency [104]. Furthermore, about two thirds of the genes repressed in Pi-deprived wild type seedlings were markedly de-repressed in Pi-starved phr1phl1 double mutants [104].

In Arabidopsis, a number of miRNA molecules have been shown to be specifically and strongly induced by Pi limitation, including miRNA399, miRNA778, miRNA827 and miRNA2111[106-108], though only the role of miRNA399 in the regulation of Pi homeostasis has been elucidated [77,109]. miRNA399 is a component of the shoot-to-root Pi-deficiency signaling pathway that moves from shoot to root via the phloem where it targets the transcripts of PHO2 [103,110]. The repression of PHO2 expression by miRNA399 causes up-regulation of root Pi transporters (e.g. PHT1;8 and PHT1;9).

The Pi-signaling network also involves IPS (Induced by Pi Starvation) genes that carry short open reading frames [111-112]. It is postulated that their transcripts operate by a mechanism called 'target mimicry' as they contain a conserved 23-bp region complementary to miRNA399 and to fine-tune the PHO$_2$–miRNA399 pathway [113]. Since homologs of Arabidopsis IPS and PHO$_2$ genes are present in numerous other plants such a mechanism is probably widespread in plants [114].

The role of miRNAs and non-coding RNAs appear to be extended to a possible role in its coordination with homeostasis of Pi and also other nutrients. For example, a hypothetical role was attributed to miRNA827, in the cross-talk between Pi-limitation and nitrate-limitation signaling pathways that affect anthocyanin synthesis [108]. In a survey on small RNAs showing differential expression, miRNA169, miRNA395, and miRNA398 were found to be suppressed in response to Pi deficiency while they also responded to other nutritional stresses [106]. Furthermore, miRNA169 and miRNA398 target genes involved in drought tolerance and oxidative stress response [115]. Suppression of miRNA395 was suggested to up-regulate the expression of APS4 and SULTR2;1 leading to increased sulfate translocation and improved utilization of sulfolipid biosynthesis under Pi-deficient conditions [115].

Taken together, the genome-wide surveys affirm the participation of miRNAs in coordination of homeostatic pathways of Pi and possible links between them and metabolic adjustment [106].

8. Limited Pi resources

Despite the existence of high amounts of P in the soil [5], the concentration of available Pi in many soil solution averages at about 1 µM and seldom exceeds 10 µM [2] which is far below the cellular Pi concentrations (5–20 mM) required for optimal plant growth and development [73]. Such a limitation often lead to reduced productivity in natural ecosystems as well as cropping systems, unless it is supplied as fertilizer [74-75]. Whilst our limited global Pi reserves are non-renewable, P in many agricultural soils is being building up. This is because 80 to 90 percent of Pi applied as fertilizer is fixed by soil particles or compounds, rendering it unavailable for plants.

9. Types of Pi mines

There are over 200 P minerals existing on the earth but only a few can be used for commercial extraction of Pi [61]. Phosphate Rocks (PR) is the commercial term applied to all

Pi bearing minerals suitable for Pi production. The primary Pi minerals in PR are *phosphorites* that include fluor-apatite ($Ca_{10}(PO_4)_6F_2$), Hydroxy-apatite ($Ca_{10}(PO_4)_6(OH)_2$, carbonate-hydroxyl-apatites, and francolite (Ca_{10-x-y} Nax $Mgy(PO_4)_6-z(CO_3)zF0.4zF_2$) which is a carbonated-substituted form of apatite mainly found in marine environments [116].

In PR industry the grade of the rocks are mostly reported as the percentage of P pentoxide (P_2O_5). Three major resources that can be profitably recovered today are as below [116-117]:

1. Sedimentary Pi deposits which are widespread throughout the world, occurring almost on all continental shelves. Francolite and apatites are deposited in layers that might cover thousands of square miles in several chemical compositions and a wide range of physical forms [116,118].
2. Igneous Pi deposits which are mostly found in continental shelf and on seamounts in the Atlantic and Pacific oceans (Russia, Canada, Brazil and South Africa). Their exploitation is economically non-feasible so that they have mostly remained untouched [118].
3. Biogenic deposits, also known as island Pi, which are mainly old bird and bat droppings built up.

About 80% of world Pi production comes from non-renewable sedimentary reserves, 15 to 19% from igneous and about 1-5% from biogenic and other deposits such as island Pi which are near total depletion due to over exploitation during the past decade [116].

10. Geographic distributions of Pi rocks

Today, PR is produced in over 40 countries, with 12 countries producing over 92 % of the world's total production [118]. US alone produce over 28% of the total production followed by China, Morocco and Russia. Among them, four major producers of PR (the United States of America, China, Morocco and Western Sahara, and the Russian Federation) produced about 72.0 percent of the world total. Twenty other countries produced the remaining 6-7 percent of world production [119].

11. Methods of Pi production

Production of fertilizers began in early 19th century when crushed bones were treated with acid and applied to soil. Since 1945 mining of PR was increased from 11.2 Mt to 145 Mt in 1999 [120]. This was translated to increasing production rate of Pi fertilizers (as P_2O_5) from 4 Mt by 1940's to over 42 Mt annually today [121]. Ever since new technologies has been evolving to maximize the production rate and purity of the fertilizers. Here a brief description of the major production methods is given.

11.1. Wet process

The wet process or hemihydrate process is the newest and dominant production method used due to its ease, lesser investment size, energy efficient and higher yield [122]. Basically, this is the process where PR is treated with sulfuric acid to hydrolyze apatite minerals. Other acids can be used to harvest Pi depending on the costs and desired final product. For example, phosphoric acid addition yields TSP as the final product. Using of nitric acid and hydrochloric acid has also been reported but never industrialized. Phosphogypsium and silicon fluorine or HF are as the main side product in this method. Sulfuric acid added to PRs forms the soluble monocalcium Pi. This product can either go to the fertilizer production (i.e. MAP or DAP) or to concentrated forms for other applications [122-123]. Concentrating the product by evaporation can yield up to 53% phosphoric acid solution. The pollution of current technology wastes (4-5 tons phosphogypsium per one ton of P_2O_5) signifies the processing costs for the industry.

11.2. Thermal process

Thermal process is an older technology that is used in only a few factories in the world. High electricity and water consumption as well as higher investment size and lower product yield makes such process less desirable for investors. This process can be performed in desired scales, but requires detailed financial feasibility estimation depending on the production site location [122]. Before heating, Pi rock is converted to 1-2 cm pellets by wet granulation and sintering to prevent blocking of furnace. Next step involves mixing with cokes (reducing agent) and SiO2 (for slag formation) before feeding into furnace. Heating at 1500°C, Pi is reduced to P4 in gaseous state which is condensed afterward. Remaining CaO combines with SiO_2 to form liquid slag that might be used for road construction [124].

11.3. Bioprocess

Bioprocessing of Pi rocks involves treatment of apatites with Pi solubilizing bacteria. This controlled fermentation process requires presence of a bacterial energy supply such as sugars which is consumed to produce organic acids to break down the PRs [125]. Optimization of this process is still under investigation of many scientific and financial institutes due to its promising outlook.

Bioprocessing of insoluble soil Pi compounds is also carried out directly in agricultural fields by the use of biofertilizers (see below) which converts a portion of precipitated fertilizers as well as natural occurring insoluble Pi. This method is extremely feasible with regards of purchasing, application and higher crop yields [126].

12. Emergence of Pi fertilizers

As mentioned earlier in this chapter, Pi has been found to be the limiting macronutrient in most agricultural soils. The first surveys that revealed such deficiency was a simple study during early 19th century in England [127]. At that time the only source of artificially added Pi to the agricultural soils was farmyard manure, which resulted in higher crop yields near manure production sites. It was only a matter of time before it was realized that crushed bones might have a positive result on crop yield on some specific types of soils. Treating the crushed bones with sulfuric acid to produce superphosphate was followed by the same treatment of PRs to produce current Pi fertilizers which was the beginning of a new era in the world agriculture history [127]. Since then, Pi fertilizers production processes, molecular types, application methods and feasibility assessments have been the subject of a wide range of researches and analyses. Table 2 summarizes popular Pi fertilizers that are currently used [128].

13. Trend of Pi usage in the world

In 2006, 167 Mt of PR were mined in the world, while China alone was responsible for 56 Mt of it. 23% of the total PR production was directly used as fertilizers which equals 12 Mt P_2O_5 per year. Only 6% was used for production of elemental P, and almost the rest went through the wet process for production of phosphoric acid, which in turn is mostly (88-95%) used up for Pi fertilizers. World's total Pi supply (P_2O_5) has been estimated to rise from 42 Mt in 2010 to 45 Mt in 2015 (Table 3) [121]. It is expected that P fertilizer supply will have an increased trend of 3.2% annually with the major Pi balance (production minus consumption) surplus in North America and Africa and deficit in Latin America and Asia. With the current rate of consumption, various estimates showed that recoverable Pi mine will be vanished in 50 to 100 years [129].

Material	P	N	K	S	P compound
Superphosphoric acid	30 to 35	-----	-----	-----	H_3PO_4
Wet Process phosphric acid	23 to 24	-----	-----	-----	H_3PO_4
Concentrated superphosphate (TSP)	20	-----	-----	1 to 1.5	$Ca(H_2PO_4)_2$
Diammonium phosphate (DAP)	20 to 21	18 to 21	-----	0-2	$(NH_4)_2HPO_4$
Monoammonium phosphate (MAP)	21 to 24	11 to 13	-----	0 to 2	$NH_4H_2PO_4$
Normal superphosphate	7 to 10	-----	-----	12	$Ca(H_2PO_4)_2$
Phosphate Rock	12 to 18	-----	-----	----	$[Ca_3(PO_4)_2]_3.CaF_x.(CaCo_3)_x.(Ca(OH)_2)_x$
Monopotassium phosphate		-----	35	-----	KH_2PO_4
Dipotassium phosphate		-----	54	-----	K_2HPO_4

Table 2. Commonly used Pi fertilizers and the percentage of each elements (quoted from Havlin and Beaton [128]).

An IFDC estimates shows that resources of unprocessed PR of varying grades could be as high as 290 Mt [130] that could become reserves once all high grade PR are depleted. Even if this takes place, the price of Pi would not be affordable for many poor farmers and still pollution because of Pi accumulation and associated wastes in the soil and water are of concerns.

Year	2010	2011	2012	2013	2014	2015
Supply	39600	42094	43966	45011	46439	47788
Demand	41700	41679	42562	43435	44245	45015
Potential balance	-2100	415	1404	1576	2194	2773

Table 3. World Pi supply, demand and balance according to FAO report (2008) [119].

It is noteworthy the extracted phosphoric acid, so-called green or wet process acid, contains upto 24% elemental P and is produced from the reaction of sulfuric acid and PR. It is used for agronomics means either by direct application to more alkaline soils or by reaction with PRs for partial or complete acidulation. Phosphoric acid is also produced by electric furnace to produce elemental P that in turn reacts with O_2 and water to make phosphoric acid. This type of phosphoric acid, called white acid, is purified to make technical and food grades which is mostly used for non-agricultural purposes such as soft drinks due to its higher purity and expensive production costs [128].

14. Effect of P fertilizers on crop yields

Soil scientists recognize three kinds of Pi: (1) deposited Pi as chemical compounds in the soil; (2) available Pi or free Pi ion in the soil; and (3) absorbable Pi or free Pi ion in vicinity of root. Even though the average use of P_2O_5 (19 kg/ha) has been less than recommended amounts in the literature (30Kg/ha) [131], it has resulted in significant increase in crop yields around the world. For example, an increase between 9 to 60% compared to control was observed in Middle East countries on barley, wheat, and chickpea. Productivity Indexes of cereal crops were as high as 11.1 when up to 90 kg/ha P_2O_5 were added. Data on oil crops shows productivity index upto 6 in groundnut and mustard. In Indonesia, upto 10 fold increase was observed after soluble Pi was added to soybeans [131]. Such examples are very easy to find in the literatures dated early 19th to 20th centuries when the application of Pi fertilizers were becoming dominant. The resultant high yield is accompanied by Pi removal of 6 kg/ha for cereals and 20 Kg/ha for other crops from soil [61].

15. The fate of Pi fertilizers in soil

Two contrasting challenges of higher crop yield and environmental concerns necessitate tracing the fate of Pi fertilizer added to soil. As the Pi fertilizer granule or droplet is added to the soil, the moisture penetrates its way into the fertilizer. This results in the release of export of Pi into the soil which might expand as far as 3 to 5 cm from an average size granule [128]. As long as the original salt remains, a saturated solution is maintained around the granule, which in turn absorbs water, increasing the soil moisture. In these areas Pi uptake can take place efficiently by plants. Depending on type of soil and its cations composition, the released Pi can re-precipitate within a few days or weeks [127-128]. This behavior indicate the proper time of Pi fertilizer application in order to reduce the chance of Pi accumulation in the soil. The use of such information also helps calculating the correct amount of Pi that should be applied for maintaining the P-level of the soil after a harvesting season which is extremely important for balanced fertilization. It has been shown repeatedly no increase in yield is detected when Pi fertilizers are over-used [124,127,128].

16. Need for Pi preservation and restoration

Historically, since mid-19th century, the use of local organic matters was replaced by other sources such as guano, deposited bird droppings in remote islands. Growing world consumption rapidly vanished guano resources by the end of 19th century shifting trades to PR. In recent years, the PR production has grown to over 140-160 Mt per year for extraction of about 25 Mt elemental P per year [132].

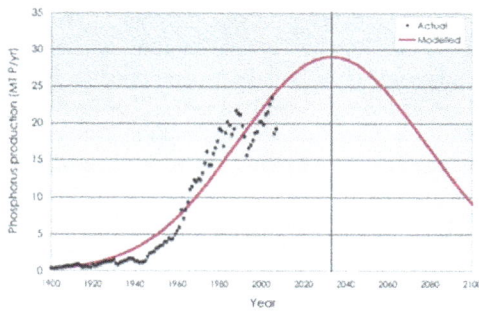

Figure 2. Predicted peak of elemental P production curve based on actual data in previous years. It illustrates global P reserves are likely to peak around year 2033 when current P production will not be economical any more (Quoted from Cordell et al. [129]).

Since PRs are non-renewable, it is estimated that the economically recoverable reserves will be depleted by the end of this century [129]. The point is that production rate will peak at a maximum when high quality and accessible Pi resources are depleted and mining and processing become quite expensive. As shown in Figure 2, the consumption of P since 19th century follows a Guassian distribution curve. Assuming the total earth reserve of elemental P is approximately 3240 Mt, then its production will peak at 29 Mt in 2033 as modeled by Cordell et al. [129] (Figure 2). Although the actual timing depends on several conditions to be met, however, this is really alarming if considering that the non-renewable Pi reserves are finite and there is no substitute for PRs. Besides, one might presume that the energy cost will be increased and the average grade of Pi will decrease in future as it was experienced to decline from 15% to 13% between 1970 to 1996 [132]. As this challenge silently threaten the food security, a global comprehensive effort is necessary. Web sites such as Pi Knowledge

Center   (http://www.GreenPi.info)   or   Global   Phosphorus   Research   Initiative (http://www.phosphorusfutures.net) could greatly help worldwide data sharing and collaborations amongst scientists and other players to seek integrative solutions.

17. Environmental concerns about Pi production

Removal of massive amounts of soil and extensive washing with huge amounts of water are major changes in the environment of the mining sites and the surrounding regions by themselves. However, the most important issue is disposal of side products that pollute the ecosystem at the stacking sites massively.

Phosphogypsum, the side product of wet process, is deposited in large land areas called gypsum stacks. For the production of 50 Mt of P_2O_5 annually, about 250 Mt of gypsum is produced, out of which only 10 percent is used for other purposes (e.g. road construction, cement and housing) and the rest remains stacked [122].

Presence of radioactive materials in side products is another issue for Pi production industry. PRs contain mainly uranium (20-300 ppm) and thorium (1 to 5 ppm) and their decay products. This would mean 0.35 kg U_2O_8 per Mt P_2O_5 or 2100 tons of unused uranium annually [122,133]. About 80 percent of this amount and their decay products are concentrated within the gypsum waste materials in wet process. These could be as high as 7 to 100 pCi/g uranium, 11 to 35 pCi/g radium and 1.7 to 12 pCi/g radon, though radioactivity is different from stack to stack depend on the original PR. The fertilizers themselves might also exhibit radioactivity which might be as high as 33 pCi/g in US [133]. These concerns are reflected in a final rule of EPA issued in 1992 stating "phosphogypsum intended for agriculture use must have a certificated average concentration of radium-226 no greater than 10 pCi/g" [133].

18. Environmental impacts of Pi fertilizers consumption

The ever growing population of the world has created an enormous pressure to cultivate larger agricultural lands and to increase production yields by improved mechanized methods and usage of fertilizers. Such practices not only alter chemical and physical properties of soils, but also affect macro and micro flora remarkably [122].

Application of Pi fertilizers to soil does not necessarily guarantee plant uptake due to their polycrystalline structure and their tendency to precipitate. An observation showed that if no crop is around, the addition of Pi fertilizer to soil increases the soluble P within the first couple of years, but it will decrease to a constant level later [127]. Studies in developed countries have also suggested that Pi application might not be as effective as it used to be in the past [125].

In addition to the effect of deposited Pi on the soil structure and microbial flora, leaching of Pi through the erosion of surface soil and organic matters to water streams and lakes cause vast eutrophication [17]. Discharged detergents sodium triphosphates as well as human or animal metabolic wastes must also be accounted for a portion of Pi inputs to surface and underground waters [122]. Eutrophication and the consequent over-enrichment of aquatic ecosystems with nutrients, mostly Pi and nitrate lead to algal blooms and anoxic events that may initiate irreversible environmental damages. Although some lakes have recovered after sources of nutrient inputs were reduced, recycling of enriched Pi from sediments causes lakes to remain eutrophic for years [122].

Recycling of Pi from sewage systems, precision farming and agricultural and industrial treatment units has been exercised in developed countries such as Japan and Germany to reduce such effects [122,124]. Replacing Pi in detergents by aluminum silicate has also been considered as option, despite its much poorer performance. Crystallization of Pi in wastewater as struvite (ammonium magnesium phosphate), separation of urine and the use of sanitized faecal matters in municipalities have been attempted to recover or reuse wasted Pi [129].

Cordell et al. [129] warned about the peak time of Pi production (Figure 2) with assumption of PRs are the sole high-value Pi reserves. Taking into consideration the massive amounts of insoluble Pi compounds accumulated in the soil, a straightforward solution could be the utilization of PSMs in the form of biofertilizers [91,126]. Simplified application method and comparable cost of Pi biofertizers on one hand and increased quantity and quality of the yields, on the other hand, have already attracted the attention of farmers.

19. Cycle of P between the physical and biological worlds

As shown in Figure 3, P like other mineral nutrients exists in soil, minerals, water as well as living organisms. Over time, geologic events bring ocean sediments to land and weathering

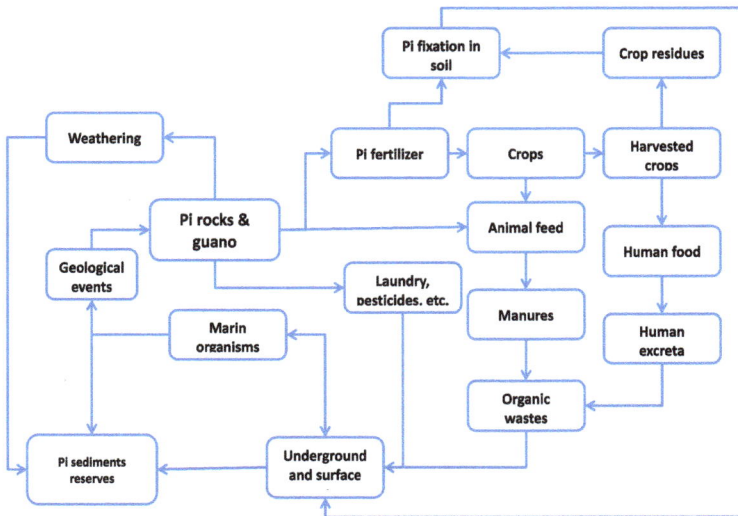

Figure 3. Pi flow through food production and consumption.

carry Pi compounds to terrestrial habitats where the land cycles occurs. Pi is absorbed by plants from soil, consumed by animals and returned to soil as organic matters that slowly released as inorganic Pi or incorporated in more stable molecules as soil organic compounds. Depending on temperature and moisture, soil microorganisms break down these compounds by secreting organic acids and APase enzymes. Pi leaching due to soil erosion or water run-off to surface or underground waters finally discharge in lakes, seas and oceans where another sub-cycle between marine organisms and Pi sediments occurs. Human action for the utilization of Pi in fertilizers, pesticides, detergents and food industries are the main sources of disturbance of these natural cycles [134].

20. Sustainable Pi usage in agriculture

The assessment of plant response to Pi level in a series of greenhouse pot trials over a wide range of soils with various characteristics established two values of soil Pi, (1) the critical value (Tc) below which it is recommended to apply Pi fertilizers in order to build up soil Pi fertility, and (2) the sufficient value (Ts) above which it is recommended not to apply it over a period of time. For Tc value, available soil Pi (measured by Olson method) ranged from 10 to15 ppm P_2O_5 whereas it ranged from 50 to 55 ppm for the Ts value [131]. Of course, it is crucial to provide a certain ratio of each nutrient to establish a proper balance. Legumes need N:P:K nutrients in a ratio of 0:1:1, 1:2:2 or 1:2:3, root crops, a ratio of 2:1:2, and other crops, a ratio of 2:1:1 or 1:1:1 [131].

An old statistics between year 1960 to 1981 shows that Pi fertilizer consumption has been well established as the share of developed countries in the world was decreased for from 75.1 % to 43.5% of the world consumption by practicing precision farming while it has increased from 6.5 to 21.3% in developing countries mainly because of the over usage [125]. The main alternative methods for Pi provision have been described below.

20.1. Direct application of PRs as Pi fertilizer

Direct application of Pi rocks as a source of Pi fertilizer has been practiced since early 19th century. Except for grinding to increase surface area in contact with soil, no other modifications are required to be made which makes it a relatively inexpensive source of Pi. The catch is, however, the lower Pi content of the rocks and required specific soils conditions such as having acidic pH (below 5.5) and low exchangeable Ca^{2+} [116]. Besides, some plants have higher capabilities in mobilization of Pi into their systems. For example, legumes and some crops of *Cruciferae* family enhance Pi solubilization and uptake by excretion of organic acids that lowers the soil pH and increases the cation exchange capacity of the soil. Acidifying the rhizosphere, secretion of APases and morphological changes are prerequisite adaptations that enable different crops to salvage Pi from PRs.

20.2. The use of composted organic materials

The increasing cost of chemical fertilizers has renewed interest of farmers across all sectors to utilize organic fertilizers as an alternative source of plant nutrients. Composted manure is simply decomposed organic material that continuously occurs in nature, often without any assistance from mankind. The rate of decomposing depends on type of material, air, moisture, and temperature and soil pH.

Manures improve the texture of both clay and sandy soil; indeed, it is the best additive to make either clay or sandy soil into rich, moisture holding, and loamy soil. They are rich in nitrogen and potassium but have very little Pi. Yet, the availability of soil Pi is enhanced due to chelating polyvalent cations by organic acids and other decay products. In addition, organic materials are rich substrates for propagation of beneficial microorganisms (see below). However, poorly available, being costly and the unbalanced levels of elements in manures have hindered the effectiveness and wide usage. In addition, these still need microbial activity to release the elements.

20.3. The use of Pi solubilizing microorganism

Pi solubilizing microorganisms (PSMs) are, as named, a group of soil fungi and bacteria involved in solubilization and mineralization of soil P from inorganic and organic pools as an integral part of the P cycle. In soil, Pi solubilizing bacteria out-number fungi with the same activity by 2–150 folds [135]. The majority of the PSMs solubilize Ca–P complexes in alkaline soils and only a few can solubilize Fe–P and Al–P in acidic soils. Nevertheless, PSMs can become effective in the latter soils when supplemented with PR.

The principle mechanism for Pi solubilization is the secretion of organic acids and APases that play major role in the mineralization of both inorganic and organic Pi compounds in surrounding soil. Several mechanisms like lowering of pH by acidulation, ion chelating, or ion exchange reactions as well as enzymatic activities in the growth environment have also been proposed to play a role in Pi solubilization by PSMs [91,126,136,137].

Through receiving energetic carbon sources from plant, PSMs facilitate the uptake of Pi and also other nutrients such as zinc, molybdenum, copper and iron. Fungi have been reported to possess greater ability to solubilize rock Pi than bacteria, though they perform better in acidic soil conditions. Species of *Aspergillus, Penicillium, Curvularia* and yeast have been widely reported to solubiliz various forms of inorganic phosphates [137]. Another mechanism which indirectly leads to increased Pi acquisition by plants is the production of phytohormones (mainly auxins) by rhizobacteria that stimulate root growth [138-139]. Therefore, it makes more sense to name such bacteria as plant growth promoting rhizobacteris (PGPR) [140].

Among the soil Pi solubilizing bacteria communities, *Pseudomonas putida, P. strata, Bacillus sircalmous* and *Pantoa agglomeranse* could be referred to as the most important strains [91]. The positive effect of Pseudomonas inoculation on plant growth has been reported in many field trials [126]. The role of Pi solubilizing bacteria and their potential capacity to restore Pi cycle processes in plant-soil system cannot be ignored [138].

Symbiotic relationship between mycorrhizal fungi and plants exists in almost all ecosystems [141]. Fungi species spread around their mycelium acquiring Pi from longer distances which is then passed to plants in exchange for carbon compounds [142]. Nevertheless, limitations in mass production and restriction of this relationship to certain plant types have been unsolved issues for wide utilizations so far.

20.4. Plant species with high Pi efficiency

Genotypic variations for tolerance to Pi deficiency has been demonstrated for many food crops [143] which provide sources for developing cultivars superior in Pi acquisition and higher yield particularly in deficient conditions through plant breeding [144]. Pi-efficient plants are defined as plants which could produce higher yields per unit of applied or absorbed nutrient compared to other plants grown under similar agroecological conditions [145]. Screening for plant species or genotypes with increased Pi absorption involves criteria such as: (i) greater root growth and, thus, extended soil exploration; (ii) the exuding of proton and organic acids that increase the solubility of Pi by decreasing pH and/or chelating; and (iii) the secretions of elevated levels of APase enzymes that break down organic Pi [146].

Most modern crop cultivars have been selected through conventional breeding approaches for better adaptation to deficient soil Pi by looking at root architectural traits which allow for more Pi acquisition from soil surface zone [147]. However, conventional plant breeding including phenotypic selection for improved root systems has proven to be difficult, prone to environmental effects, and time-consuming [148]. As tolerance to low Pi is a quantitative trait, an appropriate method to dissect its complex polygenic inheritance must be employed through quantitative trait loci (QTL) analysis. A number of QTL have been identified for tolerance mechanisms to low Pi in various food crops [149]. Molecular markers linked to the target traits are needed to be used for selection in the breeding process.

20.5. Genetic engineering for Pi acquisition and utilization

Researches on the molecular and physiological basis of Pi uptake, translocation, and internal utilization facilitate the design and generation of transgenic plants with enhanced Pi efficiency. For instance, over-expression of a Pi transporter gene in tobacco cell cultures resulted in an increased rate of Pi uptake when cells were grown under Pi limited conditions [47]. Several attempts have been made to improve soil Pi solubilization and acquisition processes in food crops through engineering of genes with bacterial, fungal or plant origins. Increasing knowledge on regulatory and signaling mechanisms involved in Pi acquisition might identify new useful genes. It is assumed that a combined overexpression of Pi-transporter and dissolving encoding genes might result in synergistic effects [150].

An overview of reported transgenic crop plants is given in Table 4. Generating plants with an enhanced metabolical Pi uptake capacity were attempted by engineered tobacco plants to produce more organic acids, specifically citrate [150]. The promising result was that these transgenic plants yielded more leaf and fruit biomass than controls when grown under Pi-limited conditions [151]. Many studies reported transgenic plants with overexpression of phytase in various food crops. Phytases are exuded into the rhizosphere and are able to hydrolyze phytates which constitutes up to 50% of the total organic Pi in soil [8]. Overexpression of phytase in potato, clover, soybean and tobacco resulted in increased Pi acquisition and content. These results point to possible manipulation of biochemical pathways that increase the ability of the plant to release organic acids or enzymes that improve Pi acquisition from soils.

20.6. Mutagenesis

Several mutant genes have been successfully introduced into commercial crop varieties that significantly enhance the nutritional value of crops. Mutants of barley, wheat, rice and soybean with low phytic acid have been released and can reduce both Pi pollution and increase bioavailability of Pi and micronutrient minerals in cereals and legumes [164]. A major QTL *pup1,* on chromosome 12 effective on Pi uptake from Pi-deficient soils was fine-mapped in rice. Liu and colleagues [165] demonstrated that chromosome 1R and 7R of rye might carry genes responsible for tolerance to Pi starvation stress, while 5R carry unfavorable genes regarding Pi starvation tolerance.

Gene Name	Gene source	Transformed crop species	Main effect under P deficiency	Reference
Secretary phytase	synthetic	Potato	Increased P content and biomass if phytate available	[152]
Phytase	*Aspergillus niger*	Clover	Increased P uptake if phytate available	[17]
Phytase	*A. niger*	Clover	Increased P uptake in sterile media (phytate available)-no effect in soil	[153]
Phytase	*A. niger*	Tobacco	Increased P uptake if phytate available	[154]
β-Propeller phytase	*Bacillus subtilis*	Tobacco	Increased P content and biomass	[155]
Phytase	*A. ficuum*	Soybean	Increased phytase activity and P content in root exudates	[156]
Phytase and acid phosphatase	*Medicago truncatula*	Clover	Increased P content and biomass	[157]
Acid phosphatase	white lupin	Tobacco	Increased P uptake and growth	[158]
Purple acid Phosphatase 18	*A. thaliana*	Arabidopsis thaliana and tobacco	Increased P uptake, P content and biomass	[158]

Gene Name	Gene source	Transformed crop species	Main effect under P deficiency	Reference
High-affinity P-transporter	Barley	Barley	No effect	[43]
High-affinity P-transporter	tobacco	Rice	Increased P uptake	[160]
Malate dehydrogenase (MDH) overexpression		Alfalfa	Increased organic acid exudation	[161]
Phosphoenolpyruvate carboxylase (PFPC) overexpression		Alfalfa	No increase in organic acid exudation	[161]
Type I H⁺ -pyrophosphatase AVP1	A. thaliana	Rice and tomato	More robust roots, higher shoot mass, higher yields	[162]
Aluminum resistance gene	wheat	Barley	Increased P uptake and grain production (acid soils)	[163]

Table 4. An overview of some reported genetic engineering in crop species

21. Conclusion

Chemical fertilizers have played a significant role in the Green Revolution to have improved yield and to combat hunger to some extent [166]. However, poorly managed consumption of resources and even poor applications of Pi fertilizers and side products have endangered our environment. These factors have encouraged researchers to seek bio-friendly, economically feasible, and replaceable sources of Pi before the world consumption rate will rich to a no-return point. In order to secure agricultural sustainability through reduction in Pi fertilizer overuse, plant and soil scientists have tried to address solutions for retrieval of Pi from P-compounds in the rhizosphere, soils in the a close proximity of root. Two major approaches in the current research and technology world are:

1. To bioengineer Pi efficient transgenic crops expressing enzymes such as phosphatases. This solution has been highly approached by many scientists (for a review see [84]). However, transgenic plants were found to be affected by the non-desired traits or epitasis among numerous APase-encoding genes. Even if a transgenic plant acquire release permission, still we are limited to use a lines variety of the transgenic crop.
2. To bioengineer rhizosphere by the use of beneficial microorganisms as biofertilizers involved in solubilization of Pi in the soil, namely *"Pi biofertilizers"*. This approach is advantageous as Pi biofertilizers could be applied to wide range of crops and their varieties as desired. Besides, signaling of sufficient availability of Pi in the rhizosphere would regulate the expression and secretion of organic acids and/or APase enzymes by

the microorganisms. Such an elegant smart mechanism leads to natural growth and development of plants.

To this end, years of researches on chemical, organic and biological Pi fertilizers have shown that none can replace the other. Practically, achieved higher yield and quality with combinations of all three in certain proportions address a call for precise farming with integrative plant nutrition programs prescribed based on crop and soil requirements.

Author details

Mohammad Ali Malboobi*, Ali Samaeian, and Tahmineh Lohrasebi
Department of Plant Biotechnology, National Institute of Genetic Engineering and Biotechnology, Tehran, I.R. Iran

Mohammad Sadegh Sabet
Department of Plant Breeding and Biotechnology, Faculty of Agriculture, Tarbiat Modares University, Tehran, I.R. Iran

Acknowledgement

We would like to express our appreciation to those who contributed to data compilation and sharing through Phosphate Knowledge Center (http://www. GreenPi.info), FAO, IFDC, IMPHOS and Global Phosphate Research Initiatives (http://www.phosphorusfutures.net).

22. References

[1] Vance CP. Symbiotic Nitrogen Fixation and Phosphorus Acquisition. Plant Nutrition in a World of Declining Renewable Resources. Plant Physiol; 2001. 127(2): 390-7.

[2] Bieleski RL. Phosphate Pools, Phosphate Transport and Phosphate Availability. Annual Review of Plant Physiology; 1973. 24: 225–252. doi: 10.1146/annurev.pp. 24.060173.001301

[3] Hinsinger P. Bioavailability of Soil Inorganic P in the Rhizosphere as Affected by Root-Induced Chemical Changes: A Review. Plant Soil; 2001. 237: 173–195.

[4] Conyers MK., Moody PW. A Conceptual Framework for Improving the P Efficiency of Organic Farming without Inputs of Soluble P Fertiliser. Crop & Pasture Science; 2009. 60: 100–104.

[5] Harrison AF. Soil Organic Phosphorus: A Review of World Literature. CAB International: Wallingford, UK; 1987.

[6] Sanyal SK., De Datta SK. Chemistry of Phosphorus Transformations in Soil. Advances in Soil Science; 1991. 16: 1–120.

[7] DalalR C. Soil Organic Phosphorus. Advances in Agronomy; 1977. 29: 85–117.

* Corresponding Author

[8]   Anderson, G. Assessing Organic Phosphorus in Soil. In: Khasawneh FE., Sample EC., Kamprath EJ. (Eds.), The Role of Phosphorus in Agriculture. American Society of Agronomy, Madison, WI, USA; 1980. p411–431.

[9]   Tate KR. The Biological Transformation of P in Soil. Plant and Soil; 1984. 76: 245–256. doi: 10.1007/BF02205584.

[10] Stewart JWB., Tiessen H. Dynamics of Soil Organic Phosphorus. Biogeochemistry; 1987. 4: 41–60. doi: 10.1007/BF02187361.

[11] Hedley MJ., Mortvedt JJ., Bolan NS., Syers JK. Phosphorus Fertility Management in Agroecosystems. In 'Phosphorus in the Global Environment: Transfers, Cycles and Management'. (Ed. H Tiessen) (Wiley and Sons: London); 1995. p59–92.

[12] Magid J., Tiessen H., CondronLM. Dynamics of Organic Phosphorus in Soils under Natural and Agricultural Ecosystems. In 'Humic Substances in Terrestrial Ecosystems'. (Ed. A Piccolo) (Elsevier Science: Amsterdam); 1996. p429–466.

[13] Frossard E., Condron LM., Oberson A., Sinaj S., Fardeau JC. Processes Governing Phosphorus Availability in Temperate Soils. Journal of Environmental Quality; 2000. 29: 15–23.

[14] Manske GGB., Ortiz-Monasterio JI., Van Grinkel M., Rajaram S., Molina E., Vick PLG. Traits Associated with Improved P-uptake Efficiency in CIMMYT's Semidwarf Spring Bread Wheat Grown on an Acid Andisol in Mexico. Plant and Soil. 2000. 221: 189–204. doi: 10.1023/A:1004727201568

[15] Simpson JR., Pinkerton A. Fluctuations in Soil Moisture, and Plant Uptake of Surface Applied Phosphate. Nutrient Cycling in Agroecosystems; 1989. 20(2): 101-108, DOI: 10.1007/BF01055434

[16] Tiessen H. Phosphorus in the Global Environment. In White PJ., Hammond JP., eds, Ecophysiology of Plant-Phosphorus Interactions. Springer, New York; 2008. p1–8.

[17] Richardson AE . Regulating the Phosphorus Nutrition of Plants: Molecular Biology Meeting Agronomic Needs. Plant and Soil; 2009. 322(1-2): 17-24, DOI: 10.1007/s11104-009-0071-5.

[18] Richardson AE., Hocking PJ., Simpson RJ., George TS. Plant Mechanisms to Optimise Access to Soil Phosphorus. Crop & Pasture Science; 2009. 60: 124–143.

[19] Barber, SA, Soil Nutrient Bioavailability. A Mechanistic Approach. Wiley-Interscience Publication, New York. 1984

[20] Gahoonia TS., Nielsen NE. Variation in Root Hairs of Barley Cultivars Doubled Phosphorus Uptake from Soil. Euphytica; 1997. 98(3): 177–182.

[21] Tinker, PB., Nye, PH. Solute Movement in the Rhizosphere. Oxford University Press, Oxford. 2000.

[22] de Willigan P., van Noordwijk M. Roots, plant production and nutrient use efficiency. PhD Thesis, Agricultural University Wageningen, The Netherlands. 1987.

[23] Raghothama KG. Phosphate Acquisition. Annu Rev Plant Physiol; 1999. 50: 665–693.

[24] Raghothama KG. Phosphate Transport and Signalling. Current Opinion in Plant Biology; 2000. 3:182–187.

[25] Mimura T. Regulation of Phosphate Transport and Homeostasis in Plants. Int Rev Cytol; 1999. 191: 149–200.

[26] Schachtman DP., Reid RJ., Ayling SM. Phosphorus Uptake by Plants: from Soil to Cell. Plant Physiol; 1998. 116: 447–453.

[27] Daram P., Brunner S., Amrhein N., Bucher M. Functional Analysis and Cell-Specific Expression of a Phosphate Transporter from Tomato. Planta; 1998. 206: 225-233.

[28] Smith SE., Read DJ. Mycorrhizal Symbiosis. Academic Press, Cambridge; 1997.

[29] Karandashov V., Nagy R., Wegmuller S., Amrhein N., Bucher M. Evolutionary Conservation of a Phosphate o` Transporter in the Arbuscular Mycorrhizal Symbiosis. Proc. Natl. Acad. Sci. USA; 2004. 101: 6285–6290.

[30] Harrison MJ. Signaling in the Arbuscular Mycorrhizal Symbiosis. Annu. Rev. Microbiol.; 2005. 59: 19–42.

[31] Bucher M. Functional Biology of Plant Phosphate Uptake at Root and Mycorrhiza Interfaces. New Phytol; 2007. 173: 11–26.

[32] Rausch C, Bucher M. Molecular Mechanisms of Phosphate Transport in Plants. Planta; 2002. 216(1): 23-37, DOI: 10.1007/s00425-002-0921-3

[33] Picault N., Hodges M., Palmieri L., Palmieri F. The Growing Family of Mitochondrial Carriers in *Arabidopsis*. Trends in Plant Science; 2004. 9: 138–146.

[34] Knappe S., Flugge UI., Fischer K. Analysis of the Plastidic Phosphate Translocator Gene Family in Arabidopsis and Identification of New Phosphate Translocator-Homologous Transporters, Classified by Their Putative Substrate-Binding Site. Plant Physiology; 2003. 131: 1178–1190.

[35] Grunwald U., Guo W., Fischer K., Isayenkov S., Ludwig-Müller J., Hause B., Yan X., Küster H., Franken P. Overlapping Expression Patterns And Differential Transcript Levels Of Phosphate Transporter Genes In Arbuscular Mycorrhizal, Pi-Fertilised And Phytohormone-Treated Medicago Truncatula Roots. Planta, 2009. 229(5): 1023-1034, DOI: 10.1007/s00425-008-0877-z

[36] Leggewie G., Wilmitzer L., Riesmeier JW. Two cDNAs From Potato Are Able to Complement a Phosphate Uptake-deficient Yeast Mutant: Identification of Phosphate Transporters from Higher Plants. Plant Cell; 1997. 9: 381.

[37] Okumura S., Mitsukawa N., Shirano Y. Shibata D. Phosphate Transporter Gene Family of *Arabidopsis thaliana*. DNA Res; 1998. 5: 261–269.

[38] Liu C., Muchal US., Mucatira U., Kononowicz AK. Raghothama KG. Tomato Phosphate Transporter Genes Are Differentially Regulated in Plant Tissues by Phosphorus. Plant Physiol.; 1998a. 116: 91-99.

[39] Liu H., Trieu AT., Blaylock LA. Harrison MJ. Cloning and Characterisation of Two Phosphate Transporters from Medicago Truncatular Roots: Regulation in Response to Phosphate and to Colonization by Arbuscular Mycorrhizal (AM) Fungi. Mol. Plant Microbe Interact.; 1998b. 11: 14- 22.

[40] Paszkowski U., Kroken S., Roux C., Briggs SP. Rice Phosphate Transporters Include an Evolutionarily Divergent Gene Specifically Activated in Arbuscular Mycorrhizal Symbiosis. Proc Natl Acad Sci USA; 2002. 99: 13324-13329.

[41] Nagy R., Vasconcelos M., Zhao S., McElver J., Bruce W., Amrhein N., Raghothama K. & Bucher M. Differential Regulation of Five Pht1 Phosphate Transporters from Maize (*Zea mays* L.). Plant Biology; 2006. 8: 186–197.

[42] Mudge SR., Rae AL., Diatloff E., Smith FW. Expression Analysis Suggests Novel Roles for Members of the Pht1 Family of Phosphate Transporters in Arabidopsis. Plant J.; 2002. 31: 341–353.

[43] Rae AL., Jarmey JM., Mudge SR. Smith FW. Overexpression of a High-affinity Phosphate Transporter in Transgenic Barley Plants Does not Enhance Phosphate Uptake Rates. Funct. Plant Biol.; 2004. 31(2): 141–148.

[44] Glassop D., Smith SE., Smith F W. Cereal Phosphate Transporters Associated with the Mycorrhizal Pathway of Phosphate Uptake into Roots. Planta; 2005. 222: 688-698.

[45] Balestrini R., Gomez-Ariza J., Lanfranco L., Bonfante P. Laser Microdissection Reveals that Transcripts for Five Plant and One Fungal Phosphate Transporter Genes are Contemporaneously Present in Arbusculated Cells. Mol Plant Microbe Interact; 2007. 20: 1055–1062.

[46] Wu Z., Zhao J., Gao R., Hu G., Gai J., Xu G., Xing H. Molecular Cloning, Characterization and Expression Analysis of Two Members of the Pht1 Family of Phosphate Transporters in Glycine max. PLoS ONE; 2011. 6(6): e19752. doi:10.1371/journal.pone.0019752.

[47] Mitsukawa N., Okumura S., Shirano Y., Sato S., Kato T., Harashima S., Shibata D. Overexpression of an Arabidopsis thaliana High-affinity Phosphate Transporter Gene in Tobacco Cultured Cells Enhances Cell Growth Under Phosphate-limited Conditions. Proc Natl Acad Sci USA; 1997. 94: 7098-7102.

[48] Shin H., Shin HS., Dewbre GR., Harrison MJ. Phosphate Transport in Arabidopsis: Pht1;1 and Pht1;4 Play a Major Role in Phosphate Acquisition from Both Low- and High-phosphate Environments. Plant J.; 2004. 39: 629–642.

[49] Marschner H. Mineral Nutrition of Higher Plants. Academic Press, San Diego, CA; 1995.

[50] Hamburger D., Rezzonico E., Petetot JMC., Somerville C. Poirier Y. Identification and Characterization of the Arabidopsis PHO1 Gene Involved in Phosphate Loading to the Xylem. Plant Cell; 2002. 14: 889-902.

[51] Jeschke W., Kirkby E., Peuke A., Pate J., Hartung W. Effects of P EFFICIENCY on Assimilation and Transport of Nitrate and Phosphate in Intact Plants of Castor Bean (Ricinus communis L.). J Exp Bot; 1997. 48: 75–91.

[52] Versaw WK, Harrisons MJ. A Chloroplast Phosphate Transporter, PHT2;1, Influences Allocation of Phosphate within the Plant and Phosphate Starvation Responses. Plant Cell; 2002. 14: 1751-1766.

[53] Williamson LC., Ribrioux SPCP., Fitter AH., Leyser HMO. Phosphate availability regulates root system architecture in Arabidopsis. Plant Physiol.; 2001. 29: 875–890.

[54] Thaler P, Pages L. Modeling the Influence of Assimilate Availability on Root Growth and Architecture. Plant Soil; 1998. 201: 307–320.

[55] Fitter A, Williamson L, Linkohr B, Leyser O. Root System Architecture Determines Fitness in an Arabidopsis Mutant in Competition for Immobile Phosphate Ions but not for Nitrate Ions. Proc. Biol. Sci.; 2002. 269: 2017–2022.

[56] Basirat M., Malboobi MA., Mousavi A., Asgharzadeh A., Samavat S. Effects of Phosphorous Supply on Growth, Phosphate Distribution and Expression of Transporter Genes in Tomato Plants. AJCS; 2011. 5(5): 537-543.

[57] Lynch JP., Brown KM. Topsoil Foraging – An Architectural Adaptation Of Plants To Low Phosphorus. Plant and Soil; 2001. 237(2): 225-237, DOI: 10.1023/A:1013324727040

[58] Neumann G., Martinoia E. Cluster Roots: An Underground Adaptation for Survival in Extreme Environments. Trends Plant Sci.; 2002. 7: 162–167.

[59] Lamont BB. Structure, Ecology and Physiology of Root Clusters: A Review. Plant and Soil; 2003. 248: 1–19.

[60] Shane MW., Lambers H. Cluster Roots: A Curiosity in Context. Plant and Soil; 2005. 274: 101–125.

[61] Lambers H., Shane MW. Role of Root Clusters in Phosphorus Acquisition and Increasing Biological Diversity in Agriculture. Wageningen UR Frontis Series; 2007. 21: 235.

[62] Peret B., Péret B., Svistoonoff S., Laplaze L.When Plants Socialize: Symbioses and Root Development. Annu. Plant Rev.; 2009. 37: 209–238.

[63] Smith SE., Smith FA. Roles of Arbuscular Mycorrhizas in Plant Nutrition and Growth: New Paradigms from Cellular to Ecosystem Scales. Annu. Rev. Plant Biol.; 2011. 62: 16.1–16.24.

[64] Forde B., Lorenzo H. The Nutritional Control of Root Development. Plant and Soil; 2001. 232: 51–68.

[65] Chevalier F. Pata M., Narcy P., Doumas P., Rossignol M. Effects of Phosphate Availability on the Root System Architecture: large-Scale Analysis of the Natural Variation between Arabidopsis Accessions. Plant Cell Environment; 2003. 26:1839–1850.

[66] Sánchez-Calderón L., López-Bucio J., Chacón-López A., Gutiérrez-Ortega A., Hernández-Abreu E., Herrera-Estrella L. Characterization of Low Phosphorus Insensitive Mutants Reveals a Crosstalk between Low Phosphorus Induced Determinate Root Development and the Activation of Genes Involved in the Adaptation of Arabidopsis to Phosphorus Deficiency. Plant Physiol.; 2006. 140: 879–889.

[67] Wu P., Ma L., Hou X., Wang M., Wu Y., Liu F., Deng X.W. Phosphate Starvation Triggers Distinct Alterations of Genome Expression in Arabidopsis Roots and Leaves. Plant Physiol.; 2003. 132: 1260–1271.

[68] Misson J., Raghothama KG., Jain A.,et al. A Genome-wide Transcriptional Analysis Using Arabidopsis thaliana Affymetrix Gene Chips Determined Plant Responses to Phosphate Deprivation. Proc. Natl. Acad. Sci. U.S.A.; 2005. 102: 11934–11939.

[69] Morcuende R. Bari R., Gibon Y., Zheng W., Pant BD., Blasing O., Usadel B.,Czechowski T., Udvardi MK., Stitt M., Schible W. Genome-wide Reprogramming of Metabolism and Regulatory Networks of Arabidopsis in Response to Phosphorus. Plant Cell Environ.; 2007. 30: 85–112.

[70] Thibaud M.C., Arrighi J.F., Bayle V., Chiarenza S., Creff A., Bustos R., Paz-Ares J., Poirier Y., Nussaume L. Dissection of Local and Systemic Transcriptional Responses to Phosphate Starvation in Arabidopsis. Plant J.; 2010. 64: 775–789.

[71] Bozzo GG., Raghothama KG., Plaxton WC. Structural and Kinetic Properties of a Novel Purple Acid Phosphatase from Phosphate-starved Tomato (*Lycopersicon esculentum*) Cell Cultures. Biochem J.; 2004a. 377: 419–428

[72] Veljanovski V., Vanderbeld B., Knowles VL., Snedden WA., Plaxton WC. Biochemical and Molecular Characterization of AtPAP26, A Vacuolar Purple Acid Phosphatase Up-regulated in Phosphate-deprived Arabidopsis Suspension Cells and Seedlings. Plant Physiol; 2006. 142: 1282–1293.

[73] Tran HT., Hurley BA. Plaxton WC. Feeding Hungry Plants: the Role of Purple Acid Phosphatases in Phosphate Nutrition. Plant Science; 2010a. 179: 14–27.

[74] Fang ZY., Shao C., Meng YJ., Wu P., Chen M. Phosphate Signaling in Arabidopsis and *Oryza Sativa*. Plant Sci.; 2009. 176: 170–180.

[75] Vance CP., Uhde-Stone C. Allan DL. P Acquisition and Use: Critical Adaptations by Plant for Securing a Non-renewable Resources. New Physiologist; 2003. 157:427-447.

[76] Jiang CF., Gao XH., Liao LL., Harberd NP., Fu XD. Phosphate starvation root architecture and anthocyanin accumulation responses are modulated by the gibberellin-DELLA signaling pathway in Arabidopsis. Plant Physiol. 2007; 145, 1460–1470.

[77] Lin WY., Lin SI., Chiou TJ. Molecular Regulators of Phosphate Homeostasis in Plants. J. Exp Bot; 2009. 60: 1427–1438.

[78] Yu B., Xu CC., Benning C. Arabidopsis Disrupted in SQD2 Encoding Sulfolipid Synthase is Impaired in Phosphate-limited Growth. Proc Natl Acad Sci USA; 2002. 99: 5732–5737.

[79] Li M., Qin C., Welti R., Wang X. Double Knockouts of Phospholipases Dz1 and Dz2 in Arabidopsis Affect Root Elongation During Phosphate Limited Growth but Do not Affect Root Hair Patterning. Physiol Plant; 2006. 140:761–770.

[80] Gaude N., Nakamura Y., Scheible WR., Ohta H., Dormann P. Phospholipase C5 (NPC5) is Involved in Galactolipid Accumulation During Phosphate Limitation in Leaves of Arabidopsis. Plant J.; 2008. 56: 28–39.

[81] Duff SMG., Moorhead GB., Lefebvre DD., Plaxton WC. Phosphate Starvation Inducible "bypasses" of Adenylate and Phosphate Dependent Glycolytic Enzymes in Brassica Nigra Suspension Cells. Plant Physiol; 1989. 90:1275–1278.

[82] Plaxton WC., Podesta FE. The Functional Organization and Control of Plant Respiration. Crit Rev Plant Sci; 2006. 25: 159–198.

[83] Gregory AL., Hurley BA., Tran HT., Valentine AJ., She YM., Knowles VL., Plaxton WC. In Vivo Regulatory Phosphorylation of the Phosphoenolpyruvate Carboxylase AtPPC1 in Phosphate-starved *Arabidopsis thaliana*. Biochem J.; 2009. 420: 57–65.

[84] Tran HT., Qian W., Hurley BA., She YM., Wang D., Plaxton WC. Biochemical and Molecular Characterization of AtPAP12 and AtPAP26: The Predominant Purple Acid Phosphatase Isozymes Secreted by Phosphate-starved Arabidopsis thaliana. Plant Cell Environ; 2010b. 33: 1789–1803.

[85] Plaxton WC., Tran HT. Metabolic Adaptations of Phosphate-Starved Plants. Plant Physiology; 2011. 20(156): 1006–1015.

[86] Ticconi CA., Abel S. Short on Phosphate: Plant Surveillance and Countermeasures. Trends Plant Sci; 2004. 9: 548–555.

[87] Nilsson L., Muller R., Nielsen TH. Dissecting the Plant Transcriptome and the Regulatory Responses to Phosphate Deprivation. Physiol Plant; 2010. 139: 129–143.

[88] Liang C., Tian J., Lam HM., Lim BL., Yan X., Liao H. Biochemical and Molecular Characterization of PvPAP3, A Novel Purple Acid Phosphatase Isolated from Common Bean Enhancing Extracellular ATP Utilization. Plant Physiol; 2010. 152: 854–865.

[89] Tarafdar JC., Claasen N. Organic Phosphorus Compounds as a Phosphorus Source for Higher Plants through the Activity Of Phosphatases Produced By Plant Roots and Microorganisms. Biol. Fert. Soils; 1988. 5: 308-312.

[90] Tarafdar JC., Rao AV., Kumar P. Role of Phosphate Producing Fungi on the Growth and Nutrition of Clusterbean (*Cyamopsis tetragonoloba* (L.) Taub.). Journal of Arid Environments; 1995. 29(3): 31-337, ISSN 0140-1963.

[91] Malboobi MA., Owlia P., Behbahani M., Sarokhani E., Moradi S., Yakhchali B., Deljou A., Morabbi K. Solubilization of Organic and Inorganic Phosphates by Three Efficient Soil Bacterial Isolates. World Journal of Microbiology Biotechnology; 2009a. 25: 1471-1477.

[92] Duff S MG., Sarath G., Plaxton WC. The Role of Acid Phosphatase In Plant Phosphorus Metabolism. Physiol. Plant.; 1994. 90: 791–800.

[93] Yan F., Zhu YY., Muller C., Zorb C., Schubert S. Adaptation of H+-pumping and Plasma Membrane H+ ATPase Activity in Proteoid Roots of white Lupin under Phosphate Deficiency. Plant Physiol; 2002. 129: 50–63.

[94] Diatloff E., Roberts M., Sanders D., Roberts SK. Characterization of Anion Channels in the Plasma Membrane of Arabidopsis Epidermal Root Cells and the Identification of a Citrate-permeable Channel Induced by Phosphate Starvation. Plant Physiol; 2004. 136: 4136–4149.

[95] Hammond JP., Bennett MJ., Bowen HC., Broadley MR., Eastwood DC., May ST., Rahn C., Swarup R., Woolaway KE., White PJ. Changes in Gene Expression in Arabidopsis Shoots during Phosphate Starvation and the Potential for Developing Smart Plants. Plant Physiol; 2003. 132: 578-586.

[96] Uhde-Stone C., Zinn KE., Ramirez-Yanez M., Li A., Vance CP., Allan DL. Nylon Filter Arrays Reveal Differential Gene Expression in Proteoid Roots of White Lupin in Response to Phosphorus Deficiency. Plant Physiol.; 2003. 131: 1064–1079.

[97] Wasaki J., Yonetani R., Kuroda S., Shinano T., Yazaki J., Fujii F., Shimbo K., Yamamoto K., Sakata K., Sasaki T., et al. Transcriptomic Analysis of Metabolic Changes by Phosphorus Stress in Rice Plant Roots. Plant, Cell and Environment; 2003b. 26: 1515–1523.

[98] Wasaki J., Shinano T., Onishi K., Yonetani R., Yazaki J., Fujii F., Shimbo K., Ishikawa M., Shimatani Z., Nagata Y., et al. Transcriptomic Analysis Indicates Putative Metabolic Changes Caused by Manipulation of Phosphorus Availability in Rice Leaves. J Exp Bot; 2006. 57: 2049–2059.

[99] Graham MA., Ramirez M., Valdes-Lopez O., Lara M., Tesfaye M., Vance CP., Hernandez G. Identification of Candidate Phosphorus Stress Induced Genes in Phaseolus Vulgaris through Clustering Analysis Across Several Plant Species. Funct Plant Biol; 2006. 33: 789–797.

[100] Wang X., Yi K., Tao Y., Wang F.,Wu Z., Jiang D., Chen X., Zhu L., Wu P. Cytokinin Represses Phosphate-starvation Response through Increasing of Intracellular Phosphate Level. Plant Cell Environ.; 2006. 29: 1924–1935.

[101] Hammond JP., Philip J. White Sugar Signalling in Root Responses to Low Phosphorus Availability. Plant Physiology; 2011. 156: 1033–1040.

[102] Muller R., Morant M., Jarmer H., Nilsson L., Nielsen TH. Genome-wide Analysis of the Arabidopsis Leaf Transcriptome Reveals Interaction of Phosphate and Sugar Metabolism. Plant Physiol; 2007. 143: 156–171.

[103] Lin SI., Chiang SF., Lin WY., Chen JW., Tseng CY., Wu PC., Chiou TJ. Regulatory Network of MicroRNA399 and PHO2 by Systemic Signalling. Plant Physiol; 2008. 147: 732–746.

[104] Bustos R., Castrillo G., Linhares F., Rubio V., Puga MI., Perez-Perez J., Solano R., Leyva A., Paz-Ares J. A Central Regulatory System Lergely Controls Transcriptional Activation and Repression Responses to Phosphate Starvation in Arabidopsis. PLoS Genet; 2010. 6(9): e1001102. doi:10.1371/journal.pgen.1001102.

[105] Rubio V., Linhares F., Solano R., Martin, AC., Iglesias J., Leyva A., Paz-Ares J. A Conserved MYB Transcription Factor Involved in Phosphate Starvation Signaling both in Vascular Plants and in Unicellular Algae Genes Dev.; 2001. 15;2122–2133.

[106] Hsieh LC., Lin SI., Shih AC., Chen JW., Lin WY., Tseng CY., Li WH., Chiou TJ. Uncovering Small RNA-mediated Responses to Phosphate Deficiency in Arabidopsis by Deep Sequencing. Plant Physiol; 2009. 151: 2120–2132.

[107] Fujii H., Chiou TJ., Lin SI., Aung K., Zhu JK. A miRNA Involved in Phosphate-Starvation Response in Arabidopsis. Curr Biol; 2005. 15: 2038–2043.

[108] Pant BD., Musialak-Lange M., Nuc P., May P., Buhtz A., Kehr J., Walther D., Scheible WR. Identification of Nutrient-responsive Arabidopsis and Rapeseed MicroRNAs by Comprehensive Real-time Polymerase Chain Reaction Profiling and Small RNA Sequencing. Plant Physiol; 2009. 150:1541–1555.

[109] Doerner P. Phosphate Starvation Signaling: A Threesome Controls Systemic Pi-homeostasis. Curr Opin Plant Biol; 2008. 11: 536–540.

[110] Pant BD., Buhtz A., Kehr J., Scheible WR. MicroRNA399 Is a Long Distance Signal for the Regulation of Plant Phosphate Homeostasis. Plant J.; 2008. 53: 731–738.

[111] Liu C., Muchhal US., Raghothama KG. Differential Expression of TPS11, A Phosphate Starvation-induced Gene in Tomato. Plant Mol Biol; 1997. 33: 867–874.

[112] Burleigh SH., Harrison MJ. A Novel Gene Whose Expression in Medicago Truncatula Roots Is Suppressed in Response to Colonization by Vesicular-arbuscular Mycorrhizal (VAM) Fungi and to Phosphate Nutrition. Plant Mol Biol; 1997. 34: 199–208.

[113] Franco-Zorrilla JM., Valli A., Todesco M., Mateos I., Puga MI., Rubio- Somoza I., Leyva A., Weigel D., Garcıa JA., Paz-Ares J. Target Mimicry Provides a New Mechanism for Regulation of MicroRNA Activity. Nat Genet; 2007. 39: 1033–1037.

[114] Zhou J., Jiao F.,Wu Z., Li Y.,Wang X., He X., Zhong W., Wu P. OsPHR2 Is Involved in Phosphate-starvation Signalling and Excessive Phosphate Accumulation in Shoots of Plants. Plant Physiol; 2008. 146: 1673–1686.

[115] Rouached H., Arpat AB. Poirier Y. Regulation of Phosphate Starvation Responses in Plants: Signaling Players and Cross-Talks. Molecular Plant; 2010. 3(2): 288–299.

[116] Van Straaten P. Rocks for Crops: Agrominerals of Sub-Saharan Africa. ICRAF, Nairobi, Kenya; 2002. p7-24.

[117] Morse D. Phosphate Rocks. US Department of the Interior Bureau of Mines Annual Report; 1992.

[118] Cisse L., Mrabet T. World Phosphate Production: Overview and Prospects; Phosphorus Research Bulletin; 2004. 15: 21-25.

[119] FAO. Current World Fertilizer Trends and Outlook to 2011/12. FAO; 2008. Rome.

[120] FAO. Use of Pi Rocks for Sustainable Agriculture. FAO; 2004. Rome.

[121] FAO. Current World Fertilizer Trends and Outlook to 2015. FAO; 2011. Rome.

[122] IMPHOS. Addressing Environmental Issues Associated with Phosphate. IMPHOS phosphate newsletter; 2009. 26. Available: www.imphos.org (accessed 18 May 2012).

[123] IMPHOS. Phosphate Trent and Outlook. IMPHOS phosphate newsletter; 2007. 25. Available: www.imphos.org (accessed 18 May 2012).

[124] Schipper WJ., Klapwijk A., Potjer B., Rulkens WH., Temmink BG., Kiestra FDG., Lijmbach ACM. Phosphate Recycling in Phosphorous Industry. Phosphorous Research Bulletin; 2004. 15: 47-51.

[125] Ivanova R., Bojinova D., Nedialkova K. Rock Phosphate Solubilisation by Soil Bacteria.J of the University of Chemical Technology and Metallurgy; 2006. 41(3): 297-302.

[126] Malboobi MA., Behbahani M., Madani H., Owlia P., Deljou A., Yakhchali B., Moradi M., Hassanabadi H. The Performance of Potent Phosphate Solubilizing Bacteria in Potato Rhizosphere. World Journal of Microbiology Biotechnology; 2009b. 25: 1479-1484.

[127] Johnston AE., Ehlert PAI., Kueche M., Amar B., Jaggard KW., Morel C. The Effect of Phosphate Fertilizer Management Strategies on Soil Phosphorus Status and Crop Yields in Some European Countries. IMPHOS; 2001.

[128] Havlin JL., Beaton JD., Tisdale SL., Nelson WL. Soil Fertility and Fertilizers: An Introduction to Nutritional Management. New Jersey: Pearson; 2005.

[129] Cordell D., Drangert J-O., White S. The Story of Phosphorus: Global Food Security and Food for Thought. Global Environmental Change; 2009. 19: 292-305.

[130] IFDC, World Phosphate Rock Reserves and Resources, 2010, IFDC report.

[131] Cisse L., Amar B. The Importance of Phosphatic Fertilizer for Increased Crop Production in Developing Countries. In: Proceedings of the A6th International Annual conference, Cairo, Egypt; 2000.

[132] Stewart W., Hammond L. Kauwenbergh SJV. Phosphorus as a Natural Resource. Phosphorus: Agriculture and the Environment, Agronomy Mono-graph No.46. Madison, American Society of Agronomy. Crop Science Society of America. Soil Science Society of America; 2005.

[133] EPA. Fertilizer and Fertilizer Production Wastes. 2012; http://www.epa.gov/radiation/tenorm/fertilizer.html (accessed 18 May 2012).

[134] Turner BL., Frossard E. Baldwin DS. Organic Phosphorus in the Environment. CAB International, Wallingford, UK; 2005. p 432.

[135] Gyaneshwar P., Kumar GN., Parekh LJ., Poole PS. *Role of Soil Microorganisms in Improving P Nutrition of Plants*. Plant and Soil; 2002. 245 83-93.

[136] Yadav KS., Dadarwal KR. In Biotechnological Approaches in Soil Microorganisms for Sustainable Crop Production (ed. Dadarwal K.R), Scientific Publishers. Jodhpur; 1997. p293–308.

[137] Kang SC., Ha CG., Lee TG. Maheshwari DK. Solubilization of Insoluble Inorganic Phosphates by a Soil-Inhabiting Fungus Fomitopsis sp. PS 102. Current Science; 2002. 82(4): 439-442.

[138] Subbarao WS. Pi Solubilizing Microorganism in: Biofertilizer in Agriculture; 1988. p133-142.

[139] Arpana N., Kumar SD., Prasad TN. Effect of Seed Inoculation, Fertility and Irrigation on Uptake of Major Nutrients and Soil Fertility Status after Harvest of Late Sown Lentil. Journal of Applied Biology; 2002. 12(1/2): 23-26.

[140] Hodge A., Berta G., Doussan C., Merchan F., Crespi M. Plant Root Growth, Architecture and Function. Plant Soil; 2009. 321: 153-137.

[141] Mehrvarz S., Chaichi MR. Effect of Pi Solubilizing Microorganisms and Phosphorus Chemical Fertilizer on Forage and Grain Quality of Barely (*Hordeum vulgare* L.). American-Eurasian Journal of Agricultural & Environmental Science; 2008. 3(6): 855-860.

[142] Shen J., Yuan L., Zhang J., Li H., Bai Z., Chen X., Zhang W., Zhang F. Phosphorus Dynamics: From Soil to Plant. Plant Physiology; 2011. 156: 997–1005.

[143] Wang X., Yan X., Liao H. *Genetic Improvement for Phosphorus Efficiency in Soybean*: A *Radical Approach. Annals of Botany; 2010.* 106: 215-222.

[144] Rengel Z., Marschner P. Nutrient Availability and Management in the Rhizosphere: Exploiting Genotypic Differences. New Phyt; 2005. 168: 305-312.

[145] Fageria NK., Baligar VC., Li YC. The Role of Nutrient Efficient Plants in Improving Crop Yields in the Twenty First Century. Journal of Plant Nutrition; 2008. 31: 1121–1157.

[146] Miyasaka C., Habte M. Plant Mechanisms and Mycorrhizal Symbioses to Increase Phosphorus Uptake Efficiency. Communications in Soil Science and Plant Analysis; 2001. 32: 1101-1147.

[147] Zhao J., Fu J., Liao H., He Y., Nian H., Hu Y., Qiu L , Dong Y., Yan X. Characterization of Root Architecture in an Applied Core Collection for Phosphorus Efficiency of Soybean Germplasm, Chin. Sci. Bull; 2004. 49: 1611–1620.

[148] Miklas PN., Kelly JD., Beebe SE.Blair MW. Common Bean Breeding for Resistance Against Biotic and Abiotic Stresses, from Classical to MAS Breeding. Euphytica; 2006. 147: 105–131.

[149] Doerge R. Mapping and Analysis of Quantitative Trait loci in Experimental Populations. Nature Reviews Genetics; 2001. 3: 43–52.

[150] Ramaekers L., Remans R., Rao IM., Blair MW., Vanderleyden J. Strategies for Improving Phosphorus Acquisition Efficiency of Crop Plants. Field Crops Research; 2010. 117: 169-176.

[151] Lopez-Bucio J., de la Vega OM., Guevara-Garcia A., Herrera-Estrella L. Enhanced Phosphorus Uptake in Transgenic Tobacco Plants That Overproduce Citrate. Nature; 2000. 18: 450–453.

[152] Zimmermann P., Zardi G., Lehmann M., Zeder C., Amrhein N., Frossard E., Bucher M. Engineering the Root–soil interface Via Targeted Expression of a Synthetic Phytase Gene in Trichoblasts. Plant Biotechnol. J.; 2003. 1: 353–360.

[153] George TS., Richardson AE., Hadobas PA. Simpson RJ. Characterization of Transgenic *Trifolium subterraneum* L. which Expresses phyA and Releases Extra-cellular Phytase, Growth and Phosphorus Nutrition in Laboratory Media and Soil. Plant Cell Environ.; 2004. 27: 1351–1361.

[154] George TS., Simpson RJ., Hadobas PA. Richardson AE. Expression of a Fungal Phytase Gene in Nicotiana tabacum Improves Phosphorus Nutrition of Plants Grown in Amended Soils. Plant Biotechnol. J.; 2005. 3: 129–140.

[155] Lung S., Chan W., Yip W., Wang L., Young EC., Lim BL. Secretion of Beta-propeller Phytase from Tobacco and Arabidopsis Roots Enhances Phosphorus Utilization. Plant Sci.; 2005. 169: 341–349.

[156] Li G., Yang S., Li M., Qiao Y., Wang J. Functional Analysis of an Aspergillus ficuumphytase Gene in Saccharomyces cerevisiae and its Root-specific, Secretory Expression in Transgenic Soybean Plants. Biotechnol. Lett.; 2009. 31: 1297–1303.

[157] Ma X., Wright E., Ge Y., Bell J., Yajun X., Bouton JH. Wang Z. Improving Phosphorus Acquisition of white Clover (*Trifolium repens* L.) by Transgenic Expression of Plant-derived Phytase and Acid Phosphatase Genes. Plant Sci.; 2009. 176: 479–488.

[158] Wasaki J., Maruyama H., Tanaka M., Yamamura T., Dateki H., Shinano T., Ito S., Osaki M. Overexpression of the LASAP2 Gene for Secretory Acid Phosphatase in White Lupin Improves the Phosphorus Uptake and Growth of Tobacco Plants. Soil Sci. Plant Nutr.; 2009. 55: 107–113.

[159] Zamani K., Sabet MS., Lohrasebi T., Mousavi A., Malboobi MA. Improved Phosphate Metabolism and Biomass Production by Overexpression of AtPAP18 in Tobacco. Biologia; 2012. 67/4:713-720.

[160] Park MR., Baek SH., de los Reyes BG. Yun SJ. Overexpression of a High-affinity Phosphate Transporter Gene from Tobacco (NtPT1) Enhances Phosphate Uptake and Accumulation in Transgenic Rice Plants. Plant Soil.; 2007. 292(1–2): 259–269.

[161] Tesfaye M., Temple SJ., Allan DL., Vance CP. Samac DA. Overexpression of Malate Dehydrogenase in Transgenic Alfalfa Enhances Organic Acid Exudation Synthesis and Confers Tolerance to Aluminum. Plant Phys.; 2001. 127: 1836–1844.

[162] Yang H., Knapp J., Koirala P., Rajagopal D., Ann Peer W., Silbart LK., Murphy A., Gaxiola R. Enhanced Phosphorus Nutrition in Monocots and Dicots Overexpressing a Phosphorus-responsive Type I H^+-pyrophosphatase. Plant Biotechnol. J.; 2007. 5: 735–745.

[163] Delhaize E., Taylor P., Hocking PJ., Simpson RJ., Ryan PR., Richardson AE. Transgenic Barley (*Hordeum vulgare* L.) Expressing the Wheat Aluminium Resistance Gene (TaALMT1) Shows Enhanced Phosphorus Nutrition and Grain Production When Grown on an Acid Soil. Plant Biotechnol. J.; 2009. 7(5): 391–400.

[164] Mohan J., Suprasanna P. Induced Mutations for Enhancing Nutrition and Food Production. Geneconserve; 2011. 40: 201–215.

[165] Liu J., Li Y., Tong Y., Gao J., Li B., Li J., Li Z. Chromosomal Location of Genes Conferring the Tolerance to Pi Starvation Stress and Acid Phosphatase (APase) Secretion in the Genome of Rye (Secale L.) Plant and Soil; 2001. 237: 267–274.

[166] Ahmed S. Agriculture-Fertilizer Interface in Asia-Issues of Growth and Sustainability. New Delhi. Oxford and IBH Publ. Co.; 1995

The Plant Nutrition from the Gas Medium in Greenhouses: Multilevel Simulation and Experimental Investigation

A.V. Vakhrushev, A.Yu. Fedotov, A.A. Vakhrushev,
V.B. Golubchikov and E.V. Golubchikov

Additional information is available at the end of the chapter

1. Introduction

It is extremely important to reduce the action of man-made factors on the environment and vegetables and crops grown in greenhouses. On the one hand, the whole year round, greenhouse technologies provide the production of vegetables and other crops, which contain micro- and macroelements necessary for the vital activity of human being. On the other hand, the use of the technologies implies the use of mineral fertilizers for extra root and foliar nutrition in order to intensify the production. However, despite the use of drop irrigation and small doses of mineral feeding, the concentrations of various heavy metal compounds, radioactive and poisonous substances in the soil of greenhouses reach the limit values after 2–3 years of its service. Then the exhausted soil should be disposed and replaced by new soil, and this is hundreds and thousands of tons.

At the same time, the problem of the neutralization of the used soil arises, since it cannot simply be stored in the open air due to the formation of the dust fractions of the above compounds. In the rain and during snow melting, water-soluble substances penetrate into soil and water sources together with run-off water. From the aforesaid it follows that searching new ways for the nutrition of plants, which allow reducing soil contamination, is urgent. In this regard, the technologies for plant nutrition from the gas phase are quite promising, since the plants receive significant portion of nutrients (up to 80%) through their foliage and stems. It should be noted that controlled gas media have already been used for quite a long time for storing vegetables and fruit. The method is based on the formation of special concentrations of nitrogen, oxygen, carbon dioxide and water vapours in order to reduce the intensity of respiration and metabolism.

Join Stocks Company Nord, (Perm, Russia) has developed a new nanobiotechnology involving nanoparticles for the production of ecologically pure vegetables in greenhouses.

The main feature of the method is the generation of a special controlled gas medium containing nanoparticles of the main macro- and microelements, which freely penetrate into the foliage and stems of crops providing their metabolic activity. The source of the gas medium is the products of self-spreading high-temperature synthesis [1–5].

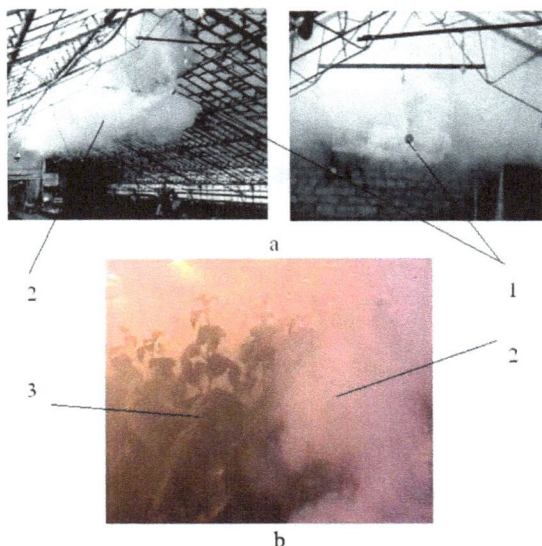

Figure 1. The diagram of the use of gas-generator (1): a) the general view of an industrial greenhouse; b) nanoaerosol gas medium (2) surrounding greenhouse plants (3).

Figure 1 demonstrates the use of a large-size gas-generator for the special controlled gas media (CGM) formation in an industrial greenhouse. CGM is complex media with specially selected concentrations of inorganic compounds in the form of nanoparticles for protecting living systems. They are used for the foliar nutrition of plants with macro- and microelements, the control of living system diseases, the protection of plants against frosts, the seed treatment, etc. The method comprises the creation of a controlled gas medium containing nanoparticles (Figure 2) which can easily penetrate into foliage and stems of plants providing their metabolic activity. The results of the application of the above technology show the absence of carcinogens, heavy metals and other dangerous substances both in foliage and fruits.

The use of this technology for growing various vegetables and crops in greenhouses of Russia, Byelorussia, the Ukraine, China and another countries shows that the controlled gas medium positively influences the metabolism of crops. It increases their productivity, improves their taste properties, reduces the content of harmful substances in them (for example, the content of nitrates is reduced by a factor of ten) and extends the vegetation period, etc. In addition, the use of such a technology cuts the amount of mineral fertilizers applied for the root systems in half. The experimental investigations show that the concentrations of useful substances in a crop leaf increase in two or more times after two-hour exposure to the "nutrient" gas medium.

It should be noted that the high efficiency of the technology of the plant nutrition from the gas medium determines a pressing need to establish certain theoretical aspects and kinetics of the processes taking place when a plant interacts with the substances of the gas medium. It is also necessary to establish the mechanisms of the nanoparticle formation and the evolution regularities for the dynamic interaction processes of condensed nanosystems and plants. In this connection, the experimental investigations and mathematical modeling of the above processes is rather urgent since it allows efficient and many-sided studying the behavior of nanosystems. The analyses of these problems were the purposes of this investigation. Research of processes of occurrence and movement of nanoparticles, being a base element of technology is especially important.

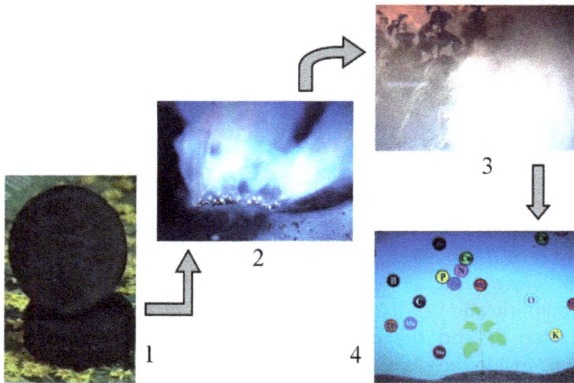

Figure 2. Main steps of biological processes of plant nutrition from controlled gas medium containing nanoparticles: 1 – generation sources, 2 – burning generation sources, 3 – the gas mixture movement from its generation source over the space where the plants are 4 – plant nutrition from gas medium.

2. The task definition. Theoretical base of multilevel simulation and experimental investigations

There are 8 main stages of the processes of the plant nutrition from the gas medium, which follow each other in time (Figure 3):

1. Combustion of a highly condensed system and gas mixture motion from a source, that generates it, in space where the plants are situated.

2. The building-up of molecular formations containing the above elements.

3. Merging atoms and molecules into nanoparticles at cooling the gas mixture to normal temperature.

4. The movement of the nanoparticles formed in the gas medium.

5. The sedimentation of the nanoparticles on the plants.

6. The penetration of the nanoparticles inside the plant from its surface.

7. The movement of the nanoparticles inside the plant along its microchannels and through its pores, and through its cellular and intercellular spaces.

8. The decomposition of the nanoparticles inside the plant into the component molecules and atoms.

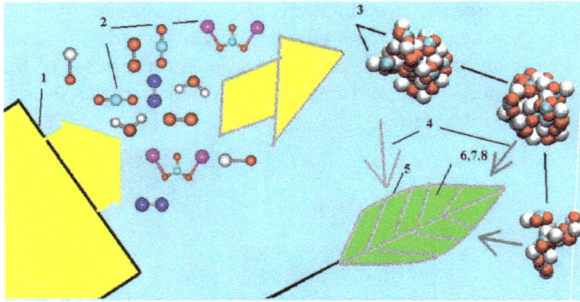

Figure 3. The main steps of the processes of the plant nutrition from the gas phase.

It should be noted that the tasks of modelling the processes of the plant nutrition from the gas phase are multilevel tasks, and each level requires special physical and mathematical approaches. Let us consider the methods for modelling the above successive processes. The calculation of the configuration of the molecular formations containing nitrogen, potassium and phosphorus (step 2) and the decomposition of the nanoparticles into component molecules and atoms inside the plant (step 8) require "ab initio" calculations. The simulation of the process of the molecule joining-up in nanoparticles at cooling the gas mixture to normal temperature (step 3) can be carried out with the help of the molecular dynamics methods. The calculation of the processes of the nanoparticle movement in the gas mixture, the sedimentation of the nanoparticles on the plant surface, the penetration of the nanoparticles inside the plant from its surface and the movement of the nanoparticles inside the plant along its microchannels, through its pores, cellular and intercellular spaces (steps 4–7) requires mesodynamics.

It is important that some of the processes could be considered within the framework of the continuum mechanics. They are the processes of the gas mixture movement over the space where the plants are (step 1) and steps 4 and 5 of the model under discussion as well. The theoretical base of multilevel simulation of the formation, movement, integration and disintegration of nanoparticles' systems in gas medium depending on the thermodynamic conditions of the medium are presented. The modelling is performed with the use of the methods of quantum mechanics, molecular dynamics and mesodynamics. As follows from the above-described process of application of nanoaerosols, nanoparticles are formed in a complex gas medium. Usually such problems are modeled by the molecular dynamics tools [6, 7]. However, their solution only within the framework of molecular dynamics requires much time and computational power. So, for example, due to a small mass of interacting atoms in order the integration scheme is stable the integration step should be taken of about 10^{-15} s, which leads to slow integration of the equations of molecular dynamics. Moreover, a collective behavior of atoms, molecules, and nanoparticles is observed at different stages of application of nanoaerosols. This causes a multilevel character of the simulation problems to which there correspond different physical and mathematical approaches [8, 9]. The main problems of such simulation are:

- multilevel related problems;
- large number of variables;
- change of scales both in space and time;
- characteristic times of the processes at different levels differ by orders of magnitude;
- change of problem variables at different levels of simulation;
- coordination of the boundary conditions in transition from one level of simulation to another with change in the problem variables;
- stochastic behavior of nanosystems.

This leads to the necessity of using different mathematical approaches and models at different levels of formation and application of nanoaerosols. Calculation of the configuration of molecular formations that constitute nanoparticles requires "ab initio" calculations, i.e., quantum-mechanical methods of simulation [10]. These methods model nanoobjects most fully and accurately, with account for quantum effects. However, they require huge computational resources. At present the application of quantum-mechanical methods of calculation to nanosystems is limited by a number of atoms that enter the nanosystem not higher than 1000–2000 atoms. The process of aggregation of molecules to nanoparticles can be calculated by the methods of molecular dynamics [11–13]. These methods allow one to consider the systems that involve 10 and more millions of atoms but they do not take into account quantum phenomena.

Calculation of the motion processes of nanoparticles in a gas mixture and their aggregation is the problem of mesodynamics [14, 15]. A characteristic feature of mesodynamics is simultaneous use of the methods of molecular and classical dynamics. One should also mention that a number of processes, especially those occurring at the completing stages of nanoaerosol technologies, can be considered within the framework of continuum mechanics. Each method mentioned has its advantages and limitations. The use of one or another method of simulation or their combination as applied to specific problems of aerosol nanotechnology depends on the required accuracy of calculations. The main reasons and aims of transition from one method of simulation to another at different stages of nanoaerosol technologies are:

1. Decrease of the number of nanosystem variables due to transition from calculation of motion of separate atoms to the analysis of motion of nanoparticles or groups of them.
2. Decrease of the number of bonds between different elements of the nanosystem due to a decrease of the number of "nearest neighbors" occurring in the region of interaction of elementary cells of the nanosystem.
3. Increase of the computational size of the nanosystem due to enlargement of the elementary reference cell.
4. Broadening of the scale of nanosystem calculation in time due to an increase of the time step of integration of the nanosystem equations.

We consider the above-presented methods of simulation step by step and indicate the methods of the correlated application of them to different stages of the problems of aerosol nanotechnologies.

2.1. Quantum-chemical methods of simulation

Quantum-chemical methods of simulation use the quantum mechanics tools and are based on the solution of the Schrödinger equation [16]. In using this method one considers a full electronic and atomic structure of objects (atoms, molecules, ions), takes into account a detailed configuration of all electron clouds. In this case, complete information on the behavior of the considered system of N particles (nuclei of atoms and electrons) in the system of coordinates x_1, x_2, x_3 is determined by the wave function Ψ that depends on $3N$ coordinates of all particles of the atomic system, projections of their spines on the axis and time t

$$\Psi = \Psi\left(x_{11}, x_{21}, x_{31}, s_{x_{31}}, x_{12}, x_{22}, x_{32}, s_{x_{32}}; \ldots; x_{1N}, x_{2N}, x_{3N}, s_{x_{3N}}, t\right). \tag{1}$$

Variation of the wave function Ψ in space and time is determined by the Schrödinger wave equation

$$\mathrm{i}\,h\frac{\partial \Psi}{\partial t} = \widehat{\mathbf{H}}\Psi, \tag{2}$$

where h is the Planck constant; $\mathrm{i} = \sqrt{-1}$;

$$\widehat{\mathbf{H}} = \sum_{k=1}^{N}\left\{-\frac{h^2}{2m_k}\nabla_k^2 + U_k\left(x_{1k}, x_{2k}, x_{3k}, s_{x_{3k}}, t\right)\right\} + $$

$$+ \sum_{k=1}^{N}\sum_{j\neq k=1}^{N} U_{kj}\left(x_{1k}, x_{2k}, x_{3k}, s_{x_{3k}} x_{1j}, x_{2j}, x_{3j}, s_{x_{3j}}, t\right) \tag{3}$$

where $\widehat{\mathbf{H}}$ is the Hamilton operator (an analogue of the classical Hamilton function) for the considered atomic system, $U_k\left(x_{1k}, x_{2k}, x_{3k}, s_{x_{3k}}, t\right)$ is the potential of the external field that acts on the kth particle, and $U_{kj}\left(x_{1k}, x_{2k}, x_{3k}, s_{x_{3k}} x_{1j}, x_{2j}, x_{3j}, s_{x_{3j}}, t\right)$ is the potential of interaction between the particles j and k.

Equation (3) holds when two conditions are met: elementary particles do not disappear and no new elementary particles appear in the evolution process of the nanosystem; the velocity of elementary particles is small compared with the velocity of light. For an aerosol nanosystem containing N^a atomic nuclei and $N^e l$ electrons the Hamilton operator (in the stationary case without account for spines of electrons) has the form

$$\widehat{\mathbf{H}} = -\widehat{\mathbf{K}}_{N^a} - \widehat{\mathbf{K}}_{N^{el}} + U_{N^a N^a} + U_{N^a N^{el}} + U_{N^{el} N^{el}}, \tag{4}$$

where

$$\widehat{\mathbf{K}}_{N^a} = -\frac{h^2}{2m_i}\sum_{i=1}^{N^a}\nabla_k^2 \tag{5}$$

is the operator of the kinetic energy of atomic nuclei,

$$\hat{K}_{N^{el}} = -\frac{h^2}{2m_{el}} \sum_{i=1}^{N^{el}} \nabla_k^2 \tag{6}$$

is the operator of the kinetic energy of electrons,

$$U_{N^a N^a} = e^2 \left(\sum_{l=1}^{N^a} \sum_{k=1, k \neq l}^{N^a} \frac{Z_k Z_l}{r_{kl}} \right) \tag{7}$$

is the potential energy of interaction of atomic nuclei,

$$U_{N^a N^{el}} = e^2 \left(\sum_{k=1}^{N^a} \sum_{i=1}^{N^{el}} \frac{Z_k}{r_{ik}} \right) \tag{8}$$

is the potential energy of interaction between the nuclei of atoms and electrons,

$$U_{N^{el} N^{el}} = e^2 \left(\sum_{i=1}^{N^{el}} \frac{1}{r_{ij}} \right) \tag{9}$$

is the potential energy of interaction of electrons; e is the electron charge, m_{el} is the electron mass, Z_k, Z_l is the number of protons in the atomic nucleus, r_{kl}, r_{ik}, r_{ij} are the distances between the atomic nuclei, between the nuclei of atoms and electrons, and between the electrons, respectively. With account for (5)–(9) the Hamilton operator (3) takes the form

$$\hat{H} = -\frac{h^2}{8\pi m_{el}} \sum_{i=1}^{N^{el}} \nabla_k^2 - \frac{h^2}{8\pi m_i} \sum_{i=1}^{N^a} \nabla_k^2 - e^2 \left(\sum_{k=1}^{N^a} \sum_{i=1}^{N^{el}} \frac{Z_k}{r_{ik}} - \sum_{i=1}^{N^{el}} \frac{1}{r_{ij}} - \sum_{l=1}^{N^a} \sum_{k=1, k \neq l}^{N^a} \frac{Z_k Z_l}{r_{kl}} \right). \tag{10}$$

In the general case, the Schrödinger equation does not have an analytical solution and is usually solved by numerical methods. Here, the main specific feature of the Schrödinger equation is that the function exists only for the entire system as a whole. An individual particle (atomic nucleus or electron) cannot be in the state which can be described by the wave function for a separate particle and the common wave function cannot be presented as a product of wave functions of separate particles. Therefore, direct solution of the Schrödinger equation requires huge computational capacities of computers. For the considered nanosystem consisting of atomic nuclei N^a and N^{el} electrons, the Schrödinger function should be determined in the configuration space $3N^a N^{el}$ dimensions. With a number of integration points over each dimension equal to 10^n, summation should be made by 10^f elements of the volume of the configuration space, where $f = 3nN^a N^{el}$. It is obvious that this is a very large number even for a small object. For example, for a nanoparticle containing 100 atoms and 100 electrons for 100 integration points over each coordinate the number of elements of the volume of the configuration space is 10^{60000}. At present the main efforts of researches are directed to the development of approximate calculation methods which will be considered during calculation of specific nanosystems.

2.2. Methods of molecular dynamics

Molecular dynamics, as applied to nanoaerosol technologies, can be successfully used for formation of the spatial structure and molecular atomic composition of nanoparticles which condense in gas mixture cooling. The molecular dynamics method was developed in the works by Hill, Dostovsky, Hughes, Ingold, Westheimer, Meyer, Alter, Older Vineyard, and other scientists in the period from 1946 to 1960 [17], and was intensely developed by the efforts of many scientists applied to different problems of simulating the processes in condensed, liquid, and gas media at the atomic level. This method is based on the concept of the Born-Oppenheimer force surface that is a multidimensional space describing the system energy as a function of the position of nuclei and atoms that form the system [17]. Thus, in the molecular dynamics method motion only of atomic nuclei of the nanosystem, but not the motion of electrons, is considered. The motion of atomic nuclei is determined by the Hamilton equations

$$\frac{d\bar{\mathbf{x}}_i}{dt} = \frac{\partial \mathbf{H}}{\partial \bar{\mathbf{p}}_i}, \tag{11}$$

$$\frac{d\bar{\mathbf{p}}_i}{dt} = -\frac{\partial \mathbf{H}}{\partial \bar{\mathbf{x}}_i}, \tag{12}$$

where

$$\mathbf{H} = \sum_{k=1}^{N} \left\{ \frac{p_k^2}{2m_k} + U_k\left(x_{1k}, x_{2k}, x_{3k}, t\right) \right\} +$$

$$+ \sum_{k=1}^{N} \sum_{j \neq k=1}^{N} U_{kj}\left(x_{1k}, x_{2k}, x_{3k}, x_{1j'2j}, x_{3j}, t\right) - \sum_{k=1}^{N} \alpha_k \bar{\mathbf{p}}_k \bar{\mathbf{x}}_k \tag{13}$$

is the Hamilton function, $\bar{\mathbf{x}}_i$ is the vector of the coordinates (x_{1i}, x_{2i}, x_{3i}), and $\bar{\mathbf{p}}_i$ is the vector of moment $\left(m_i \frac{dx_{1i}}{dt}, m_i \frac{dx_{2i}}{dt}, m_i \frac{dx_{3i}}{dt}\right)$.

Having substituted (13) into Eqs. (11) and (12), we obtain the equations of the motion of nanosystem atoms in the form of the Newton equations

$$m_i \frac{d\bar{\mathbf{V}}_i}{dt} = \sum_{j=1}^{N^a} \bar{\mathbf{F}}_{ij} + \bar{\mathbf{F}}_i^g(t) - \alpha_i m_i \bar{\mathbf{V}}_i, i = 1, 2, .., N^a, \frac{d\bar{\mathbf{x}}_i}{dt} = \bar{\mathbf{V}}_i, \tag{14}$$

where m_i is the mass of the ith atom, $\bar{\mathbf{F}}_{ij}$ are the forces of interatomic interaction, α_i is the coefficient of "friction" in the atomic system, $\bar{\mathbf{F}}_i^g(t)$ are the external forces, and t is the time.

As a rule, a random force $\bar{\mathbf{F}}_i^g(t)$ that affects the ith atom and is specified by the Gauss distribution. The Gauss distribution is the δ-correlated in time Gaussian random process with the following properties:

a mean value of the random force is 0

$$\left\langle \bar{\mathbf{F}}_i^g(t) \right\rangle = 0; \tag{15}$$

$\bar{\mathbf{F}}_i^g(t)$ does not correlate with the velocity $\frac{d\bar{\mathbf{x}}_i}{dt}$ of the atom under consideration, thus

$$\left\langle \bar{\mathbf{F}}_i^g(t) \frac{d\bar{\mathbf{x}}_i}{dt} \right\rangle = 0; \tag{16}$$

$$\left\langle \bar{\mathbf{F}}_i^g(t) \bar{\mathbf{F}}_i^g(0) \right\rangle = 2k_B T_0 \alpha_i m_i \delta(t), \tag{17}$$

where k_B is the Boltzmann constant, $\delta(t)$ is the Dirac delta function, and T_0 is the temperature. Thus, the interaction of the atomic molecular system with the external medium (the heat reservoir), which consists of the following two parts: the systematic friction force

$$\sum_{j=1}^{N^a} \bar{\mathbf{F}}_i^f(t) = -\alpha_i m_i \bar{\mathbf{V}}_i, \tag{18}$$

and the random force $\bar{\mathbf{F}}_i^g(t)$ (noise) is described. In this case, the equations of motion are called the Langevin equations and the molecular dynamics method of calculation by these equations is given the name of the Langevin dynamics method.

Nanoparticles start formation from the gas medium that at the initial instant consists of the atoms of different materials and molecules. In cooling, due to condensation the gas medium is supplemented with time by nanoparticles. This process occurs in a macro-volume including a large number of atoms and molecules. Simulation of such system by the methods of molecular dynamics is impossible due a large number of variables, therefore simulation occurs in a calculated cell occupying nano- or microvolume with the specific boundary conditions of the Born-Karman surface of the cell [7, 17]. In this case, the molecular dynamics tools allow obtaining of not only general characteristics of the system, but also make it possible to follow the trajectories of each atom and nanoparticle. The essence of these boundary conditions is explained in Figure 4. The space modeled is divided to 27 equal cells. The central cell is the calculated cell, other cells are the "images" of it. In this case, simulation is made in the central cell only. All other cells-images contain the same set of atoms, molecules and nanoparticles as the calculated cell has. The total number of the images of the calculated cell is 26. The molecular dynamics equations (14) are written only in the calculated cell. The trajectories of motion of atoms, molecules, and nanoparticles in the images of the calculated cell are specified by absolutely similar trajectories of their motion in the calculated cell according to Eqs. (14).

When any atom, molecule, or nanoparticle from the inner space crosses any boundary, a similar atom, molecule, or nanoparticle with the same properties and velocity appears in the calculated cell from the opposite boundary. When an atom, molecule, or nanoparticle approaches the inner boundary of the cell the atoms, molecules, and nanoparticles of the images begin to affect them. Thus, the entire calculated space is presented in the form of the set of equal calculated volumes and motion of microobjects and occurrence of any processes in them are taken to be identical and the effect of the calculated cell edges is eliminated. Thanks to the hypothesis on the periodicity of the modeled region, the periodic boundary conditions allow a decrease of the calculated volume and, thus, a considerable decrease of the computation expenses in simulation. We consider a mathematical formulation of the Born-Karman periodic boundary conditions. A set of images of the ith atom with the

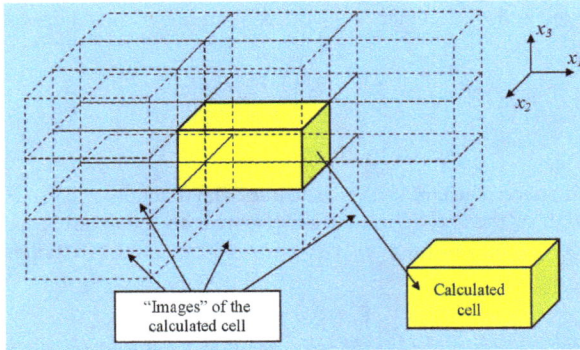

Figure 4. Periodic boundary conditions.

coordinate $\bar{\mathbf{r}}_i$ is specified as follows:

$$\bar{\mathbf{r}}_{i,image(k_1,k_2,k_3)} = \bar{\mathbf{r}}_i + L_j \cdot \bar{\mathbf{i}}_j, \ k_1, k_2, k_3 = 1, 2, 3, \tag{19}$$

where L_j is the length of the calculated cell along the corresponding coordinate, k_1, k_2, k_3 are the numbers of cells-images in all directions, $\bar{\mathbf{r}}_{i,image(k_1,k_2,k_3)}$ is the radius-vector of the image of ith atom, and $\bar{\mathbf{i}}_j$ is the transfer vector;

$$\bar{\mathbf{i}}_j = \begin{pmatrix} i_1 \\ i_2 \\ i_3 \end{pmatrix}, \ i_1, i_2, i_3 = -1,0,1, i_1 \neq i_2 \neq i_3 = 0, \tag{20}$$

In such a manner we create all 26 images of the ith atom. As the atom reaches the boundary of the calculated cell it enters the cell from the opposite side with the coordinates $\bar{\mathbf{r}}_i = \bar{\mathbf{r}}_i^* + L_j \cdot \bar{\mathbf{I}}_j$, where

$$\bar{\mathbf{I}}_j = \begin{pmatrix} l_1 \\ l_2 \\ l_3 \end{pmatrix}, \ l_1, l_2, l_3 = -1,0,1, l_1 \neq l_2 = \pm 1, l_2 \neq l_3 = \pm 1, l_1 \neq l_3 = \pm 1, \tag{21}$$

$\bar{\mathbf{r}}_i^*$ is the radius-vector of the ith atom at the boundary of the calculated cell and $\bar{\mathbf{I}}_j$ is the transfer vector. The atom preserves the motion parameters:

$$\bar{\mathbf{V}}(\bar{\mathbf{r}}_i) = \bar{\mathbf{V}}(\bar{\mathbf{r}}_i^* + L_j \cdot \bar{\mathbf{I}}_j). \tag{22}$$

The presented mathematical formulation of the periodic boundary conditions eliminates the effect of edges and allows one to accurately describe interactions occurring in the calculated region. To calculate the problem of nanoparticles formation in the calculated cell by the molecular dynamics method we use the structure of molecules constituting the gas mixture, which is obtained at the first stage of simulation, and specify the coordinates and velocities of atoms of all molecules at the time instant $t = 0$

$$\bar{\mathbf{x}}_i = \bar{\mathbf{x}}_{i0}, \frac{d\bar{\mathbf{x}}_i}{dt} = \bar{\mathbf{V}}_i = \bar{\mathbf{V}}_{i0}, \ t = 0, \bar{\mathbf{x}}_i \subset \Omega, \tag{23}$$

where Ω is the volume of the calculated cell. The initial coordinates of atoms and molecules $\bar{x}_{i0}$ are specified proceeding from the uniform distribution of them in the gas mixture and random mixing within the calculated cell. Modules of the initial velocities of atoms and molecules are calculated according to the Maxwell distribution [17] proceeding from the initial temperature of the gas mixture T_0. The relation between the initial temperature T_0 and initial velocities $\hat{V}_{i0}$ is determined by the expression

$$T_0 = \frac{1}{3Nk_B} \sum_{i=1}^{N_a} m_i \left(\hat{V}_{i0}\right)^2, \tag{24}$$

where N is the total number of the degrees of freedom of the atoms in the gas mixture and N_a is the number of atoms in the gas mixture. The Maxwell distribution for the velocity vector $\hat{V}_0 = \left(\hat{V}_{x0}, \hat{V}_{y0}, \hat{V}_{z0}\right)$ is the product of the distributions for each of three directions:

$$f_V \left(\hat{V}_{x0}, \hat{V}_{y0}, \hat{V}_{z0}\right) = f_V \left(\hat{V}_{x0}\right) f_V \left(\hat{V}_{y0}\right) f_V \left(\hat{V}_{z0}\right), \tag{25}$$

where the distribution along one direction is determined by the normal distribution:

$$f_V \left(\hat{V}_{x0}\right) = \sqrt{\frac{m}{2\pi k_b T_0}} \exp\left(-\frac{m(\hat{V}_{x0})^2}{2k_b T_0}\right). \tag{26}$$

Integrating the system of equations (14) with respect for time using the presented initial conditions, at the time instant t we obtain the main parameters of nanoaerosol:

a mean kinetic energy of the system

$$E(t) = \frac{\sum\limits_{i=1}^{N} m_i \left(\hat{V}_i(t)\right)^2}{2}, \tag{27}$$

an instantaneous value of temperature

$$T(t) = \frac{1}{3Nk_B} \sum_{i=1}^{N} m_i \left(\hat{V}_i(t)\right)^2. \tag{28}$$

The temperature is obtained by averaging of the instantaneous values $T(t)$ over some time range

$$T(t) = \frac{1}{3Nk_B\tau} \int_{t_0}^{t_0+\tau} \sum_{i=1}^{n} m_i \left(\hat{V}_i(t)\right)^2 dt. \tag{29}$$

Under real conditions the molecular system usually exchanges energy with the surrounding. Special algorithms - thermostats - are used to take into account such energy interactions. The use of the thermostat allows one to calculate molecular dynamics at a constant temperature of the medium or, on the contrary, to change the medium temperature according to a certain law.

In the general case, the thermostat temperature does not coincide with the temperature of the molecular system. At a fixed temperature of the thermostat the molecular system temperature can change due to different reasons. In the case of the established equilibrium, the thermostat temperature and the mean temperature of the molecular system should coincide. The simplest way to maintain a constant temperature of the thermostat is scaling of velocities. Scaling is done using the expression

$$\bar{\mathbf{V}}_j^{new}(t) = \bar{\mathbf{V}}_j^{old}(t) \cdot \sqrt{3Nk_B T_t \bigg/ \left\langle \sum_{i=1}^{N} m_i \left(\bar{\mathbf{V}}_i^{old}(t) \right)^2 \right\rangle}, \tag{30}$$

where T_t is the thermostat temperature. Averaging of a value of the total moment within the time interval between scaling of velocities is denoted in angular brackets. Another important factor is the effect of the initial temperature on the distribution of molecule velocities. As is shown above, in the molecular dynamics problem the velocity field at the initial instant of time is usually selected according to the Maxwell distribution. This distribution has a form of normal distribution. As should be expected, for gas at rest a mean velocity in any direction is zero. It is of interest to know the distribution of the velocities of molecules or atoms not over the projections but over the absolute value of velocities. The velocity modulus V is determined as

$$V = \sqrt{V_{x_1}^2 + V_{x_2}^2 + V_{x_3}^2}. \tag{31}$$

Therefore, the velocity modulus will always be larger than or equal to zero. Since all V_j are distributed normally, V^2 will have the $f(V)$ distribution with three degrees of freedom. If $f(V)$ is the probability density function for the velocity modulus, then it has the form:

$$f(V)\, dV = 4\pi V^2 \left(\frac{m_i}{2\pi k_B T} \right)^{3/2} \exp\left(\frac{-m_i V^2}{2k_B T} \right) dV. \tag{32}$$

The following expression is used for calculation of the nanosystem pressure:

$$P(t) = \frac{1}{3W} \left[\sum_{i=1}^{N} m_i \left(\bar{\mathbf{V}}_i(t) \right)^2 - \sum_{i,j;i<i} \left(\bar{\mathbf{r}}_{\bar{j}}(t) - \bar{\mathbf{r}}_i(t) \right) \bar{\mathbf{F}}_{i\bar{j}}(t) \right], \tag{33}$$

where W is the volume occupied by the nanosystem. The first term in (33) depends on the energy of motion of atoms or molecules and the second term is determined by pairwise interaction of atoms. Along with the pair of ith and jth atoms, all images of the jth atom are considered and interaction between the closest image $\bar{j}$ and the ith atom is calculated. The function $\bar{\mathbf{F}}_{i\bar{j}}(t)$ characterizes a value of interaction between the atoms. The use of the barostat algorithms allows simulation of the behavior of the system at constant pressure. The simplest of them is the Berendsen barostat where a value of pressure is maintained constant by scaling the calculated cell. The position of the particles in the system at each time step is modified according to the scaling coefficient of the Berendsen barostat μ:

$$\bar{\mathbf{r}}_i(t) \rightarrow \mu\bar{\mathbf{r}}_i(t),\ i = 1, 2, ..., N. \tag{34}$$

The scaling coefficient is determined by the expression

$$\mu = \sqrt[3]{1 - \frac{\Delta t}{\tau_P}\,(P - P_b)},$$ (35)

where Δt is the integration step, τ_P is the time of barostat implementation, P is the current pressure, and P_b is the barostat pressure. Transformation of the position of particles by formula (34) leads to the change of the calculated cell size and volume and thus to the change of pressure. The primary problem of the molecular dynamics method is calculation of the forces of interaction between atoms (molecules). Forces of this interaction are potential and are determined from the expression

$$\bar{F}_{ij} = -\sum_{i=1}^{N} \frac{\partial U(\bar{r})}{\partial \bar{r}_i},$$ (36)

where $\bar{r} = \{\bar{r}_1, \bar{r}_2, ..., \bar{r}_N\}$; $\bar{r}_i$ is the radius vector of the ith atom, is the potential of intramolecular interaction that depends on the mutual position of all atoms. This method is based on the concept the Born-Oppenheimer force surface that is the multidimensional space describing the system energy as a function of the position of the nuclei of atoms that form it [17]. The potential $U(\bar{r})$, in the general case, is specified in the sum of several components that correspond to different types of interaction:

$$U(\bar{r}) = U_b + U_\theta + U_\varphi + U_{ej} + U_{LJ} + U_{es} + U_{hb},$$ (37)

where the terms correspond to the following types of interactions: the change of the bond length U_b, the change of bond angle U_θ, the torsion angles U_φ, the plane groups U_{ej}, the van-der-waals interactions U_{LJ}, the electrostatic interactions U_{es}, and the hydrogen bonds U_{hb}. The indicated terms have different functional forms [11].

2.3. Simulation of nanosystems by mesodynamics methods

Calculations of controlled gas media by the molecular dynamics method are effective at the initial stage of formation of nanoparticles. However, simulation of processes that occur in the gas medium by molecular dynamics at the atomic level requires huge computer resources and time. This stipulates the topicality of the development of economic methods of calculation. In this section, we suggest a technique that is based on the mesodynamics methods [18]. We note that mesodynamics is the development of method of particles. The essence of the method is as follows. As atoms and molecules merge to nanoparticles, the larger number them manifest a collective behavior. Atoms and molecules making the nanoparticle move together with small oscillation near the equilibrium position within the nanoparticle structure. This allows one to decrease a number of simulated objects and to use another simulation method – mesodynamics. Mesodynamics is based on a collective behavior of atoms and uses force parameters that are calculated by the molecular dynamics methods. In this case, motion of nanoparticles is studied by the methods of classical mechanics. We consider main stages of application of mesodynamics.

The first stage is the calculation of interaction of the pair of nanoparticles. We give the problem formulation for symmetric nanoparticles and then for nanoparticles of an arbitrary shape. With this in mind, we consider a nanoparticle that consists of N^a atoms occupying the region Ω at the time instant $t = 0$ (Figure 5). The position of each ith atom of the nanoelement is specified by the coordinates x_{i1}, x_{i2}, x_{i3}. Atoms interact with each other. Figure 6 shows the interaction forces $\bar{F}_{ik}$ (the force of interaction between atoms i and k) and $\bar{F}_{ij}$ (the force of interaction between atoms i and j). The force of interaction between two atoms is directed along the line that connects their centers. Moreover, each ith atom of the nanoelement is affected by the external force $\bar{F}_i^b$. The direction and value of this force is determined by the type of interaction between the nanoelement and the surrounding medium. Atoms of a nanoparticle move under the action of the system of these forces.

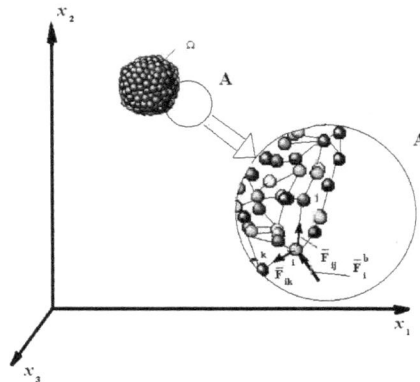

Figure 5. Nanoparticle: A) Magnified image of the part of a nanoparticle.

The motion of atoms that form a nanoparticle is determined, according to the molecular dynamics method, by the system of differential equations (14) which is supplemented with forces caused by interactions of atoms with the surrounding medium

$$\bar{F}_{bi} = - \sum_{1}^{N} \frac{\partial U_b(\rho_{bi})}{\partial \rho_{bi}} \, \bar{e}_{ib}, \tag{38}$$

where ρ_{bi} is the distance between atom i and atom b from the surrounding medium, $\bar{e}_{ib}$ is the unit vector directed from atom i to atom b, and $U_b(\rho_{bi})$ is the potential of interaction between nanoparticle atoms and atoms from the surrounding medium. We consider two symmetric nanoparticles lying at a distance S from each other (Fig. 4). In this case, Eq. (14) takes the form

$$m_i \frac{d^2 \bar{x}_i}{dt^2} = \sum_{j=1}^{N_1 + N_2} \bar{F}_{ij} + \bar{F}_i(t) - \alpha_i m_i \frac{d\bar{x}_i}{dt}, \ i = 1, 2, .., (N_1 + N_2), \tag{39}$$

at the boundary conditions

$$\bar{x}_i = \bar{x}_{i0}, \bar{V}_i = \bar{V}_{i0}, \ t = 0, \bar{x}_i \subset \Omega_1 \bigcup \Omega_2, \tag{40}$$

where N_1 and N_2 is the number of atoms in the first and second nanoparticles, respectively, Ω_1 and Ω_2 are the regions occupied by the first and second nanoparticles, respectively. Solution of (39) at the boundary conditions (40) allows calculation of the trajectories of motion of atoms of each nanoparticle and, consequently, of nanoparticles as a whole. In this case, the total forces of interaction between the particles will be determined by the relation

$$\bar{F}_{b1} = -\bar{F}_{b2} = \sum_{i=1}^{N_1} \sum_{j=1}^{N_2} \bar{F}_{ij}, \qquad (41)$$

where i and j are the atoms of the first and second nanoparticles, respectively.

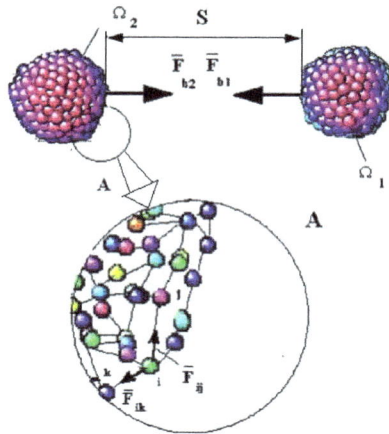

Figure 6. Schematic of interaction of nanoparticles: A) Magnified image of the part of a nanoparticle.

In the general case, the force of interaction of nanoparticles $\bar{F}_{bi}$ can be written as the product of functions dependent on the size of nanoparticles and the distance between them:

$$|\bar{F}_{bi}| = \Phi_{11}(S_c) \cdot \Phi_{12}(D). \qquad (42)$$

The direction of the vector $\bar{F}_{bi}$ is determined by the director cones of the vector that connects the centers of masses of nanoparticles. Of course, the forces of interaction between the particles change in time with small oscillations about a mean value. Therefore, Eq. (42) determines a mean value of the interaction force of nanoparticles. We consider two interacting nonsymmetric nanoparticles lying from one another at a distance S_c between their centers of masses and oriented at certain specified angles relative to one another (Figure 7). In contrast to the previous problem interaction of atoms entering into nanoparticles leads not only to relative displacement, but to rotation of the latter. Thus, in the general case, the sum of all forces of interaction of atoms of the nanoelements is reduced to the resultant vector of forces

$\overline{F}$ and the principal moment $\overline{M}$

$$\overline{F} = \overline{F}_{b1} = -\overline{F}_{b2} = \sum_{i=1}^{N_1} \sum_{j=1}^{N_2} \overline{F}_{ij}, \tag{43}$$

$$\overline{M} = \overline{M}_{b1} = -\overline{M}_{b2}, \tag{44}$$

where i and j are the atoms of the first and second nanoparticles, respectively.

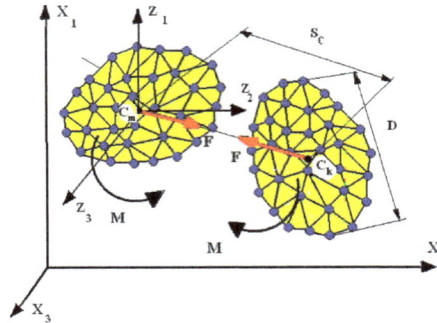

Figure 7. Interacting nanoparticles: $\overline{M}, \overline{F}$ are the principal moment and resultant vector of forces.

The main aim of this stage of calculation is the construction of the dependences of forces and momenta of interaction of nanoparticles on the distance between their centers of masses S_c, angles of mutual orientation of nanoparticles $\Theta_1, \Theta_2, \Theta_3$ (shape of a nanoelement), and their characteristic dimension D. In the general case, these dependences can be presented in the form

$$\overline{F}_{bi} = \tilde{\Phi}_1(S_c, \Theta_1, \Theta_2, \Theta_3, D), \tag{45}$$

$$\overline{M}_{bi} = \tilde{\Phi}_2(S_c, \Theta_1, \Theta_2, \Theta_3, D). \tag{46}$$

For spherical nanoparticles the angles of mutual orientation do not influence the force of their interaction, therefore the moment in Eq. (39) is identically equal to zero. In the general case, functions (45) and (46) can be approximated, by analogy with (42), as a product of the functions of $S_c, \Theta_1, \Theta_2, \Theta_3, D$, respectively. In the study of the evolution of the system of interacting nanoparticles we consider motion of each nanoparticle as a comprehensive whole. In this case the translatory motion of the center of masses of each nanoparticle is specified in the system of X_1, X_2, X_3, and rotation of a nanoparticle is described in the system of coordinates Z_1, Z_2, Z_3 related to its center of masses (Figure 8).

This transition allows one to pass to other variables (from the coordinates and velocities of atoms to the coordinates and velocities of the center of masses of nanoparticles) and to decrease the number of them. For a nanosystem consisting of N^{np} nanoparticles each of which contains N^a atoms the number of mesodynamics variables, compared to the molecular dynamics method, decrease k_x times, which is calculated by the formula [15]

$$k_x = \xi_x \frac{N^a N^{np}}{N^{np}} = \xi_x N^a, \tag{47}$$

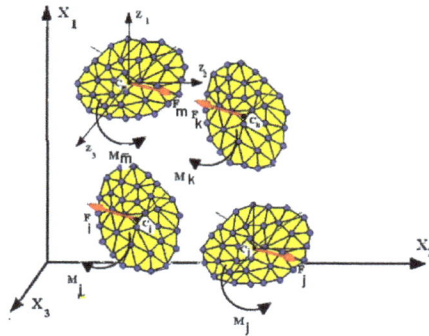

Figure 8. System of interacting particles.

where $\xi_x = 1$ for force interaction of nanoparticles, $\xi_x = 0.5$ when both forces and momenta of interaction of nanoparticles are taken into account. The system of equations that describes motion of interacting nanoparticles has the form

$$
\begin{cases}
M_k \dfrac{d^2 X_1^k}{dt^2} = \sum\limits_{j=1}^{N_e} F_{X_1}^{kj} + F_{X_1}^{ke}, \\[2mm]
M_k \dfrac{d^2 X_2^k}{dt^2} = \sum\limits_{j=1}^{N_e} F_{X_2}^{kj} + F_{X_2}^{ke}, \\[2mm]
M_k \dfrac{d^2 X_3^k}{dt^2} = \sum\limits_{j=1}^{N_e} F_{X_3}^{kj} + F_{X_3}^{ke}, \\[2mm]
J_{Z_1}^k \dfrac{d^2 \Theta_1^k}{dt^2} + \dfrac{d\Theta_2^k}{dt} \cdot \dfrac{d\Theta_3^k}{dt}(J_{Z_3}^k - J_{Z_2}^k) = \sum\limits_{j=1}^{N_e} M_{Z_1}^{kj} + M_{Z_1}^{ke}, \\[2mm]
J_{Z_2}^k \dfrac{d^2 \Theta_2^k}{dt^2} + \dfrac{d\Theta_1^k}{dt} \cdot \dfrac{d\Theta_3^k}{dt}(J_{Z_1}^k - J_{Z_3}^k) = \sum\limits_{j=1}^{N_e} M_{Z_2}^{kj} + M_{Z_2}^{ke}, \\[2mm]
J_{Z_3}^k \dfrac{d^2 \Theta_3^k}{dt^2} + \dfrac{d\Theta_2^k}{dt} \cdot \dfrac{d\Theta_1^k}{dt}(J_{Z_2}^k - J_{Z_1}^k) = \sum\limits_{j=1}^{N_e} M_{Z_3}^{kj} + M_{Z_3}^{ke},
\end{cases}
\tag{48}
$$

where X_i^k, Θ_i are the coordinates of the centers of masses and the orientation angles of the principle axes Z_1, Z_2, Z_3 of nanoparticles; $F_{X_1}^{kj}, F_{X_2}^{kj}, F_{X_3}^{kj}$ are the interaction forces of nanoparticles calculated by formulas (42) or (43); $F_{X_1}^{ke}, F_{X_2}^{ke}, F_{X_3}^{ke}$ are the external forces acting on nanoparticles; N_k is the number of nanoparticles; M_k is the nanoparticle; $M_{Z_1}^{kj}, M_{Z_2}^{kj}, M_{Z_3}^{kj}$ are the momenta of interaction forces of nanoparticles calculated by formulas (46); $M_{Z_1}^{ke}, M_{Z_2}^{ke}, M_{Z_3}^{ke}$ are external momenta acting on nanoparticles; $J_{Z_1}, J_{Z_2}, J_{Z_3}$ are inertia momenta of nanoparticles. The initial conditions for the system of equations (48) have the form

$$
\bar{\mathbf{X}}^k = \bar{\mathbf{X}}_0^k, \Theta^k = \Theta_0^k, \bar{\mathbf{V}}^k = \bar{\mathbf{V}}_0^k, \frac{d\Theta^k}{dt} = \frac{d\Theta_0^k}{dt}.
\tag{49}
$$

During the motion of nanoparticles a pair of nanoparticles can merge. In this case, a number of nanoparticles in the system decreases by unity and, correspondingly, the number of equations in the system (48) decreases by six. Thus, the parameters of a new, merged, nanoparticle is calculated and a new reduced system of equations is integrated. It should be noted that if the momenta of interaction of nanoparticles with each other and with the surrounding medium are zero, only the first three equations remain in the system of equations (48). Equations of the motion of nanoparticles (48) allow for all interactions between nanoparticles and their interactions with the surrounding medium. However, in the gas medium where nanoparticles are formed the distance between nanoparticles is large and interaction between nanoparticles is short-range. For this case, the system of equations (48) can be written as

$$
\begin{cases}
M_k \dfrac{d^2 X_1^k}{dt^2} = F_{X_1}^{ke}, \\
M_k \dfrac{d^2 X_2^k}{dt^2} = F_{X_2}^{ke}, \\
M_k \dfrac{d^2 X_3^k}{dt^2} = F_{X_3}^{ke}, \\
J_{Z_1}^k \dfrac{d^2 \Theta_1^k}{dt^2} + \dfrac{d\Theta_2^k}{dt} \cdot \dfrac{d\Theta_3^k}{dt} (J_{Z_3}^k - J_{Z_2}^k) = M_{Z_1}^{ke}, \\
J_{Z_2}^k \dfrac{d^2 \Theta_2^k}{dt^2} + \dfrac{d\Theta_1^k}{dt} \cdot \dfrac{d\Theta_3^k}{dt} (J_{Z_1}^k - J_{Z_3}^k) = M_{Z_2}^{ke}, \\
J_{Z_3}^k \dfrac{d^2 \Theta_3^k}{dt^2} + \dfrac{d\Theta_2^k}{dt} \cdot \dfrac{d\Theta_1^k}{dt} (J_{Z_2}^k - J_{Z_1}^k) = M_{Z_3}^{ke}.
\end{cases}
\tag{50}
$$

If the moment and rotation of nanoparticles is disregarded, the system of equations (50) takes the form

$$
\begin{cases}
M_k \dfrac{d^2 X_1^k}{dt^2} = F_{X_1}^{ke}, \\[2mm]
M_k \dfrac{d^2 X_2^k}{dt^2} = F_{X_2}^{ke}, \\[2mm]
M_k \dfrac{d^2 X_3^k}{dt^2} = F_{X_3}^{ke}.
\end{cases}
\tag{51}
$$

The forces of interaction of nanoparticles with the gas medium can be presented as

$$
\bar{\mathbf{F}}_{X_i}^{ke}(t, \bar{\mathbf{r}}(t)) = -M_i g + \bar{\mathbf{f}}_i(t) - m_i b_i \dfrac{d\tilde{\mathbf{X}}_i(t)}{dt}, \quad i = 1, 2, ..., n,
\tag{52}
$$

where $\bar{\mathbf{f}}_i(t)$ is the random force acting on the ith nanoparticle from the side of the gas medium, b_i is the coefficient of "friction" in the nanoparticles-gas medium system. The random force $\bar{\mathbf{f}}_i(t)$ is similar to the random force in the Langevin dynamics. We note that the random force $\bar{\mathbf{f}}_i(t)$ reflects the effect of the gas phase molecules on nanoparticles moving in it. It is determined from the Gauss distribution with the following properties: a mean value of the random force $\bar{\mathbf{f}}_i(t)$ is zero and it correlates with the velocity $\hat{\mathbf{V}}_i(t)$ of the considered nanoparticle such that

$$
\left\langle \bar{\mathbf{f}}_i(t) \hat{\mathbf{V}}_i(t) \right\rangle = 0, \quad \left\langle \bar{\mathbf{f}}_i(t) \bar{\mathbf{f}}_i(0) \right\rangle = 2 k_B T_0 b_i m_i \delta(t).
\tag{53}
$$

In order to model the random force fthe Box-Müller transformation is used in Eq. (52). Let x and y be independent random quantities uniformly distributed on the section $[-1, 1]$. We determine $R = x^2 + y^2$. In the case when $R > 1$ or $R = 0$, the values of x and y should be generated anew. As soon as the condition $0 < R \leqslant 1$ is met, z_0 and z_1 are calculated. z_0 and z_1 are independent random quantities satisfying the standard normal distribution

$$z_0 = x\sqrt{\frac{-2\ln R}{R}}, \; z_1 = y\sqrt{\frac{-2\ln R}{R}}. \tag{54}$$

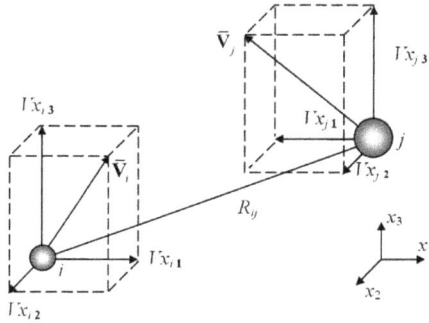

Figure 9. Mutual position of the ith and jth particles.

On obtaining the standard normal random quantity z, we can easily pass to the quantity $\xi\,(\mu, \sigma^2)$ distributed normally with the mathematic expectation μ and the standard deviation σ by the formula $\xi = \mu + \sigma z$. According to the data on the random force $\bar{\mathbf{f}}_i(t)$ in Eq. (52), it has the mathematical expectation $\mu = 0$ and standard deviation

$$\sigma = \sqrt{2k_B T_0 b_i m_i \delta\,(t)}. \tag{55}$$

The processes of condensation of nanoparticles in the gas medium are stipulated by the presence of potentials of interaction between the atoms. In the gas medium with nanoparticles two main factors affect the processes of their merging: the distance between the interacting particles and the direction and value of velocities. We consider two nanoparticles that are at a distance R_{ij} from each other at an arbitrary instant of time (Figure 9). When R_{ij} is small, the condition of "sticking" of nanoparticles is met. The second factor that affects condensation of nanoparticles is determined by the value and direction of velocities. It is obvious that very "fast" particles can overshoot one another even in contact. The angle α between the velocity vectors (Figure 10) that determines the direction of the motion of particles also substantially affects merging of particles.

The choice of an adequate condition of sticking of nanoparticles determines the processes of formation of new nanoparticles and the dynamics of their motion. The sticking criterion of nanoparticles can be presented in the form

$$\Phi(|R| - R_{ij}; \bar{\mathbf{V}}_i - \bar{\mathbf{V}}_j) = 0. \tag{56}$$

In the problem of merging of non-interacting nanoparticles of great importance is the choice of the correct time step of integration of Eqs. (51) in both analytical and numerical solution of them. The choice of a small time step increases the time of problem calculation. A very

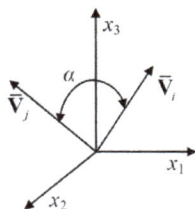

Figure 10. Velocity vectors of the ith and jth particles.

large integration step can lead to "overshoot" of nanoparticles despite condition (54). Thus, the problem of selecting an optimum time step arises. Solution of this problem in the present paper is based on the position of nanoparticles and values of their velocities. Figure 11 shows a pair of nanoparticles at some instant of time. Nanoparticles have the velocities $\bar{\mathbf{V}}_i$ and $\bar{\mathbf{V}}_j$ and their position is determined by the current radius vectors $\bar{\mathbf{r}}_i$ and $\bar{\mathbf{r}}_j$ respectively. The projections of the velocities of the ith and jth particles are determined according to the formulas

$$pr_{R_{ij}}\bar{\mathbf{V}}_i = \bar{\mathbf{V}}_i \cdot \bar{\mathbf{R}}_{ij}/R_{ij}, \ pr_{R_{ij}}\bar{\mathbf{V}}_j = \bar{\mathbf{V}}_j \cdot \bar{\mathbf{R}}_{ji}/R_{ij}. \tag{57}$$

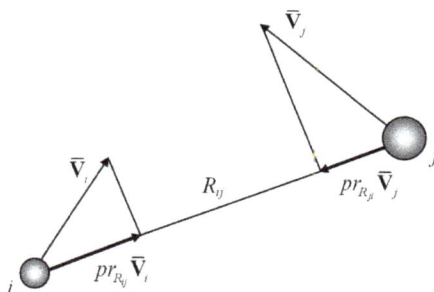

Figure 11. Two nanoparticles moving at the velocities $\bar{\mathbf{V}}_i$ and $\bar{\mathbf{V}}_j$.

The vectors $\bar{\mathbf{R}}_{ij}$, $\bar{\mathbf{R}}_{ji}$ and the distance between the particles R_{ij} are calculated from the relations

$$\bar{\mathbf{R}}_{ij} = \bar{\mathbf{r}}_j - \bar{\mathbf{r}}_i, \ \bar{\mathbf{R}}_{ji} = \bar{\mathbf{r}}_i - \bar{\mathbf{r}}_j, \ R_{ij} = \left|\bar{\mathbf{R}}_{ij}\right| = \left|\bar{\mathbf{R}}_{ji}\right| = \left|\bar{\mathbf{r}}_j - \bar{\mathbf{r}}_i\right|. \tag{58}$$

The time interval in which collision of the ith and jth particles becomes possible is directly proportional to the distance between the particles and inversely proportional to the velocity projections:

$$\Delta t_{ij} = \frac{R_{ij}}{pr_{R_{ij}}\bar{\mathbf{V}}_i + pr_{R_j}\bar{\mathbf{V}}_j}. \tag{59}$$

Using formula (59), by selection of all possible pairs of particles, one finds the smallest positive value Δt_{ij} and the integration step is calculated

$$\Delta t = \frac{1}{m} \cdot \min_{i,j,\ dt>0}\left(\Delta t_{ij}\right); i = 1, 2, \ldots, n; j = i, i+1, \ldots, n, \tag{60}$$

where m is the integer number that determined which part of the period of the fastest particles is the time step. Thus, the choice of the integration step is, first of all, affected by motion of the fastest particles. For the problem solution not to be delayed by additional calculations, it is more reasonable to select the integration step not in each iteration, but in a certain number of them. An important issue of multilevel simulation of the problems of aerosol nanotechnologies is gradual increase of the space scale during calculation. This possibility is stipulated by the fact that during formation of nanoparticles in a gas mixture the number of atoms and molecules decreases. They combine to nanoparticles and then demonstrate a collective behavior. Thus, the number of variables of the simulation problem becomes smaller.

This process occurs at the stage of gas system simulation by the molecular dynamics method. As nanoparticles become larger, their concentration in the calculated cell under consideration decreases rapidly. Then the gas phase is no longer the source of nanoparticles and nanoparticles enlarge only due to sticking of smaller nanoparticles. Therefore, from now on it is not expedient to calculate the nanosystem by the molecular dynamics method, it is necessary to use methods of mesodynamics. Further enlargement of nanoparticles can lead to the situation when particles within the calculated cell virtually do not interact and their trajectories do not intersect. However, if we take into account the effect of nanoparticles from neighboring cells, the condensation process will continue. Thus, adequate investigation of the condensation problem of nanoparticles requires timely increase of the space scale of the cell by uniting several calculated cells into one. Since the problem is solved using the periodic boundary conditions, the space scale can be increased by symmetric mapping of atoms, molecules, and nanoparticles on the neighboring calculated cells. This problem was considered in detail in [19].

2.4. The software package

In conclusion we consider the common algorithm of simulation of the problems of aerosol nanotechnologies and indicate in which way the solutions at different structural levels agree. The methods of solution of aerosol nanotechnologies presented above allow simulation of the proceses of formation and motion of n anoparticles in different space and time scales. Each method makes it possible to simulated the system at different structural levels and demonstrates an increase of computational capacities when mathematical description of the nanosystem changes. The general scheme of calculation by different methods is presented in Figure 12.

Quantum-chemical methods of simulation allow construction of the wave function Ψ. This function is used for calculation of the structure and dimensions of molecules entering into the gas mixture. Molecular dynamics, using this information, makes it possible to calculate velocities and coordinates of atoms (both free and joined into molecules) $\bar{V}_i$, $\bar{x}_i$, and variation of these quantities with time. The result of calculation is the shape and structure of nanoparticles formed by molecules and atoms and the forces of interaction between the atoms. Simulation by the mesodynamics methods on the basis of the calculated data by the molecular dynamics method allows one to calculated linear and angular velocities of motion of nanoparticles, coordinates of the center of masses of nanoparticles $\bar{V}_i$, Θ_3^k, $\bar{x}_i$, and variation of these parameters with time. The shape and spatial structure of nanostructures formed from nanoparticles is calculated based on these data.

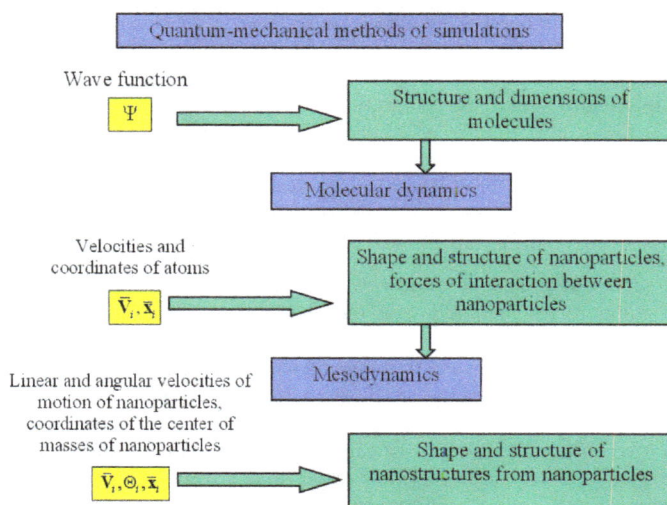

Figure 12. General scheme of the solution of problems of aerosol nanotechnologies.

It should be noted that the presented methods of calculation are not isolated from each other. In solving a number of problems they may be used simultaneously. For example, the trajectories of motion of atoms and molecules can be calculated by the molecular dynamics methods and the forces of interaction between the atoms and molecules at each calculated time step by the methods of quantum mechanics. Motion of nanoparticles in the gas phase can be treated as motion of supermolecules the trajectory of which is calculated by the mesodynamics methods and the motion of atoms and molecules of the gas mixture is calculated by the molecular dynamics methods.

The software package was created for the calculation of the above methodology (Figure 13), designed for the study of the formation of heterogeneous nanoparticles. The software package allows to multi-level mathematical simulation of the formation of nanoparticles. It consists of the initial data preparation section, the computational unit, data analysis and visualization unit and the section of data reconciliation. The initial data preparation section determines the initial conditions for simulation and generates input files for the computer unit. Computing init includes sections for quantum-mechanical calculations, molecular dynamics simulation program, and calculate the motion of nanoparticles mesodynamics methods. The simulation analysis of the system is performed in the unit of analysis and visualization. The unit solves the problem of analyzing the structure, properties and characteristics of the formed nanochatits, as well as visualization of the results. This unit supports the output files for quantum mechanics calculations, molecular dynamics and mezodinamike, and is able to reproduce both static and dynamic state of the system being modeled. Algorithms constitute unit structure, such as the establishment of the atoms are grouped in the nanoparticles, the definition of uniform composition formed from a mixture of atoms, molecules and nanoparticles, and nanoparticle properties: radius, shape, volume, bulk quantities of the

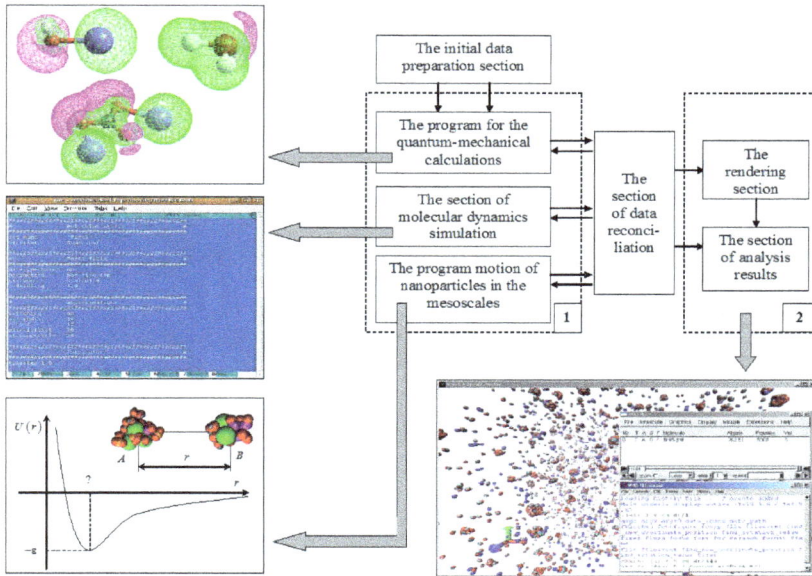

Figure 13. The structure of the software system, where 1 – compute unit 2 – unit of analysis and visualization.

surface. In addition, this module allows to determine the chemical composition and ratios of the source of the chemical elements contained in the nanoparticles, to calculate the fraction of atoms and molecules condensed into nanoparticles, to build the internal structure of nanoparticles. A detailed description of the software system is presented in the paper [20].

2.5. Technique of experimental research

Experimental investigations were carried out on two main areas:

- The study of the formation of nanoparticles in the gas phase.
- Investigation of the effect of the gas phase with nanoparticles on plants.

The experimental investigations were carried out according to the following technique.

1. The laboratory glass was prepared: it was washed, dried and degreased.
2. The glass was fixed on a holder at the distance of 150 mm from a solid-fuel grain in the laboratory cabinet.
3. The grain was ignited.
4. The laboratory cabinet was sealed.
5. The sample was held in the gas atmosphere for 5 minutes.

6. The laboratory cabinet was unsealed.

7. The sample was unfixed from the holder.

8. The sample was placed in the microscope Bikmed-1.

9. With the use of the digital camera Canon Power Shot A95, the sample was photographed at different magnifications.

10. The digital information obtained was entered into the computer Samsung Q30 and processed.

11. Then the sample was placed in the atom force microscope NTEGRA Maximus.

12. The optical image of the sample surface was built up.

13. An area on the sample surface, which was free from microparticles, was selected.

14. The above area was scanned.

15. The information in the digital form was entered into the computer and processed.

3. Results of simulation and experimental research

3.1. The analysis of the calculation results

Modelling is carried out in three steps (Figure 14). At the first step, the structures and shapes of the initial molecules are calculated using the method of quantum mechanics. In this case, the basic data are the chemical formula of the molecule, the number of bonds between the atoms and their lengths, the electrostatic charge of the atoms in the molecule, the angles between the bonds in the molecule (for molecules containing no less than 3 atoms) and some other information related to the spatial arrangement of the molecule atoms relative to each other. The second step of the calculation of the processes in the gas mixture is realized by the molecular dynamics method. The third step of the calculation is realized by the mesodynamics. The calculation of the structures and shapes of nanoparticles is carried out. The stability of nanoparticles and nanoparticles' systems in the process of their static or dynamic interaction is analyzed. The effect of the composition, shape and size of nanoparticles on their movement processes in gas medium (Brownian movement, agglomeration, sedimentation) is studied. The processes of spatial sedimentation of nanoparticles on plants are analyzed.

The second step of the calculation of the processes in the gas mixture is realized by the molecular dynamics method. The investigated gas mixture 1 is admitted inside a cylindrical calculation cell C_1 of length $L = 220$ nm (nanometres) and radius $R = 32$ nm. The gas mixture consists of 18 molecules of different types with a certain ratio of mass portions. For the calculations it is sufficient to take into account only six components of the gas mixture, since their mass makes 99 percent of the total mass of the system. They are the following molecules: O_2, CO_2, K_2CO_3, H_2O, N_2, MgO. The system under study contains 8850 atoms joined into 3500 molecules. The number of molecules of different types in the gas composition is determined in proportion to their mass portion in the gas mixture.

To start the calculation by the molecular dynamics method at the moment $t = 0$, we use the structures of the gas mixture molecules, which were obtained at the first step and then specify

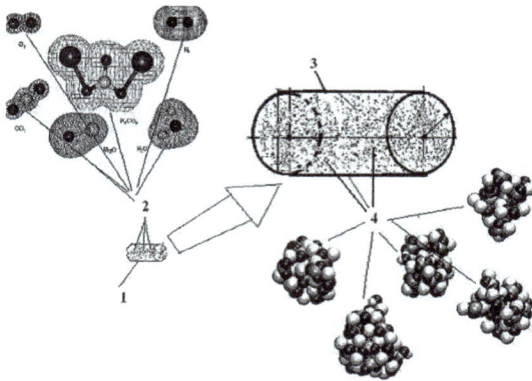

Figure 14. The settlement circuit: 1 – calculation cell C_1: 2 – molecules; 3 – calculation cell C_2; 4 – the nanoparticles.

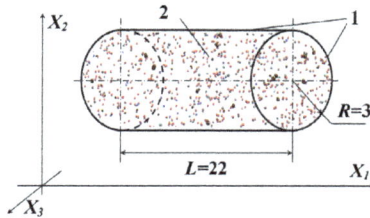

Figure 15. The calculation scheme for modeling the process of the nanoparticle formation and movement: 1 – gas mixture containing nanoparticles; 2 – cylindrical calculation cell (the sizes are in nanometers).

Molecular formula	The number of molecules
O_2	1600
CO_2	800
H_2O	600
N_2	200
MgO	150
K_2CO_3	150

Table 1. The molecular composition of the gas mixture.

the coordinates and velocities of the atoms of all the molecules (23). The initial coordinates of the molecules $\bar{x}_{i0}$ are given based on the uniform distribution of the gas mixture molecules and their random intermixing within the calculation cell. The modules of the molecule velocities are calculated in accordance with Maxwell distribution and at an initial temperature $T_0 = 600K$. The initial temperature and the initial velocities of the molecules are determined by the relation (24).

For the velocity vector, Maxwell distribution is the product of the distributions for each of the three directions (25)–(26). Then the gas mixture is gradually cooled down to the temperature

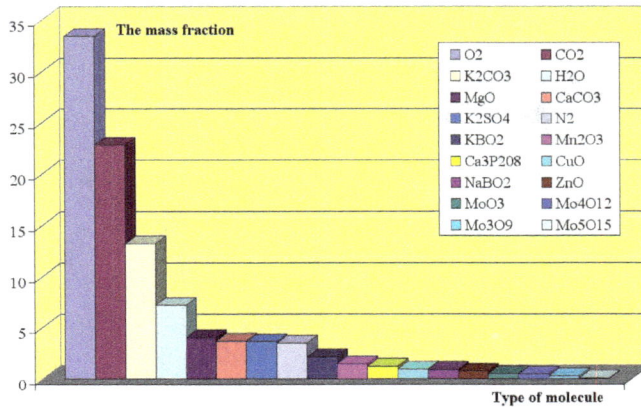

Figure 16. Mass portions of the molecules of the calculated gas mixture.

$T_p = 300K$ for 1.1 ns (nanosecond). The system temperature is further kept constant at the above level. The calculations are carried out based on Verlet scheme, with a step of integration with respect to the time 10^{-15} s. Thus, the period of 1 ns is calculated over one million steps of integration with respect to the time, which is quite sufficient for a detailed calculation of the main parameters of the molecular dynamic processes in the calculated gas mixture. For the interaction of the molecules with each other, the main types of forces between atoms and molecules are taken into account.

The calculation shows that nanoparticles form in the gas mixture at cooling. The analysis of the qualitative composition of the condensed nanoparticles shows that their main component is potassium. As it follows from Figure 17, the main mass of potassium (95%) transits in the condensed phase and only 5% of this element remains in gas phase.

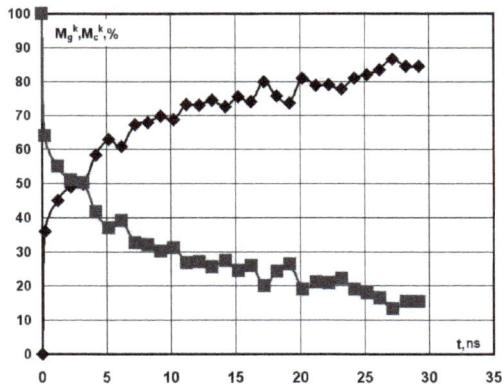

Figure 17. The variation of the condensed potassium mass and the mass of potassium in the gas phase.

Figure 18 shows the variation of volumetric density of the nanoparticles in unit of volume $Nr = N^{np}/\Omega$ with time, which form during the cooling process of the initial gas mixture. From the plot, one can see that at the initial period (about 3 ns), an intensive formation of nanoparticle takes place in the gas mixture. Their maximal density of the nanoparticles observed at $t = 4.3$ ns. It is necessary to note that during this period a noticeable variation of the number of nanoparticles is taking place. This is indicative of the presence of simultaneous and competing processes of the formation and decomposition of nanoparticles. By the 55th nanosecond, these processes subside; new particles do not form, and previously formed nanoparticles consolidate. Therefore, the density of the nanoparticles gradually reduces to $10^{22,5}$ with time. However, despite the decrease in the quantity of the nanoparticles, their total volume Ω_S is increasing with the time (Figure 19). From this figure, it follows that the volume of the nanoparticles is actively increasing during the initial period, and then the rate of the nanoparticle volume growth is decreasing. This also indicates that the formation of new nanoparticles is subsiding. By the 75th nanosecond, the nanoparticle volume growth practically stops.

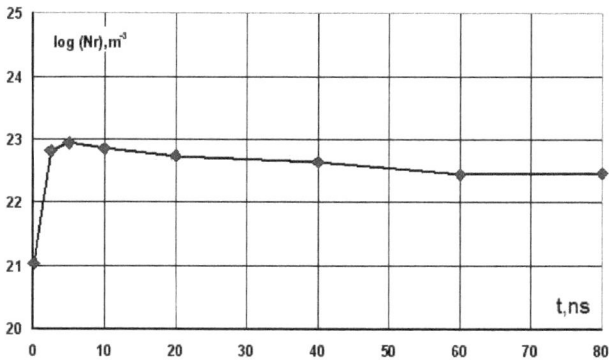

Figure 18. The variation of the number of nanoparticles in the unit of volume with time.

Similar dependences are observed when the mass components of the system calculated are considered. The content of the gas phase is decreasing, and the content of the condensed phase is increasing; further the ratio of the phases remains constant. Hence, only about 15% of the system substance turns into condensate and 85% remains in the gas phase. Then the gas phase is not the nanoparticle source any more and the nanoparticle enlargement takes place solely due to the reason that small nanoparticles stick together. From this moment, therefore, it is expedient to calculate a nanosystem with the use of the mesodynamics methods rather than the molecular dynamics method. This is the last stage of the simulation.

The presence of the nanoparticle consolidating process is confirmed by an increase in the number of coarse nanoparticles with time. Figure 20 shows histograms of the distribution of the nanoparticles by sizes at different moments: $t = 1.3; 30.3; 75.3$ ns. The maximum of this distribution does not practically shift with time relatively to the size of the nanoparticles; with a decrease in the number of the nanoparticles, the quantity of the coarser nanoparticles is growing.

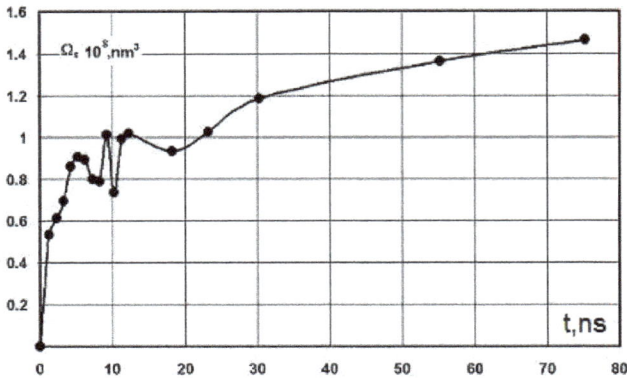

Figure 19. The variation of the total nanoparticle volume Ω_S with time.

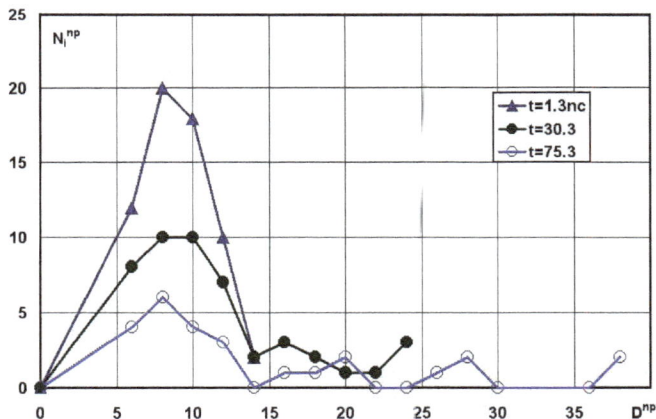

Figure 20. The variation of the distribution of nanoparticles by sizes with time.

For the simulation, a calculation cell C_2 is used (Figure 14-3), and the nanoparticle movement (Figure 14-4) is calculated without taking into consideration the movement of the gas phase molecules. The calculation cell C_2 is five times larger than a cell C_1. Consequently, the volume of the calculation cell C_2 is 125 times larger than the volume of the cell C_1 and, hence, the number of particles in it is 125 times larger than that in the cell C_1. For this calculation stage, the initial conditions are determined in accordance with the relations

$$t_2^0 = t_1^*, \, \tilde{\mathbf{X}}_i \subset \Omega_2, \, \tilde{\mathbf{V}}_\mathbf{i} = \frac{1}{M_i} \sum_{j=1}^{N_i} m_j \tilde{\mathbf{V}}_j, \tag{6-}$$

where $\bar{X}_i$ are coordinates of the nanoparticle centre of mass; m_j and $\bar{V}_j$ are masses and velocities of atoms contained in a nanoparticle, respectively; M_i and m_j and $\bar{V}_i$ are masses of nanoparticles and velocities of the nanoparticles' centres of mass, respectively; t_1^* and t_2^0 are the time of the completion of the second stage and the time of the beginning of the third stage of the calculations, respectively. The mesodynamics calculations show that with time nanoparticles combine into larger ones. Their number decreases down to 16, and their average size increases fivefold approximately. This allows a sevenfold increase of the calculation cell. In this case, in comparison with the calculation cell C_2, its volume increases by 343 times and the number of nanoparticles in it reaches 5488. The step of integration with respect to the time is 10^{-13} s.

Thus, in the example considered, the use of the mesodynamics method allowed increasing the calculation cell volume and the volume of the modelled space by 42875 times. The integration step increased hundredfold, and at the same time, the number of the variables remains almost unchanged. The structures and compositions of the formed nanoparticles are shown in Figure 21. The calculations show that the nanoparticles mainly consist of the K_2CO_3 molecules with small inclusions of the molecules of water, carbon dioxide and oxygen.

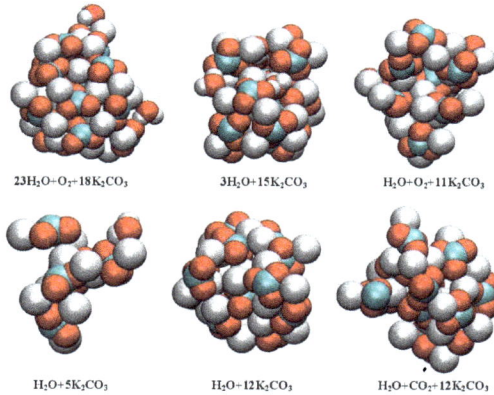

$23H_2O+O_2+18K_2CO_3$ $3H_2O+15K_2CO_3$ $H_2O+O_2+11K_2CO_3$

$H_2O+5K_2CO_3$ $H_2O+12K_2CO_3$ $H_2O+CO_2+12K_2CO_3$

Figure 21. The structures and compositions of the nanoparticles.

The investigation of the movement of nanoparticles is important since it determines the character of the interaction of the nanoparticles and the surface. Below, the calculated movement paths for a massive nanoparticle (Figure 22-a) and a particle of a smaller size (Figure 22-b) are given.

From the figures, it is clear that the nanoparticles are moving in a random and complex way. Moreover, a massive nanoparticle passes a shorter path than a particle with a smaller mass. Its path is "smoother". Judging by the shapes of the paths, one can assume that the particles are in Brownian motion, moving in different directions in space. Consequently, it can be suggested that the process of their sedimentation occurs on the surfaces that are randomly oriented in space, in other words, on all the surfaces of a plant.

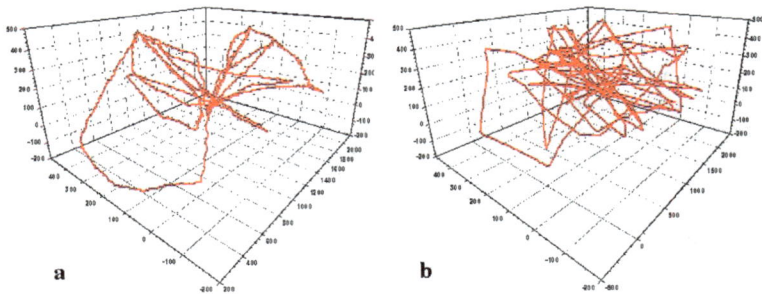

Figure 22. Brownian motion path of a massive nanoparticle (a) and Brownian motion path of a nanoparticle with a smaller mass (b).

3.2. The analysis of the experimental results of nanoparticle formation

The maximal size of the optically-investigated sample area was 7125x5400 μm and the minimal size was 95x72 μm. The accuracy of the determination of the microparticle boundary was 0.2 μm in this case. The size of the area scanned with the use of the power atomic microscope was 1.5x1.5 μm. The accuracy of the determination of the nanoparticle boundary was 2 nm.

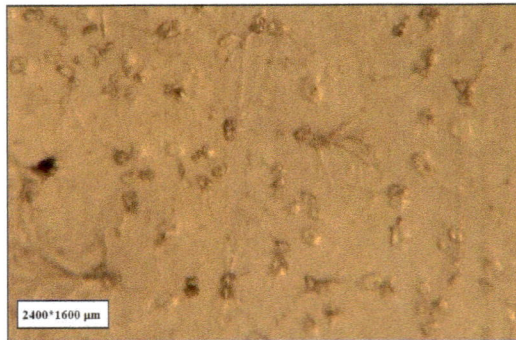

Figure 23. Optical image of particles deposited on the glass (75x magnification).

Let consider the experimental investigation results on the forms and distribution of the sizes of particles deposited on a glass surface from the gas phase. The investigations conducted show that the sizes of the particles deposited on the glass in accordance with the above technique lie in a wide range: from tens of microns to tens of nanometers. In Figure 23, a 75x enlarged view of the sample is presented. It is seen from the figure that there are particles of 100 μm in size on the glass. However, the smaller particles are seen as well. A more enlarged image presented in Figure 24 (187.5x magnification) allows a reliable identification of a particle with the size of about 10 μm.

In Figure 25 an optical-digital image of the particles deposited on the glass (5375x magnification) shows that the particles with smaller sizes (1 micron and less than 1 micron)

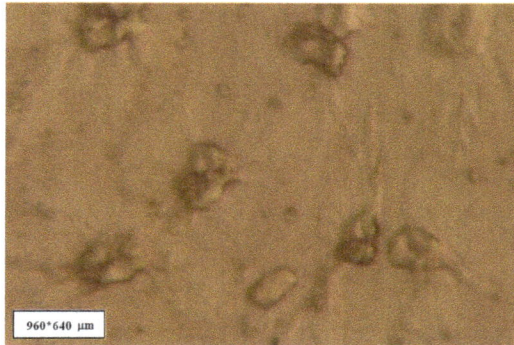

Figure 24. Optical image of particles deposited on the glass (185.5x magnification).

are also observed. It should be noted that further digital magnification leads to a no distinct image and does not allow establishing the sizes of particles reliably.

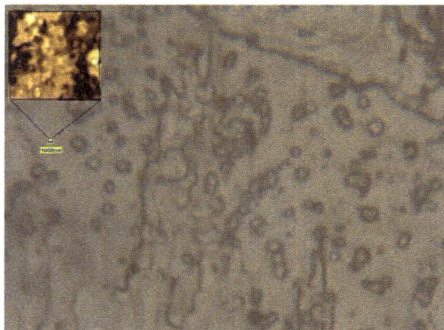

Figure 25. Optical-digital image of particles (5375x magnification, image size: 95 μm x 72 μm).

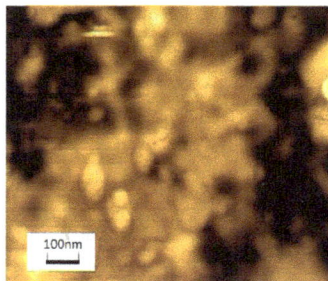

Figure 26. Image of nanoparticles deposited on the glass, which was obtained using the power atomic microscope (image size: 1000 nm x 1000 nm).

In Figure 25, one also can see an area of 1000 nm (1 μm) that was investigated with the use of the atomic-force microscope. The pattern of scanning is presented on an enlarged scale in

The fruits of the plant	The control plants	Plants treated with aerosol
Cucumbers	1.00	1.80
Beet	3.07	3.45
Radish	0.73	2.10

Table 2. The content of potassium in vegetables in milligrams.

Figure 26. The pattern of scanning shows that nanoparticles have precipitated on the glass. The size distribution histogram for nanoparticles is displayed in Figure 27. It follows from the plot that most nanoparticles lie in the range from 15 nm to 45 nm.

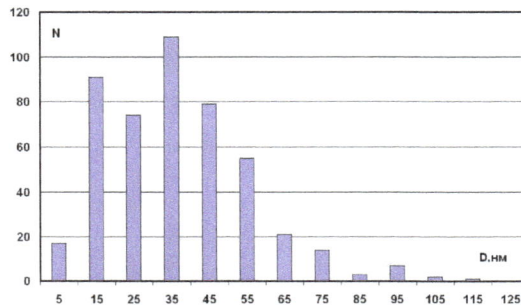

Figure 27. Size distribution histogram for nanoparticles.

3.3. Research of influence of the gas phase with nanoparticles on plants

Experimental researches included: structural botany researches, studying of activity of the photosynthetic processes, the biochemical analysis of ferment activity. Also studying of nanoaerosol action on regulation of a potato growth and dynamics of growth and development of salad cultures were carried out.

Let's compare amount of mineral fertilizers traditionally used at cultivation of plants in hothouses with amount of fertilizers at aerosol nutrition of plants. In the closed soil need of plants for nitric fertilizers (urea, ammoniac saltpeter) is 1.1 kg/m^2, and need for potash fertilizers is 1.4 of kg/m^2 [21]. As mineral fertilizers are brought in the form of water solution with concentration from 0.5% to 0.8%, the total amount of solution brought in the soil is 80 80 liters per m^2. The amount of mineral fertilizers thus is 6.97 kg/m^2. Processing by means of nanoaerosols is carried out one time in 5–7 days. For all period of growth of plants the fertilizer expense is only 8–14 g/m^2. On the average, it approximately in 500 times is lower than amount of the mineral fertilizers brought in the traditional way in the soil. Respectively and harmful substances it is brought less. First of all change of the maintenance of various elements (bohrium, manganese, zinc, cobalt, magnesium, potassium, calcium and iron) in samples of plants after processing by an aerosol was investigated. Experiments confirmed increase of the maintenance of elements of a nutripion in various parts of plants (Table 2–3).

Let's consider the results of studies on the effects of nanoparticles regenerants of potato. The work was carried out jointly with the Institute of Experimental Botany, National Academy of

Plant	Number of samples	Type of treatment	The content of elements, %				
			B	Cu	Mn	Mo	K
Tomatoes	1	Control	37.5	12.4	43.2	0.8	6.8
		Aerosol	74.0	43.0	60.0	11.9	7.5
	2	Control	33.3	11.5	35.5	0.6	8.8
		Aerosol	72.6	50.8	45.9	11.9	7.5
Cucumbers	1	Control	45.0	9.3	17.5	4.7	5.2
		Aerosol	67.2	29.1	14.3	14.0	4.0
	2	Control	30.0	8.3	15.9	0.8	3.2
		Aerosol	70.5	24.0	27.0	4.3	3.8

Table 3. The content of elements in the leaves of plants.

Belarus. The dynamics of growth in regenerants of potato micropropagation in the prolonged culture of morphometric characteristics was considered. Before each aerosol treatment were measured morphometric parameters and monitoring of rooting cuttings explants. The data showing a clear lead in the growth processes of regenerants of potato varieties Dolphin processed aerosol to 12 days of observation was obtained (Figure 28).

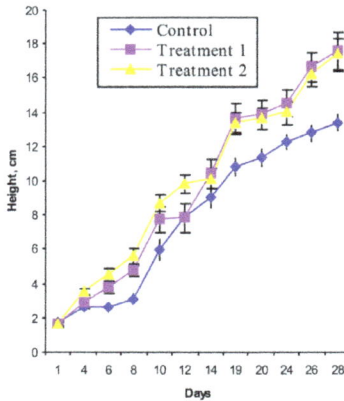

Figure 28. Growth dynamics of potato varieties Dolphin from the experimental data.

As can be seen from Figure 29 experimental plants are actively developing, exceeding the height of control plants at 1.22 and 1.19 times (treatment 1 and 2, respectively). A similar plant development was supported by Figure 29, which displays the daily activity of the growth of plants as they develop.

For plants, left for further growth and development in plastic containers, we observed for 42 days. In Table 4 was shown that lead to the growth of experimental plants and stored at the age of 42 days: plants with processing options 1 and 2 at 1.45 and 1.44 times higher than controls at the same height on the formation of leaves and internodes.

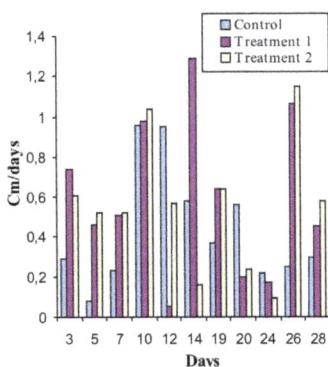

Figure 29. The growth rate of potato varieties Dolphin for the period of observation from the experimental data.

Version	Height, cm	Number of leaves, items	Number of internodes, items	Wet weight, g		
				leaves	stems	roots
Control	22.47±1.04	11.06±0.66	15.75±0.75	2.69±0.50	1.77±0.22	0.47±0.13
Treatment 1	32.69±1.45	10.87±0.59	16.12±0.39	2.78±0.51	2.76±0.37	0.94±0.23
Treatment 2	32.44±4.43	11.44±0.76	16.69±0.81	2.68±0.53	2.59±0.32	0.72±0.14

Table 4. Morphological characteristics of potato cultivar Dolphin at harvest at age 42 days.

The development of potato varieties Lazurit is slightly different from class Dolphin on the dynamics of growth. As follows from Figure 30 plants in a series of treatment 2 exceed the height of the control and experimental plants (treatment 1) to 28 days by 38% and 32% respectively. The plants left in the plastic containers for further growth and development, as well as in the case of grade Dolphin, were conducted further follow-up.

As shown in Table 5 advantage of the growth experienced in the plant varieties Lazurit age of 42 days over the control is saved. Plants with treatment 1 and 2 were of 1.87 and 1.96 times higher than the control plants in height and 1.13 and 1.17 times the number of internodes The mass of the leaves advantage of control plants, with an equal number of it's with the experimental data shows about the formation of a dense leaf. At a height equal to the control plants and the option of processing 1 plants have experienced a significant differences in the accumulation of fresh weight of leaves and roots of 2.2 and 3.4 times. At the age of 42 days in 40% of plant varieties Lazurit (treatment 2) were formed side shoots on the average length of 4.5 cm. In control plants and the processing 1plants the side shoots are not detected. We can assume that a often treatment of options 1 and 2 (one day) was contributed to the formation of young growing aerial parts of plants, thus extending the growing season.

Results of studies of dry matter content in some parts of potato plants Lazurit (Table 6) showed that frequent drug treatment "Greenhouse" leads to a "rejuvenation" of plants, reducing the synthesis of organic matter, such as starch.

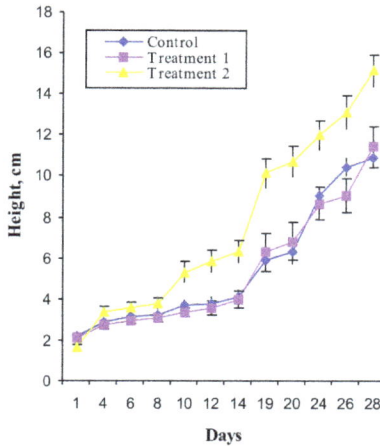

Figure 30. Growth dynamics of potato varieties Lazurit for different conditions experience.

Version	Height, cm	Number of leaves, items	Number of internodes, items	Wet weight, g		
				leaves	stems	roots
Control	18.17±2.17	10.83±1.18	11.83±0.60	4.93±0.78	2.48±0.20	1.15±0.16
Treatment 1	34.03±1.25	9.00±0.59	13.37±0.56	2.21±0.50	2.53±0.34	0.34±0.09
Treatment 2	35.63±1.01	10.87±0.54	13.80±0.39	3.15±0.70	3.11±0.70	1.01±0.29

Table 5. Morphological characteristics of potato varieties Lazurit at the age of 42 days.

Version	Part of absolutely dry mass, %			The absolutely dry mass, g		
	leaves	stems	roots	leaves	stems	roots
Control	22.47±1.04	11.06±0.66	15.75±0.75	2.69±0.50	1.77±0.22	0.47±0.13
Treatment 1	32.69±1.45	10.87±0.59	16.12±0.39	2.78±0.51	2.76±0.37	0.94±0.23
Treatment 2	32.44±4.43	11.44±0.76	16.69±0.81	2.68±0.53	2.59±0.32	0.72±0.14

Table 6. Characterization of potato varieties Lazurit at the age of 42 days.

Consider the results of a study of the photosynthetic activity of the processes of photosynthesis. About the functioning of the photosynthetic transport chain in plants can talk on the basis of photoinduced changes in absorption or fluorescence. As is known, light green plants is accompanied by increased fluorescence yield. Fluorescence undergoes complex transient phenomena before reaching a constant level. It is assumed that the induction of fluorescence due to the variable part of fluorescence of chlorophyll a, the output of which depends on the state of the active center of photosystem and the rate of electron transfer. The accurate registration of the fluorescence changes for small periods of time is required to obtain a complete picture of fluorescence induction. To characterize the photosynthetic processes determine the rate of decay of variable chlorophyll fluorescence was used, reflecting the photochemical activity (FHA).

The systematic study of the change dynamics in the decay rate of variable fluorescence of chlorophyll per day processing plants nanoaerosol was carried out (after 2, 4, and 6 hours after treatment). Analyzing the results, you may notice that the substance treatment of "Greenhouse" vegetative regenerants causes changes in the studied parameters as the Dolphin class and a grade Lasurit. Changes in the direction of activation of photosynthetic activity strongly manifested at the Dolphin class. The second version of the experience with pre-treatment plants before meristem cuttings was more sensitive to changes in photosynthetic activity of plants. The additional introduction of CO_2 in the nanoaerosol processing, gives rise to a new round of metabolic processes, especially evident after 16–19 treatments when the plants are in the process of budding and early flowering. The relative values of the parameter Ft/Fm, reflecting the velocity of the electron transport chain of chloroplast membranes is in the range 0.3–0.4, which corresponds to the high photosynthetic activity. In the leaf tissue homogenates regenerantnyh plants varieties Lasurit and Dolphin processed aerosol "Greenhouse", and in the control group the content of peroxidase have been studied, which is an indicator of physiological stress of plants. With an increase in enzyme activity can be stated that the action of a factor caused a chain of physiological and biochemical processes that lead to the response to stress.

Such a reaction may have immune activating effect. As a result of the immune enzyme analysis with antibodies to the enzyme peroxidase, it was found that the drug "Greenhouse" causes a change in the activity of this enzyme. Moreover, there are reliable changes in activity (significance level 0.01), depending on the multiplicity of processing. The results of determination of peroxidase activity are shown in Figure 31.

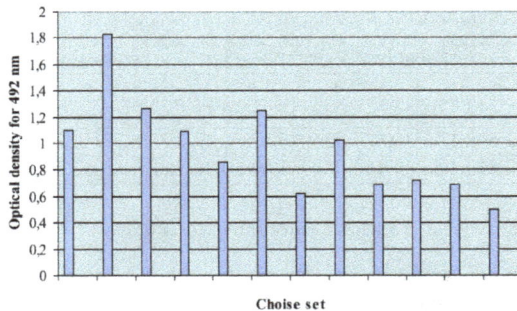

Figure 31. Peroxidase activity in membrane fractions of regenerated potato varieties Dolphin and Lazurit.

As a result of the experiments, significant differences in the value of this indicator depending on the variety were observed. In the variety Dolphin pretreatment before the cuttings a similar reaction regenerates the stress was caused that was manifested in an increase in peroxidase activity. In the grade Lazurit option 3, on the contrary, was the stress is not sensitive (Table 7).

#	Version	Treatment 1	Treatment 2	Treatment 3	Treatment 4	Treatment 5
		Grade Dolphin – medium early				
1	Control	0.776	0.806	0.854	0.760	0.836
2	Variant 2	0.846	0.906	0.917	0.940	0.909
3	Variant 3	1.351	1.390	1.441	1.387	1.366
		Grade Lazurit – early				
4	Control	1.358	1.433	1.488	1.505	1.554
5	Variant 2	1.981	2.031	2.097	2.121	2.073
6	Variant 3	0.949	0.979	1.003	0.996	0.979

Table 7. Change in optical density at $\lambda = 492$ under the action of peroxidase enzyme spray "Greenhouse" in regenerant leaves of potato varieties Dolphin and Lazurit.

We studied the effects on the biological properties of nanoaerosol seeds: vegetables – carrots, beets, kale, forage grasses – goat grass, clover, plants in greenhouses: cucumber, sweet pepper, tomato, parsley, lettuce. Work was carried out with seeds of varying quality. Germination of control samples in different batches ranged from 27% to 99%. Some of the seeds before the experiments were struck by bacterial diseases. The results of the studies have shown high efficiency of this method of improving the biological properties of the seeds. In all cultures were obtained by increasing the energy of germination up to 18% and 20% germination. For control group of seed germination was 67% and 71% germination, and in an experimental batch exposed to a special gas environment, these figures were 87% and 93%, respectively. The study of the nanoaerosol treatment on growth and development of salads also confirmed the high efficiency of the proposed technology. Comparison of germination of lettuce seeds "Gribovsky curly" grade is shown in Figure 32.

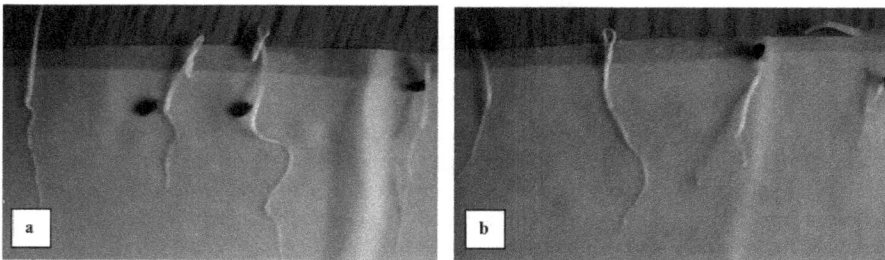

Figure 32. Germination of lettuce seeds "Gribovsky curly" grade (treatment of "Greenhouse" - a, control - b).

It is evident that the seeds after nanoaerosol treatment developed intensively. In addition – treated seeds during storage, preparation for the planting and growing season were significantly more resistant to various fungal diseases. Observations over three years for the treated seeds showed that the rate of germination and vigor are in time at a high level. Reducing of germination does not exceed 1–1.5% per year. Pilot studies according to productivity of potatoes, carrots, a tomato and cucumbers showed that the crop increases on the average by 20–30% [5].

4. Conclusion

1. In the present paper, the multilevel mathematical model is developed for solving the problem of formation, movement, hashing, and condensation of nanoparticles by methods of quantum mechanics, molecular dynamics and mesodynamics, describing the behavior of nanoaerosols for studying the processes of condensation of nanoaerosols for the nutrition of plants.

2. The numerical calculations have showed:
   * the growth of nanoparticles and their interactions depend on time and on the number of free molecules that actively form nanoparticles;
   * the speed of association of molecules into nanoparticles depends on temperature and pressure and molecules concentration in the gas medium;
   * the molecules of potassium carbonate and water are into nanoparticles and therefore they can be used for the nutrition of plants.

3. Experiments have confirmed that in aerosols nanoparticles are formed.

4. In experimental studies influence nano- and the microparticles generated by specially solid propellant composition, on biological objects (plants, fruits and seeds) is established. A number of positive effects which are widely used already today hothouse and the farms working with plants in hothouses were received.
   * Productivity increase to 30%, increase in ovaries of fruits, development of more powerful stalk and leaves and uniformity of maturing of fruits was established.
   * Plants become more resistant to diseases.
   * Early fructification and increase in term of vegetation is observed.
   * Germination of various plants seeds increases to 100%.
   * From treated seeds produced high-quality embryos, superior on all counts of untreated seed embryos.
   * Stability of sprouts to fungous diseases raised.

5. Method of foliar spray fertilizer plant almost does not require any material costs and specialized equipment. This method significantly reduced the complexity of foliar feedings. On average, holding a spray fertilizer on the area of 1000 m^2 spent 0.3 person/hour, and overall costs are reduced by 40% in recalculation per m^2, while increasing the productivity of more than 20%.

6. Technology makes it possible to make fertilizer to the requirements of the manufacturer, to add or to remove from it those nutrients that are necessary to use soil or certain climatic regions.

7. The executed theoretical and experimental studies allowed use this technology for growing various vegetables and crops in greenhouses of Russia, Byelorussia, the Ukraine and China. Practical value of the received results is very good.

Acknowledgements

Ltd. Nord supported this work. The work was supported by the Russian Foundation for Basic Research (grant 10–01–96044–p_Ural_a), by the Presidium of the Ural Branch of

Russian Academy of Sciences as part of a research project for young scientists "Investigation of the interaction of nanoparticles with a stream of gas and solid surfaces" (grant 11-1-NP-50) and by the program of the Presidium of the Russian Academy of Sciences "Nanosystems: fundamental correlations of nano- and macroparameters". Calculations are executed in Joint Supercomputer Center of the Russian Academy of Sciences.

Author details

Vakhrushev A.V., Fedotov A.Yu. and Vakhrushev A.A.
Institute of Mechanics, Ural Branch the Russian Academy of Sciences, Izhevsk, Russia

Golubchikov V.B and Golubchikov E.V.
Join Stocks Company Nord, Perm, Russia

5. References

[1] Alikin V.N, Vakhrouchev A.V, Golubchikov V.B, Lipanov A.M and Serebrennikov S.Y (2010) Development and Investigation of the Aerosol Nanotechnology. Moscow: Mashinostroenie. 196 p.

[2] Golubchikov V.B, Sibiriakov S.V, Levin D.G, Alikin V.N (2006) The Regulated Gas Medium for Preseeding Processing Seeds of Vegetables. Hothouses of Russia j. 1: 55.

[3] Vakhrouchev A.V, Golubchikov V.B (2006) Numerical Investigation of the Dynamics of Nanoparticle Systems in Biological Processes of Plant Nutrition. Abstracts of Conference on Nanoscience and Technology ICN&T. Basel, Switzerland: 209.

[4] Vakhrouchev A.V, Golubchikov V.B (2007) Numerical Investigation of the Dynamics of Nanoparticle Systems in Biological Processes of Plant Nutrition. Journal of Physics, Conference Series. 61: 31–35.

[5] Vakhrouchev A.V, Golubchikov V.B (2006) The Regulated Gas Medium with a Set of Nanoparticles Generated by the Solid Fuel Composition as a Method for Producing Eco-vegetables. NanoBiotech World Congress, Boston, MA, USA: 16–17.

[6] Chushak Y (2001) Molecular Dynamics Simulations of the Freezing of Gold Nanoparticles. Eur. Phys. J.D. 16: 43–46.

[7] Vakhrouchev A.V, Fedotov A.Y, Vakhrushev A.A, Golubchikov V.B, Givotkov A.V (2011) Multilevel Simulation of the Processes of Nanoaerosol Formation. Part 1. Theory Foundations. Nanomechanics Science and Technology: An International Journal. Vol. 2, issue 2: 105–132.

[8] Cagin T, Che J, Qi Y, Zhou Y, Demiralp E, Gao G, Goddard III W (1999) Computational Materials Chemistry at the Nanoscale. J. 1: 51–69.

[9] Steinhauser O.M (2008) Computational Multiscale Modelling of Fluids and Solids. Theory and Application. Berlin: Springe. 428 p.

10] Marx D, Hutter J (2008) Ab Initio Molecular Dynamics: Basic Theory and Advanced Methods. Cambidge: University Presse. 567 p.

[11] Brooks B.R, Bruccoleri R.E, Olafson B.D, States D.J, Swaminathan S and Karplus M (1983) CHARMM: A Program for Macromolecular Energy Minimization, and Dynamics Calculations. J. Comput. Chemistry. Vol. 4, issue 2: 187–217.

[12] Gauss J (2000) Molecular Properties, Modern Methods and Algorithms of Quantum Chemistry. Proceedings, Second Editions. 3: 541–592.

[13] Heermann D.W (1986) Computer Simulation Methods in Theoretical Physics. Berlin: Springer-Verlag. 148 p.

[14] Holian B.L (2003) Formulating Mesodynamics for Polycrystalline Materials. Europhysics Letters. 64: 330–33.

[15] Vakhrushev A.V (2009) Modeling of the Nanosystems Formation by the Molecular Dynamics, Mesodynamics and Continuum Mechanics Methods. Multidiscipline Modeling in Materials and Structures. Vol. 5, issue 2: 99–118.

[16] Landau L.D, Lifshits E.M (1972) Quantum Mechanics. Moscow: Science. 368 p.

[17] Burkert U, Allinger N.L (1982) Molecular Mechanics. Washington D.C.: ACS Monograph. 339 p.

[18] Imry Y (2002) Introduction to Mesoscopic Physics. Oxford: University Press. 236 p.

[19] Vakhrushev A.V, Fedotov A.Y, Vakhrushev A.A (2011) Modeling of Processes of Composite Nanoparticle Formation by the Molecular Dynamics Technique. Part 1. Structure of Composite Nanoparticles. Nanomechanics Science and Technology: An International Journal. Vol. 2, issue 1: 9–38.

[20] Fedotov A.Y, Vakhrushev A.V (2011) The Software Package for Multi-level Modeling of the Formation of Heterogeneous Nanoparticles ComplexDyn v.5.0. Certificate of Registration of electronic resource number 17335, Russian Federation: 1–6.

[21] Golubchikov V.B, Sharupich V.P (2002) Foliar spray plants as a way of saving in greenhouses. Greenhouses of Russia. 2: 15–24.

Chapter 3

Enzymatic Antioxidant Responses of Plants in Saline Anthropogenic Environments

Piotr Kamiński, Beata Koim-Puchowska, Piotr Puchowski, Leszek Jerzak, Monika Wieloch and Karolina Bombolewska

Additional information is available at the end of the chapter

1. Introduction

Plants under natural conditions are frequently exposed to combined stressors including drought stress and desiccation, salt stress, chilling, heat shock, heavy metals, ultraviolet, radiation, air pollutants such as ozone and SO_2, mechanical stress, nutrient deprivation, pathogen attack and high light stress [1]. A common result of most abiotic and biotic stresses is an increased production of reactive oxygen species (ROS), which frequently result in oxidative stress [2], [3]. The production of ROS results from pathways such as photorespiration, from the photosynthetic apparatus and from mitochondrial respiration [1].

The reduction of molecular oxygen to H_2O yields to intermediates reactive oxygen species (superoxide anion ($O_2^{\bullet-}$), hydroxyl radical ($OH^\bullet$), hydroperoxyl radical ($HO_2^\bullet$), hydrogen peroxide (H_2O_2), singlet oxygen ($^1O_2^{\bullet-}$), which are potentially toxic because they are relatively reactive compared with O_2 [4]. Reactive oxygen species may lead to the unspecific oxidations of proteins, membrane lipids or DNA injury. Malonodialdehyd (MDA) content, as a product of lipid peroxidation, has been considered as an indicator of oxidative damage in the cell membrane, resulting in disruption of metabolic function and loss of cellular integrity [5], [6]. Many reports confirm lipid peroxidation has been associated with damages provoked by a variety of environmental stresses [7], i.e. with salt stress [8], [9], [10] and heavy metals [11], [12].

However, there is evidence that ROS also plays important roles in the plant's defence system against pathogens and any pathogenic factors, mark certain development stages such as tracheary element formations, lignification and other cross linking process in the cell wall and act as intermediate signalling molecules to regulate the expression of genes [4]. Thus it is very important for cells to control of level of reactive oxygen species but not eliminate them completely. Defence mechanisms against free radical-induced oxidative stress involve: (1) preventive mechanisms, (2) repair mechanisms, (3) physical defences, and

(4) antioxidant defences [13]. The plants defend against these ROS by the induction of activities of certain antioxidative enzymes such as catalase (CAT), glutathione peroxidase (GPx), glutathione reductase (GR), and superoxide dismutase (SOD), which scavenge reactive oxygen species [14], [15].

Superoxide dismutase catalyses the reaction of dismutation of superoxide radicals: $2O_2^{\bullet-}$ + $2H^+ \rightarrow H_2O_2 + O_2$. SOD is a metaloprotein with metals Cu, Zn, Mn, and Fe as co-factor [16], Stroiński 1999). Cu-Zn-SOD (homodimer; 31-33 kD) is present in cytosol and chloroplasts. Mn-SOD (tetramer 90kD) was found in the matrix mitochondrium [17], [18], whilst Fe-SOD, commonly occurring in *Procaryota*, was detected only in some plants [19], [18]. On the other hand catalase catalyses the reaction: $2 H_2O_2 \rightarrow 2H_2O + O_2$. This enzyme contains three isoenzymatic forms (CAT-1, CAT-2, and CAT-3). CAT-1 and CAT-2 occur mainly in glyoxisomes, peroxysomes, and cytosol, while CAT-3 in mitochondria and cytosol [20], [18].

Another important factor decomposing H_2O_2 is peroxidase of ascorbic acid (APOX) (m.w. 28-34 kD) which contains protoheme as a prosthetic group and 4 cysteines responsible for sensitivity of the enzyme to compounds blocking a thiol group [21], [18]. It occurs in chloroplasts and cytosol [22], [18], vacuole [23], [18], and in apoplasts [24], [18]. APOX, in particular the chloroplast isoenzyme, becomes labile if the concentration of ascorbic acid drops significantly and it decomposes under the influence of its own radical product. Under such circumstances both monodehydroascorbate (MDHAA) and dehydroascorbate (DHAA) are easily reduced along the Halliwell-Asada path [18], [25]. The cell defensive mechanism also depends on small-cellular antioxidants such as glutathion (GSH), proline (Pro) or ascorbic acid (AA) which also react with radicals generated in the oxidative-stress [18].

The origin of oxidative stress under salinity conditions is well documented in leaves, where the inhibition of Calvin cycle results in over-reduction of oxygen and formation of superoxides [26]. Salt stress limits also gas exchange and thereby CO_2 supply to the leaf. One consequence is the over-reduction of photosynthetic-chain electron transport. This induces the generation of ROS [15]. Little is known about the origin of oxidative stress in roots, although salt stress-related impairment of mitochondrial function is likely to be involved [27]. Recent studies have revealed that salt-induced oxidative stress occurs in root mitochondria in a wild salt-tolerant tomato species as indicated by increased concentrations of H_2O_2 and MDA [28], [27]. It is now widely accepted that superoxide radicals produced during respiration, in response to salt stress, are the main precursors of mitochondrial H_2O_2 and make an important contribution to the oxidative load experienced by the cell [26]. Root mitochondrial ROS production is increased as a result of high salinity as constraints are imposed on electron transport through mitochondrial complexes I and II [27].

Calcium as macroelement and important physiological chemical element reacts with heavy metals and is required for various roles in plant cell [29], [30]. It is implicated in the movement of cellular organelles such as the spindle apparatus and secretory vesicles, and may play a key role in integrating plant cell metabolism [31], [32]. However, the homeostasis of Ca is maintained principally by the action of extrusion proteins. The cytosol is strongly buffered against high concentrations of calcium by numerous ranges of calcium-binding proteins such as calmodulins and calmodulin-binding proteins [2]. Reports showed close interaction between intracellular H_2O_2 and cytosolic calcium in response to biotic and abiotic stresses -increase in cytosolic calcium boosts the generation of H_2O_2. The protein calmodulin binds to activates plant catalases in the present of calcium. It indicated a dual function of Ca in regulating H_2O_2 homeostasis [30].

Metals are involved in the direct or indirect generation of free radicals (FR) and reactive oxygen species in the following ways: 1. direct transfer of electron in the single electron reduction; 2. disturbance of metabolic pathways resulting in an increase in the rate of FR and ROS formation; 3. inactivation and down regulation of the enzymes of the antioxidative defence system, and 4. depletion of low molecular weight of antioxidants [33]. Microelements such as Fe, Zn, Cu and Mn fulfil various roles in the metabolism of plant organism and are necessary for the regularity of physiological processes, however the excess and deficiency of these elements leads also to disturbance of ionic homeostasis [34]. Fe, Mn, Cu, and Zn as transition metals have frequently unpaired electrons and they are, therefore, very good catalysts of oxygen reduction. In aqueous solutions at neutral pH, $O_2^{\cdot-}$ can generate H_2O_2, which can subsequently decompose to produce $^{\cdot}OH$ by the Haber-Weiss reaction in which Cu and Fe being involved: $O_2^{\cdot} + H_2O_2 \rightarrow O_2 + OH^- + {}^{\cdot}OH$. When iron is the transition metal in the Haber-Weiss reaction, it is called the Fenton reaction [35], [33].

Toxic heavy metals Cd an Pb are considered as not essential metals for plant metabolism and stimulate formation of free radicals and reactive oxygen species [36], [37], [38]. The ROS formation by toxic metals is indirectly rather than directly [39]. The toxicity of Cd may result from its binding to sulfhydryl groups (S-H) of proteins leading to inhibition of activity and disruption of structure, due to perturbations in the nutrient balance and disturbance of cellular redox control 40]. Due to a Cd inhibitory effect on the Calvin cycle, there is a decrease in NADPH utilization, resulting in the one-electron reduction of a large number of oxygen molecules on the reducing side of PSI [39]. The reports of [41], [42], [43], [44], [45], [46], [47], and [48] have concluded that an oxidative stress could be involved in Cd toxicity, by either inducing oxygen free radical production, or by decreasing enzymatic and non-enzymatic antioxidants.

The subject of numerous studies for at least the past four decades has been physiological responses of plants for salinity and heavy metals in the controlled laboratory conditions. However, many of these determinations are concentrated to one type of pollutant, whilst plants in the natural conditions are subjected to many stressful differentiated ecophysiological sources and factors. Therefore there still remains a need for research on the interdependencies of plants with multiple biotic and abiotic factors in their natural habitats, their adaptation mechanisms and responses. The aim of this paper was thus to investigate the enzymatic antioxidant mechanisms and responses in plants subjected to destabilization of chemical elements management in the natural conditions. We thus studied antioxidant enzymes SOD, CAT, and APOX, and the content of malondialdehyde (MDA) variations in different ecological groups of glycophytes Creeping thistle *Cirisium arvense*, Common nettle *Urtica dioica*, Yarrow *Achillea millefolium*, and Burdock *Arctium lappa* in various types of environments: salted and alkaline anthropogenic environments, agricultural environments and also pollution free environments of the Pomeranian region of Poland. We also investigated the halophyte Common glasswort *Salicornia europaea* from only salted environments. Simultaneously, we examined the levels of chemical elements Na, Ca, Fe, Zn, Cu, Mn, Cd, and Pb in roots and green parts of plants as probably the factor generated reactive oxidant species and thus activated enzymatic antioxidant mechanisms. We compare environmental and ecophysiological determinations of plant groups under salinity and acidity and their adaptation strategies. These studies were planned in order to understand whether the consideration of particular relationships between destabilization of free radicals homeostasis are connected with the excess of Na and Ca and pro-antioxidant balance.

2. Study area

The study area is situated in the Pomeranian Kujawy region (52-53°N, 18-20°E, central Poland) connected with Permian rock-salt uplifted in the form of salted domes and associated salt springs and saline ground waters, which had influence on the development of a sodium industry in districts of Janikowo and Inowrocław and non-polluted and industrialized area of the Tuchola Forestry complex (54-53°N and 17°30′-18°30′ E); Fig. 1.

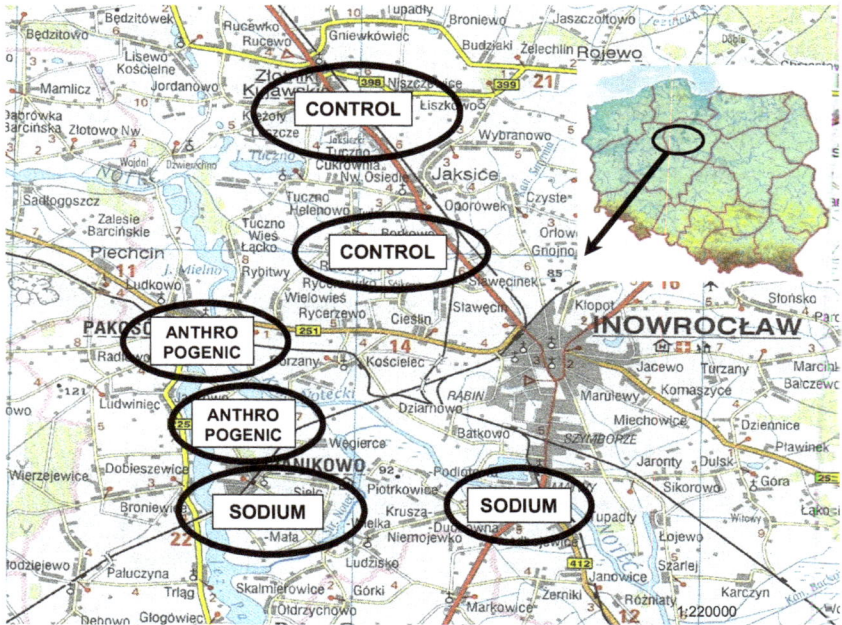

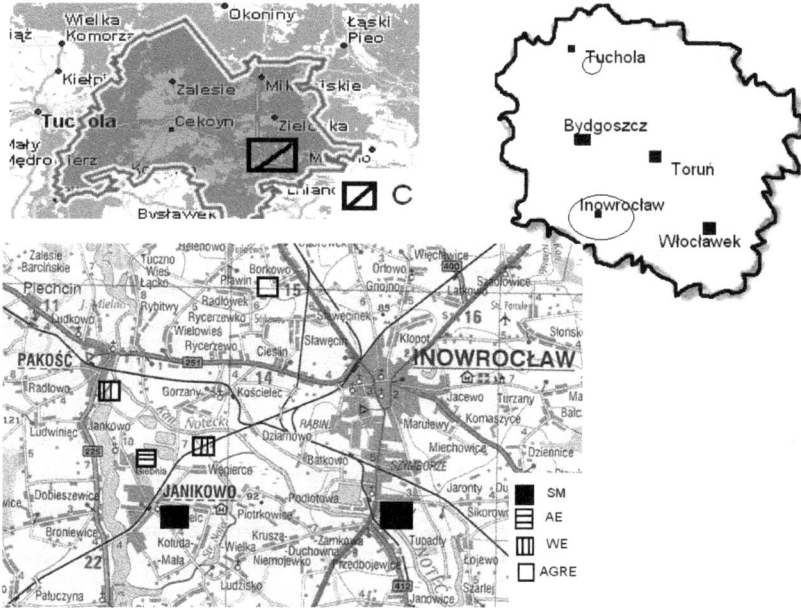

Figure 1. Study area. Environments: C – control, AGRE – agricultural, WE – wetlands, AE – anthropogenic, SM – sodium manufactures (http://www.maps.google.pl/kujawsko-pomorskie, modified).

The mainly human management caused secondary salinity and consequently alkalization of Kujawy areas as a result of inappropriate tightening of sediment traps causing wastes to infiltrate into the soil and grassland irrigation. Our research was done in selected habitats with a variety of anthropogenic disturb: 1) "Sodium manufactures" (SM) - Inowrocław and Janikowo district, closely by sodium factories; 2) "Anthropogenic" (AE) waste dumping sites near Giebnia and neighboured polluted areas; 3) "Wetlands" (WE) floodplains near Noteć Channel and Pakoskie Lakeland; 4) "Agriculture" (AGRE) – agriculture areas near Borkowo and 5) "Control" – in Tuchola Forestry named "Bory Tucholskie" (see Fig. 1).

3. Material and methods

Field studies were carried out in 2008-2010 during summer (May-June-July). Plant species were estimated on the basis of the work by [49] and were divided onto glycophytes - Creeping thistle *Cirisium arvense*, Common nettle *Urtica dioica*, Yarrow *Achillea millefolium*, Burdock *Arctium lappa* and obligatory halophyte: Common glasswort *Salicornia europea*. Common glasswort appears only in salted habitats and is known as succulent [50]. It is one of the most saline tolerant plants in general and is capable to grow under highly saline conditions in the lowest part of salt marshes [51], [52]. Samples of roots and green parts of plants: leaves from glycophytes (No of samples N=220), and shoots of Common glasswort *Salicornia europea* (No of samples N=150), and samples of soils (No of samples N=300) from rhizosphera were collected from all studied environments (glycophytes) and only from anthropogenic salted areas (halophytes), because of their natural distribution.

Samples of soil were air-dried to a constant mass up to 65^0C, homogenized and sieved through 1 mm mesh. The electrolythical conductivity (expressed as mS or μS) of soil (Ec) was measured by conductivity meter (Elmetron CC-401). Instead the soil acidity (pH) was determined in bidistilled water at soil solution ratio of 1:2.5 with a potentiometric glass electrode by used pH-meter (Elmetron C-501).

Samples were collected from randomly selected plants at midday, and were kept cool in freezer bags for transfer to the laboratory. Green parts samples were then washed three times with deionized water; roots were gently separated from soils, then abundantly washed with tap water to eliminate soil particles, and also rinsed three times with deionized water. Root and washed shoot samples were divided into two parts. One was immediately stocked at −80°C until further biochemical analysis, while a second was cut into small pieces and dried in oven for 24 h [53], [54]. Mineralized samples of soil and plants were analyzed for Na, Ca, Fe, Zn, Cu, Mn, Cr, Ni, Cd, and Pb using inductively coupled plasma mass spectrometry ICP-MS (AGILENT 7500 CE; plasma ICP-MS spectrophotometer from Agilent Technologies Inc. (Palo Alto, CA, USA). The results were interpreted comparatively with standard well-known concentration, i.e. in relation to the analysis of reference materials. Parallel measurements were taken in the blind trials. The results were given in mg*kg^{-1} of dry weight.

To determine the levels of superoxide dismutase (SOD) and catalase (CAT) roots and green parts of plants (1g) were homogenized in a 0.05 M phosphate buffer (pH 7.0) containing 1mM EDTA (3 ml). During homogenization polyvinyl-pyrrolidone (PVP) (0.25 g) was added (modified by [40]). The homogenate was centrifugated twice for 10 min first at 10 000 x g and second at 20 000x g. For ascorbate peroxidase (APOX) assay the roots and green parts (1g) were homogenized according to [54], i.e. by adding 0.05 M phosphate buffer (pH 7.5) containing 1 mM EDTA, 1 mM sodium ascorbate, 1 mM DTT and 4.0% (w/v) polyvinyl-pyrrolidone (PVP), 1 mM EDTA (5 ml). The homogenate was centrifugated at 15 000x g for 20 min. Sodium ascorbate was added only in case of green parts. All assays were conducted at 4° C. The protein content in the supernatant was measured according to [55].The activity of all enzymes was expressed in units of mg^{-1} of protein.

The activity of SOD was assayed by measuring its ability to inhibit the photochemical reduction of NBT, adopting the method of [56]. The reaction mixture consist of 50 mM phosphate buffer (pH 7.8) containing 0.1 mM EDTA, 120 μM riboflavine, 97.5 mM methionine, 2.25 mM NBT. After the addition of plant homogenate switching on a UV lamp for one minute the start of reaction is begin. The absorbance at 560 nm of wave length was measured. Fifty percent reduction in color was considered as one unit of enzyme activity.

The activity of CAT was assayed according [57], [58] in reaction solution (3 ml) composed of 0.1 M phosphate buffer (pH 7.0) to which 30% (w/v) H_2O_2 was added until reaching an absorbance at 240 nm between 0.520 and 0.550. The reaction was initiated by the addition of 10 μl of plant extract and the activity of CAT was monitored by the decreasing absorbance at 240 nm at for 2 min. One unit of enzyme was the amount necessary to decompose 1 μmol of H_2O_2 per min at 25°C (extinction coefficient 43.6 $M^{-1}*cm^{-1}$).

The activity of APOX was measured according method of [59]. The soluble protein extract (25 μl) was added to 1.975 ml reaction mixture of 0.05 M phosphate buffer (pH 7.0), 0.5 mM sodium ascorbate, 0.1 mM H_2O_2 and the decrease absorbance of ascorbate at 290 nm for 2 min was used to calculate APOX activity. Enzyme activity was expressed in enzyme unit of mg^{-1} protein. One unit of enzyme was the amount necessary to decompose 1 μmol of substrate per min at 25°C (extinction coefficient 2.8 $mM^{-1}*cm^{-1}$).

The intensity of lipid peroxidation was estimated following the method by [60]. Approximately 0.5 g of frozen plant tissue samples was cut into small pieces, homogenized with 2.5 ml of 5% trichloroacetic acid, and then centrifuged at 10000 g for 15 min. at room temperature. The equal volumes of supernatant and 0,5% thiobarbituric acid in 20% trichloroacetic acid were added in a new tube and incubated in 96°C for 25 minutes and then quickly cooled in an ice bath. After centrifugation at 8 000 g for 5 min, the absorbance of supernatant was recorded at 532 and 600 nm. The value for non-specific absorption at 600 nm was subtracted. The concentration of MDA was calculated using coefficient of absorbance 155 mM-1 cm-1 and was expressed as μmol g^{-1} FW.

3.1. Statistical analysis

Arithmetic means and descriptive statistics of activation of antioxidant enzymes: SOD, CAT, APOX and concentrations of Na, Ca, Fe, Zn, Cu, Mn, Pb, and Cd in roots and green parts of plants were calculated. We also calculated arithmetic mean of concentrations of the same elements in soils from root zone of plants. The data did not show normal distribution, so non parametric tests were used. We used ANOVA Kruskal-Wallis test, followed by multiple Kruskal-Wallis comparison and U-Mann-Whitney test to estimate the significance of differences in the activation of antioxidant enzymes and concentrations of elements in roots and green parts from different environments, between organs and also groups of plants from the same environments (significance level at $p<0.05$), and also the differences between concentrations of metals in soil in particular environments. The dependence of activity of antioxidant enzymes and concentrations of elements in the roots and green parts of plants from different environments were calculated by correlation coefficient (r) according to the ranks of Spearman test (significance level at $p<0.05$) [61].

This work required permits from General and Regional Nature Conservation Dpts. These were obtained and had the following respective numbers: DOPozgiz-4211/I-14/995/09/ep, RDOS.PN.6631/2/08/KLD, and RDOS.04.PN.6631/37/09/KLD.

4. Results

4.1. Soil parameters

As shown in Table 1, the soils at Kujawy region were in the various degree of alkalinity (sodium manufactures; strong alkaline), anthropogenic environments, wetlands, agriculture environment (moderate alkaline), whilst pH of soils from control was acid. The electrical conductivity of soils was higher in sodium manufactures and anthropogenic environments than in control (Table 1).

As soils collected in SM and AE environments mainly consisted of tailings from nearby sodium factory and AGRE environment near field used agriculturally, heavy metal contents in the sample of soils were high and varied greatly in comparison with control and wetlands. The concentration of Na, Ca, Fe, Zn, Cu, and Mn were higher in soils at disturbed than control environments but in case of the same elements also from wetlands. However, concentration of toxic metals Pb, and Cd were higher in Tuchola Forestry than in Wetlands and did not differ with concentrations in soils from remaining environments (Table 1).

4.2. Chemical elements in roots and green parts of glycophytes and Common glasswort *Salicornia europaea*

4.2.1. Glycophytes

We found lower concentrations of Na, Ca, Fe, Cu, and Pb in roots in control in compare with environments in Kujawy region in opposite to level of Zn (higher in control than in agricultural and did not differ in the remaining environments). Similar to Zn concentration, Cd level was also higher in non-polluted environment (control) in comparison with factory disturbed sodium manufactures and anthropogenic (Table 2). Simultaneously, sodium level in glycophytes from anthropogenic environments was higher than in those from remaining areas.

Concentrations of Fe and Zn were lower in leaves from control but only in comparison with wetlands. Simultaneously, Pb level was lower in control areas than in wetlands and agriculture fields. It is interesting that the level of Cd, in opposite to Pb, was higher in the unpolluted environment (C) compared with all environments studied in Kujawy region, except anthropogenic areas. Similar concentration of Mn was also higher in the control (C) as compared with industrial disturbed SM and AE and also wetlands (Table 2). We did not find significant differences in the concentrations of Mn in roots and the level of Ca, and Cu in leaves among investigated environments (Table 2).

4.2.2. Common glasswort Salicornia europaea

We found only higher concentrations of Na in roots and Mn in leaves from anhropogenic than from sodium manufactures. Simultaneously, the roots and green parts of plants had higher concentrations of Pb in sodium manufactures than in the anthropogenic. Concentrations of Ca, Fe, Zn, Cu and Cd did not differ between both organs and amongst environments studied (Table 3).

	SM (N=35)		AE (N=55)		WE (N=85)		AGRE (N=50)		C (N=75)		H	p
	Mean	SD	Mean	SD	Mean	SD	Mean	SD	Mean	SD		
pH	8,78	1,713	8,25	0,509	7,84	0,181	7,83	0,114	5,00	1,518	24,66	0,000 C<SM, AE, WE
EC	33,90	28,625	7,55	8,504	1,68	1,156	1,45	0,354	1,31	0,980	14,30	0,006 C<SM, AE
Na	821,81	823,663	451,04	412,291	52,02	33,097	67,00	4,282	51,77	42,593	19,67	0,001 C, WE<AE
Ca	83778,73	104267,017	40552,16	38357,177	5472,83	3456,820	32691,09	4396,396	2121,80	1854,719	29,39	0,000 C<SM, AE, AGRE; WE<AE
Fe	5724,14	1541,227	9312,59	2272,352	4955,36	2448,168	7211,14	221,338	2035,80	272,179	28,94	0,000 C<AE, AGRE; WE<AE
Zn	42,79	32,909	72,82	46,669	14,65	5,187	31,21	1,201	17,42	7,450	30,39	0,000 C, WE<AE; WE<AGRE
Cu	8,79	5,318	20,28	16,444	3,33	1,368	9,21	0,085	1,90	0,820	33,83	0,000 C<SM,AE,AGRE; WE< AE, AGRE
Mn	200,83	33,028	220,05	45,325	158,46	70,534	287,31	28,798	138,75	98,082	19,42	0,001 C,WE<AGRE
Pb	17,78	10,227	59,20	39,063	8,95	3,424	13,64	0,787	13,23	3,849	30,31	0,000 WE<AE,C
Cd	0,13	0,074	0,16	0,098	0,06	0,025	0,09	0,001	0,14	0,058	11,75	0,019 C>WE

Table 1. Soil characteristics of sodium manufacture (SM), anthropogenic environments (AE), wetlands (WE), agricultural environments (AGRE) and control (C); mean ± SD. Significant differences in the concentrations of Na, Ca, Fe, Zn, Cu, Mn, Pb, and Cd in soil among environments (p<0.05).

	SM (N=34) Mean	SD	AE (N=49) Mean	SD	WE (N=72) Mean	SD	AGRE (N=34) Mean	SD	C (N=31) Mean	SD	H	p	
roots													
SOD	83,83	52,06	119,19	79,51	96,25	64,43	143,73	87,06	96,80	37,29	13,54	0,009	AGRE> SM, WE
CAT	0,35	0,49	0,23	0,24	0,19	0,13	0,21	0,19	0,19	0,10	1,12	0,891	
APOX	34,09	22,73	41,42	37,50	46,72	51,18	48,73	32,82	38,04	20,87	4,99	0,289	
MDA	11,40932	6,507260	11,12850	5,715258	10,59186	6,015241	5,86504	2,785663	11,12856	8,162991	25,68	0,000	AGRE<SM, AE, WE, C; C<SM, AE ; WE<SM, AE
Na	2028,65	2193,52	2249,29	2193,05	572,37	726,67	797,09	693,15	663,97	981,73	42,29	0,000	
Ca	13322,53	8226,41	9033,28	5182,81	8 175,14	3716,05	9141,37	7099,62	6098,64	2384,96	28,76	0,000	SM>AE, WE, AGRE, C
Fe	675,88	713,57	461,23	327,59	579,35	411,30	600,60	464,30	364,68	346,99	10,99	0,027	C<WE
Zn	24,67	10,64	23,72	10,89	18,51	11,59	15,05	3,40	26,27	18,09	29,32	0,000	AGRE<SM, AE, C; WE<AE, SM
Cu	13,67	9,60	10,45	12,93	9,57	8,98	7,37	5,06	9,72	11,14	12,98	0,011	SM>WE, AGRE, C
Mn	40,91	29,92	40,66	37,82	45,90	41,03	57,59	30,33	77,61	75,52	10,01	0,402	
Pb	4,15	4,15	3,89	6,55	1,54	1,48	1,16	0,91	1,18	0,74	39,67	0,000	AE>AGRE, C, WE; SM>AGRE, C, WE
Cd	0,17	0,11	0,15	0,11	0,19	0,24	0,10	0,04	0,25	0,22	16,31	0,003	AGRE<C, SM
green parts													
SOD	120,21	89,97	140,63	104,34	73,02	42,80	169,19	161,18	123,77	90,81	29,86	0,000	WE<SM, AE, AGRE, C
CAT	0,34	0,34	0,38	0,64	0,19	0,16	0,25	0,18	0,46	0,64	8,72	0,069	
APOX	57,52	39,03	45,00	36,42	55,74	52,06	59,92	40,74	48,03	47,10	6,90	0,141	
MDA	9,51898	3,440164	14,53614	4,554324	10,01070	4,226993	9,79719	4,854250	11,04482	5,097336	34,35	0,000	AE>SM, WE, AGRE, C; C<SM, AE, WE ;
Na	171,24	209,12	1423,17	2379,17	193,29	371,57	147,09	538,10	80,85	122,73	78,24	0,000	AGRE<SM, AE, WE ; AE>SM, WE
Ca	21754,94	14162,79	18801,86	11008,24	20 838,65	12124,47	25354,39	14019,17	21597,36	14022,93	3,69	0,449	
Fe	178,51	97,44	217,27	164,10	224,47	173,91	185,20	116,29	152,36	99,02	10,80	0,029	C<WE
Zn	33,27	21,09	27,46	13,26	24,06	10,40	26,96	15,56	35,12	20,39	11,15	0,025	C<WE
Cu	7,08	4,31	7,46	2,97	7,88	6,03	7,44	2,77	7,25	3,59	3,11	0,540	
Mn	37,38	46,00	54,90	74,88	72,79	112,25	53,04	24,37	128,35	197,60	48,60	0,000	SM<AGRE, WE; C>SM, AE, WE
Pb	0,82	0,73	0,87	0,50	0,68	0,38	0,54	0,34	0,51	0,53	30,07	0,000	C<AE, WE ; AE>AGRE
Cd	0,06	0,05	0,07	0,05	0,17	0,43	0,06	0,04	0,19	0,22	13,53	0,009	C>SM, WE, AGRE

Table 2. Elements concentrations, activity of antioxidant enzymes, and degree of lipoperoxidation in organs of populations of glycophytes in sodium manufacture (SM), anthropogenic environments (AE), wetlands (WE), agricultural environments (AGRE) and control (C); mean ± SD. Significant differences among environments (p<0.05).

	SM (N=70)		AE (N=80)			
	Mean	SD	Mean	SD	Z	p
roots						
SOD	152,34	207,76	127,13	104,47	0,00	1,000
CAT	0,25	0,30	0,17	0,14	0,03	0,975
APOX	81,59	42,20	98,08	77,09	-0,03	0,975
MDA	5,19	1,73	5,08	0,91	-0,10	0,922
Na	10266,67	6637,63	18772,32	7713,88	-2,14	0,032
Ca	6112,00	5671,56	3900,34	1462,37	-0,09	0,926
Fe	467,42	391,22	470,88	274,67	-0,59	0,555
Zn	19,49	13,36	17,39	7,22	0,28	0,780
Cu	5,80	5,09	6,33	3,01	0,28	0,780
Mn	30,46	19,87	24,72	7,81	0,96	0,335
Pb	2,64	1,71	1,01	0,29	3,13	0,002
Cd	0,16	0,24	0,14	0,09	-0,84	0,401
green parts						
SOD	110,04	100,51	171,65	146,89	-1,93	0,053
CAT	0,39	0,32	0,15	0,14	1,88	0,061
APOX	34,44	34,77	21,68	17,38	0,55	0,583
MDA	2,48	0,69	4,04	0,98	-3,64	0,000
Na	44235,06	32261,31	49327,20	22546,83	-0,78	0,436
Ca	22766,03	20946,63	25326,90	12844,86	-0,49	0,624
Fe	378,85	482,21	239,69	159,86	-0,38	0,707
Zn	19,98	10,44	13,38	6,51	1,82	0,069
Cu	8,25	18,29	4,07	1,62	-0,89	0,371
Mn	36,92	36,70	127,42	84,06	-3,20	0,001
Pb	1,32	0,55	0,27	0,08	4,13	0,000
Cd	0,20	0,29	0,20	0,18	-0,49	0,624

Table 3. Elements concentrations, activity of antioxidant enzymes, and degree of lipoperoxidation in the organs of populations of Common glasswort *Salicornia europaea* in sodium manufacture (SM), and anthropogenic environments (AE); mean ± SD. Significant differences among environments at p<0.05.

4.3. Antioxidant enzymes activity and lipoperoxidation

Our results indicated significant differences only in the activity of SOD (glycophytes) in both organs. SOD activity was higher in roots from agricultural environments comparatively with sodium manufactures and wetlands. Simultaneously, SOD activity was lower in leaves from wetlands in comparison with remaining environments. We also found lower MDA content in roots from agricultural environments than from remaining examined areas. Instead the highest degree of lipid peroxidation was in the green parts of plants from anthropogenic environments (Table 2). We did not found differences in the biochemical parameters in organs of Common glasswort *Salicornia europaea* except of MDA content in the green parts of plants (higher level in the AE than in the SM) from different environments (Tables 4, 5).

4.4. The impact of metals on the antioxidant enzymes activity in roots and leaves of plants

Spearman coefficient analyses were performed with concentrations of each of the metals studied in roots and green parts of glycophytes and Common glasswort *Salicornia europaea* versus the activity of antioxidant enzymes and degree of lipoperoxidation. The activity of enzymes was also correlated with MDA content. However, these relationships were depending on the type of environment (Tables 4-6).

4.4.1. Glycophytes

We found more relationships between activity of antioxidant enzymes and metals in the green parts than in roots, depending on the type of environment. However, concentrations of Na (SM), Fe (AE, WE), Cu (SM, AE, WE, C) and Pb (AE) were positively correlated with SOD activity. Instead, CAT activity had also positive relationships with Na, Cu (AE) and Zn, Mn (C). Simultaneously, the activity of another enzyme, APOX, was positively related with level of Na (SM, AGRE), Zn, and Mn (AE) and negatively – with Cu (C), and Cd (AGRE, C).

Our results indicated positive correlations of SOD activity with concentrations of Zn (SM, AGRE), Cu (SM) and Cd (SM, WE, AGRE) but also negative with Mn (AGRE, C) and Pb (AGRE). Concentrations of Na were also related with SOD activity, but were positive (AE) or negative (C). We found negative relationships between CAT activity and concentrations of Ca (WE), Fe (SM, WE, AGRE), Mn (SM), and Pb (WE, AGRE) but also positive with Zn (AGRE). The level of Cu was positive related with activity of CAT in the leaves of plants from sodium manufactures but also negative in the leaves from anthropogenic areas. APOX activity was positively correlated with Ca concentrations in the leaves of plants from all environments and with Fe in the leaves collected at wetland and agriculture environments. However, the activity of this enzyme was negatively related with Na, Zn (C) and Cd (SM, AE, C). Concentrations of Cu had both positive (WE) and negative (AE) relationships with APOX in the leaves.

We found positive relations between concentrations of Na (AE), Cu, (AE, C), Pb (WE) and Cd (WE, C) and degree of lipoperoxidation processes in roots. However, we also found correlations between Na, Zn, Mn, and Cd but in the control environment (Na), AE, WE (Ca), AE (Zn, Cd), and SM (Mn). It should be emphasized that we did not state relation between concentrations of Na, and also Mn, and Pb and MDA content in the leaves. Concentrations of Ca were positively correlated with level of MDA (AE, WE) in the green parts in opposite to metals – Zn and Cd (AE, AGRE). Iron and copper stimulated lipoperoxidation (Fe in AGRE, Cu in WE) and was mutually negatively correlated (SM, SM, C) with content of MDA but it depend on the type of environment (Tables 4, 5, Figs. 2, 3). MDA content was positively correlated with CAT activity in roots (AE, AGRE) and APOX in leaves (AE, WE, AGRE). On the other hand, we found negative relations between MDA and the activity of APOX in roots (AE) and SOD, and CAT in leaves (AGRE); Table 4.

SM (N=34)			AE (N=49)			WE (N=72)			AGRE (N=34)			C (N=31)		
relations	R	p	relations	R	p	relations	R	p	relations	R	p	relations	R	p
roots														
SOD & Cu	0,43	0,011	SOD & Fe	0,46	0,001	SOD & Fe	0,29	0,015	APOX & Cu	-0,37	0,031	SOD & Cu	0,364	0,044
SOD & Na	0,61	0,000	SOD & Cu	0,59	0,000	SOD & Cu	0,42	0,000	APOX & Cd	-0,51	0,002	APOX & Cu	-0,565	0,001
APX & Na	0,36	0,036	SOD & Pb	0,34	0,016	MDA & Cu	0,37	0,001	APOX & Na	0,61	0,000	APOX & Cd	-0,360	0,040
MDA & Mn	-0,35	0,040	APX & Mn	0,32	0,024	MDA & Cd	0,39	0,001	CAT & MDA	0,63	0,000	CAT & Mn	0,421	0,040
			APX & Zn	0,32	0,025	MDA & Pb	0,33	0,005				CAT & Zn	0,528	0,040
			CAT & Cu	0,43	0,002	MDA & Na	0,32	0,006				MDA & Cu	0,498	0,040
			CAT & Na	0,34	0,018	MDA & Ca	-0,27	0,021				MDA & Cd	0,720	0,040
			MDA & Cu	0,35	0,014							MDA & Na	-0,510	0,040
			MDA & Zn	-0,43	0,002									
			MDA & Cd	-0,32	0,023									
			MDA & Na	0,49	0,000									
			MDA & Ca	-0,41	0,003									
			APOX & MDA	-0,51	0,000									
			CAT & MDA	0,55	0,000									
green parts														
SOD & Cu	0,34	0,047	SOD & Na	0,35	0,013	SOD & Cd	0,42	0,000	SOD & Mn	-0,41	0,017	SOD & Mn	-0,521	0,003
SOD & Zn	0,38	0,025	APOX & Cd	-0,45	0,001	APOX & Fe	0,34	0,003	SOD & Zn	0,60	0,000	SOD & Na	0,411	0,022
SOD & Cd	0,35	0,044	APOX & Ca	0,59	0,000	APOX & Cu	0,29	0,014	SOD & Cd	0,57	0,000	APOX & Zn	-0,473	0,007
APOX & Cd	-0,36	0,036	CAT & Cu	-0,35	0,012	APOX & Ca	0,31	0,008	SOD & Pb	-0,48	0,004	APOX & Cd	-0,649	0,000
APOX & Ca	0,41	0,017	MDA & Zn	-0,43	0,002	CAT & Fe	-0,33	0,004	APOX & Fe	0,39	0,024	APOX & Ca	0,545	0,002
CAT & Mn	-0,50	0,002	MDA & Cd	-0,43	0,002	CAT & Pb	-0,31	0,007	APOX & Ca	0,38	0,029	MDA & Cu	-0,628	0,000
CAT & Fe	-0,40	0,020	MDA & Ca	0,50	0,000	CAT & Ca	-0,23	0,049	cat & Fe	-0,47	0,005	MDA & Zn	-0,378	0,036
CAT & Cu	0,38	0,028	APOX & MDA	0,54	0,000	MDA & Cu	0,24	0,040	CAT & Zn	0,50	0,002			
MDA & Fe	-0,46	0,006				MDA & Ca	0,40	0,000	MDA & Mn	0,34	0,048			
MDA & Cu	-0,60	0,000				APOX & MDA	0,42	0,000	MDA & Fe	0,59	0,000			
									MDA & Zn	-0,52	0,002			
									MDA & Cd	-0,61	0,000			
									MDA & Cd	-0,61	0,000			
									SOD & MDA	-0,49	0,003			
									APOX & MDA	0,51	0,002			
									CAT & MDA	-0,63	0,000			

Table 4. Concentration of element–activity of enzyme, content of MDA interactions and content of MDA–activity of enzyme interaction related changes of elements level and antioxidant enzymes activity, the level of lipoperoxidation in roots and green parts of glycophytes in different environments.

SM (N=70)			AE (N=80)		
Relations	R	p	relations	R	p
Roots					
SOD & Fe	0,66	0,036	CAT & Fe	-0,77	0,002
SOD & Cu	0,64	0,046	CAT & Cu	-0,76	0,003
SOD & Zn	0,74	0,015	CAT & Zn	-0,62	0,023
APOX & Cd	-0,82	0,004	CAT & Cd	-0,67	0,012
APOX & MDA	-0,64	0,048	CAT & Na	-0,71	0,006
green parts					
SOD & Zn	0,65	0,023	ns		
SOD & Cd	0,82	0,001			
APOX & Fe	0,59	0,044			
APOX & Cu	0,65	0,023			
APOX & Cd	0,79	0,002			
APOX & Ca	0,60	0,041			
CAT & Pb	0,60	0,039			

Table 5. Concentration of element–activity of enzyme, content of MDA interactions and content of MDA–activity of enzyme interactions related changes of elements level and antioxidant enzymes activity, level of lipoperoxidation in roots and green parts of Common glasswort *Salicornia europaea* in different environments.

Cu R = 5,2464+0,1295*x-0,0003*x^2
R²=0,1521
Na R = -1173,3968+50,287*x-0,1049*x^2
R²=0,4079

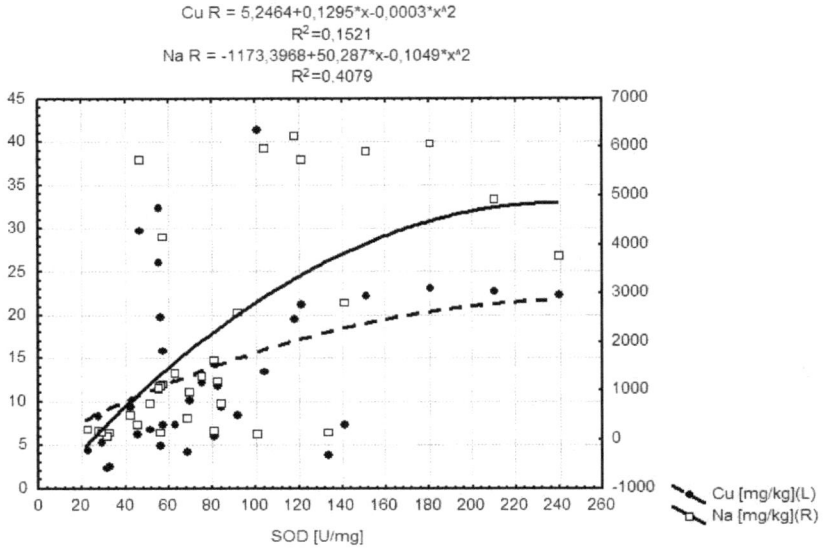

Figure 2. Interrelationships between SOD activity and the concentration of Cu, and Na (mg*kg⁻¹) in roots of glycophytes from sodium manufacture.

Ca R = 8518,9211+320,442*x-1,2445*x^2
R²= 0,3189
Cd R = 0,0832+9,1622E-5*x-4,935E-6*x^2
R²= 0,2569

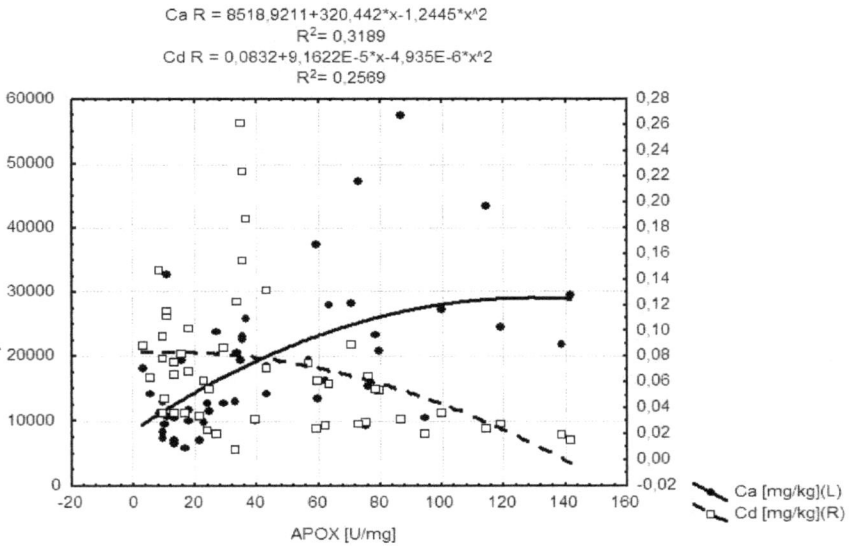

Figure 3. Interrelationships between APOX activity and the concentration of Ca, and Cd in (mg*kg⁻¹) the green parts of glycophytes from anthropogenic environment.

4.4.2. Common glasswort Salicornia europaea

Our results indicated relationships between activity of antioxidant enzymes and concentrations of metals in roots of plants in both environments and only in shoots from sodium manufactures. Concentrations of Fe, Cu, and Cd were positively related with SOD activity in the roots of plants from sodium manufactures in opposite to activity of CAT in roots of plants from anthropogenic areas. However, Na and Cd concentrations were also negatively correlated with CAT activity in shoots of plants from anthropogenic environments. We also found negative relations between concentrations of Cd and APOX activity in roots of plants from sodium manufactures. On the other hand, the activity of antioxidant enzymes was positively correlated with concentrations of Zn, Cd (SOD), Pb (CAT), Ca, Fe, Cu, and Cd (APOX) in shoots of plants. We did not find relations between MDA content and chemical elements concentration in both studied organs of Common glasswort Salicornia europaea (Tables 4, 5, Figs. 4, 5). However, we found negative relations between MDA content and APOX activity in the roots of plants from sodium manufactures.

$$Cu\ R = 9.3822-15.4595*x-9.5462*x^2$$
$$R^2=0.8476$$
$$Zn\ R = 23.6814-25.0323*x-44.5043*x^2$$
$$R^2=0.8028$$

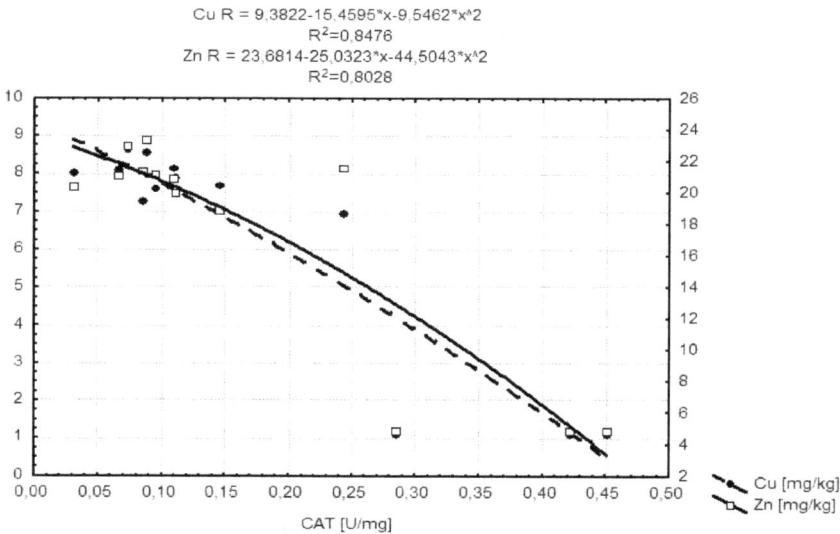

Figure 4. Interrelationships between CAT activity and the concentration of Cu, and Zn in roots (mg*kg^{-1}) of Common glasswort Salicornia europaea from anthropogenic environment.

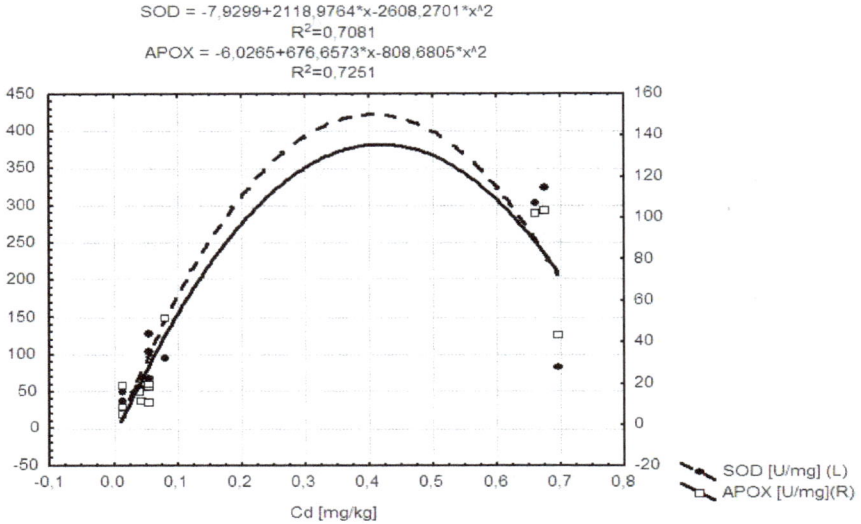

Figure 5. Interrelationships between Cd concentration (mg*kg⁻¹) and SOD, and APOX (U/mg) activity
in the green parts of Common glasswort *Salicornia europaea* from sodium manufacture.

4.5. Differences between glycophytes and Common glasswort *Salicornia europaea* in their natural environments

We found significant higher activity of APOX in roots of glycophytes than in Common
glasswort *Salicornia europaea* from both examined environments (SM, AE) in opposite to
green parts. CAT activity was similarly higher in the green parts of glycophytes than in the
halophyte (Common glasswort *Salicornia europaea*) but only in the anthropogenic
environments. Instead, the level of MDA was higher in the glycophytes except green parts
in the anthropogenic areas. Common glasswort *Salicornia europea* cumulated more sodium in
both organs in all environments studied and also had higher level of Zn, Cu, Mn, and Cd in
the green parts from AE than glycophytes. However, Cu and Pb concentrations were also
higher in the green parts of this halophyte in sodium manufactures. On the other hand,
glycophytes accumulated more calcium in roots from both environments, and more copper
in roots from sodium manufactures and also Pb in both organs from anthropogenic
environments than Common glasswort *Salicornia europaea*. The differences between Fe
concentration and SOD activity in the organs of examined plants from two various
environments did not differ (Table 6).

	SM		AE		SM		AE	
	roots				green parts			
	Z	p	Z	p	Z	p	Z	p
SOD	ns		ns		ns		ns	
CAT	ns		ns		ns		-2,02	*
APX	3,23	**	3,54	***	-2,14	*	-2,06	*
MDA	-3,25	**	-4,00	***	-5,20	***	-5,82	***
Na	4,19	***	5,24	***	5,09	***	5,32	***
Ca	-2,53	*	-3,86	***	ns		ns	
Fe	ns		ns		ns		ns	
Zn	ns		ns		ns		-3,66	***
Cu	-2,53	*	ns		-2,51	*	-3,53	***
Mn	ns		ns		ns		3,86	***
Pb	ns		-3,10	***	3,01	**	-5,00	***
Cd	ns		0,14		ns		2,22	***

Table 6. Differences among biochemical parameters: activity of enzymes (SOD, CAT, APOX), content of MDA, concentrations of elements between root and green part of glycophytes and Common glasswort *Salicornia europaea* from different environments (p<0.005).

4.6. Differences between enzymes activity, lipoperoxidation and concentrations of elements in plants

We noted higher activity of SOD and APOX in the green parts of glycophytes from sodium manufactures. However MDA level was significantly higher only in green parts of glycophytes than in roots from AE and AGRE. On the other hand, APOX activity in shoots of Common glasswort *Salicornia europaea* was lower, similarly to MDA level of these plants from SM, but the intensity of lipoperoxidation was significantly higher in roots also at AE. Concentrations of Na were higher in roots of glycophytes in opposite to halophyte. Simultaneously, Ca concentrations were higher in leaves of both glycophytes and Common glasswort in opposite to level of Fe, which was rather cumulated in roots (not significant at SM in Common glasswort *Salicornia europea*). Simultaneously, Zn was found in higher concentrations only in glycophytes but in non salted environments (WE, AGRE, C). Cu and Mn were rather mobile metals but we found significant higher Cu concentrations in roots of glycophytes (SM) and – for both elements – in roots of Common glasswort (AE). Lead, as toxic metal was accumulated in roots of plants. Glycophytes also accumulated Cd in roots but it is interesting that we did not find the differences among organs in plants from control environments. Instead, concentrations of Cd were not differs amongst organs of Common glasswort *Salicornia europea* from both environments (Table 7).

glycophytes						Common glasswort	
	SM	AE	WE	AGRE	C	SM	AE
SOD	-1,98 *	ns	ns	ns	ns	ns	ns
CAT	ns	ns	ns	ns	ns	ns	ns
APX	-2,46 *	ns	ns	ns	ns	2,67 *	3,78 ***
MDA	ns	-3,26 **	ns	-4,02 ***	ns	3,73 ***	2,38 *
Na	5,61 ***	3,091 **	6,41 ***	6,38 ***	4,55 ***	-3,13 **	-2,75 **
Ca	-2,59 **	-5,492 ***	-7,79 ***	-5,01 ***	-5,14 ***	-2,14 *	-4,22 ***
Fe	5,17 ***	5,372 ***	5,86 ***	4,63 ***	3,01 **	ns	2,09 *
Zn	ns	ns	-4,25 ***	-4,94 ***	-2,25 *	ns	ns
Cu	3,05 **	ns	ns	ns	ns	ns	2,26 *
Mn	ns	ns	ns	ns	ns	ns	-4,16 ***
Pb	5,83 ***	5,642 ***	4,26 ***	2,39 *	3,63 ***	2,27 *	4,22 ***
Cd	5,15 ***	4,860 ***	4,37 ***	3,72 ***	ns	ns	ns

Table 7. Differences among the content of biochemical parameters: MDA, proline and concentration of elements in organs of plant groups from the same environment ($p<0.005$).

5. Discussion and conclusions

The presence of a sodium factory and industrialization of Kujawy region affected the salinity, alkalinity and higher content of Na, Ca but also Fe, Cu, and Zn in the environments of this region, especially neighbored with sodium manufactures and waste dumping sites. On the other hand, higher concentration of microelements in the agricultural environments than in controls was probably reflected the use of fertilizers in the agricultural practices. It should be emphasized that we did not find differences in the concentration of toxic elements (Pb, Cd) in plants and soils between control and disturbed environments (SM, AE); Table 1. However, concentration of chemical elements studied in plants differed among environments but relations were not directly similar to those occurred in soils (Tables 2, 3). It probably depends on the environmental conditions. We should consider that bioavailability of chemical elements by plant is affected by numerous basal environmental characteristics of soil: pH, red-ox potential, salinity, the content of organic matter, etc. [62], [63]. Simultaneously, different factors than edaphic conditions of soils, i.e. seasonal physiology, the condition of plants, species-species capacities for uptake, translocation and compartmentalization, may contribute to the differential bioaccumulation of elements [64]. The roots of glycophytes accumulated about 3 fold higher concentrations of Na than collected at control environments, whilst green parts of plants – even 17 fold higher (Table 2). We can thus conclude that at sodium manufactures, especially anthropogenic excess of sodium affects ionic balance and primarily impacts upon metabolic processes [65]. Instead, calcium concentration in the organs of plants do not exceed average concentrations recommended by [66]; 0.2-5% d.w. However, the higher Ca concentrations, especially in salted environments, could be related with higher level of this element in these environments. On the other hand, increased intake of calcium could be connected with excess of sodium and mobilization of defence mechanisms [67]. Simultaneously, our results indicated that concentration of Zn and Cu in roots and leaves of examined plants do not exceed the phytotoxic range recommended by [68] and [69], i.e., 100-400 mg*kg^{-1}; 30 mg*kg^{-1}, respectively). Similar level of Pb, and Cd in plant's organs in our study was below ranges of 50-300 mg*kg^{-1}, and 5-30 mg*kg^{-1}, respectively, considered by [68] as phytotoxic. However, iron and manganese concentration in the range of 40-500 mg*kg^{-1} and 50-500 mg*kg^{-1}, respectively, according to [70] are found as toxic for plants, so our result indicate that Fe level is high in both organs of plants, but high concentrations of Mn in roots and green parts of glycophytes was connected with environmental factors.

Stressful factors (acidity, salinity, toxic heavy metals) adversely affects the associated ecological balance and increases the level of free radicals and reactive oxygen species (ROS), which is related with changes in activities of antioxidant enzymes [71], [72], [73]. Most studies show that the higher activities of these enzymes were positively correlated with plant stress-resistance [74], [6], [73]. However, the changes of enzyme activities also depend on plant genotypes and stress intensity [74], [75].

Based on the results obtained by us, we can conclude that physiological activity of important antioxidant enzymes can maintain oxidative homeostasis and also reduce the membrane damages in the plant organism (mainly by lipoperoxidation). Moreover, SOD, CAT, and APOX is related with concentration of examined elements (Na, Ca, Fe, Zn. Cu, Mn, Pb, Cd), depending on the status of the environment. These relations were positive and negative, and they depend on the ecological group of plants, the type of organ and on stress intensity (various degree of anthropogenic impact). Glycophytes studied by us were subjected on many stressors at disturbed environments in Kujawy region. These plant species strive to maintain their oxidative homeostasis and develop efficient antioxidative responses. Thus we found positive relationships between sodium concentrations and the content of MDA as indicator of oxidant injury, but also between CAT activity in roots of plants from anthropogenic environments, and also similar between calcium concentrations and the level of lipoperoxidation and APOX in green parts of plants from anthropogenic and wetland environments. Copper concentration was also related with APOX activity and MDA level in the wetlands. Instead, Fe was correlated with MDA and APOX in the agricultural areas (Table 4). We can thus concluded that Na, Ca, Cu and also Fe stimulated lipoperoxidation processes but also mobilization antioxidant mechanisms of glycophytes in Kujawy region. Furthermore, negative relation of Cd with APOX in both organs from controls and also positive correlations with degree of lipoperoxidation in roots (Table 4) indicate that discrimination of one of important enzymes involved in antioxidant mechanism and impact on ROS generation. Similar to Cd, also Cu stimulated higher level of MDA and negatively affect the activity of APOX in opposite to activity of SOD in roots of glycophytes in control sites (Table 4). In the organs of Common glasswort *Salicornia europea* we did not find these relations (Table 5). Reports of [8], [76], and [6] showed correlations between stress level and MDA content. Moreover, lipid peroxidation and H$_2$O$_2$ levels, and also SOD, CAT, APOX and GR activities increased in pea roots and leaves under Cd stress [77], whilst APOX and CAT decreased at high Cd concentrations [78]. Also [47] reported that Cd-induced oxidative stress in *Arabidopsis* is due to H$_2$O$_2$ accumulation.

It should be emphasized that we found more correlations between concentrations of chemical elements and studied biochemical parameters (Tables 4, 5), which indicate the complexity of antioxidative processes and simultaneously the participation of chemical elements in stimulation and modification of lipoperoxidation processes and SOD, CAT or APOX activities. Simultaneously, the same elements could impact upon the activation of many oxidative pathways in organs of glycophytes (Fig. 5) and it depends on the environmental conditions. In disturbed sodium manufactures the level of Na stimulated activity of SOD and APOX, but in the other environments it stimulated the activity of CAT. Similarly, copper concentration is positively correlated with CAT in roots but negatively in the green parts of plants from anthropogenic environments. Furthermore, Na and Ca concentrations stimulated activity of antioxidant mechanisms, especially in disturbed environments. On the other hand, calcium is also related with APOX activity in all studied environments. Simultaneously, concentrations of transition metals could impact the increase or decrease of antioxidant enzymes activity (Figs. 2-5). We suggest it depends on the interactions amongst elements, the intensity of stressors or impact of other not examined factors. However, more analysis will require for estimation these problems. We also should consider that Cu, Zn, Mn, and also Fe are cofactors of metalloprotein, i.e. SOD. Simultaneously, cadmium and lead as toxic metals especially defected the activity of APOX and CAT, respectively, in the organs of glycophytes, which is in opposite to SOD activity (Tab. 4). Similarly, Fe, Zn, and Cu stimulate activity of SOD in sodium manufactures in roots and shoots of Common glasswort *Salicornia europea*. Moreover, the same metals decrease CAT activity in roots from anthropogenic environments. Cadmium and lead concentration are also defected the activity of APOX and CAT and impacted the increase of SOD. We also found correlations between APOX and Ca concentrations in shoots of Common glasswort *Salicornia europaea* in sodium manufactures, alike as in the green parts of glycophytes. Moreover, sodium concentrations influence the decrease of CAT in roots of this halophyte (Table 5). Similarly, CAT activity is also higher in green parts of glycophytes than in halophytes. This indicate that glycophytes are better adaptation to stressful conditions, however there are many indistinct questions required to estimation.

There is much evidence that salinity and heavy metals enhanced or decreased activity of antioxidant enzymes but usually under control conditions. Both increase as well as decrease in the activity of SOD has been reported in plants in response to salinity stress [8], [79]. It appears that the activity of SOD under salinity varies depending upon plant species, organ analyzed as well as upon the level of salinity [80], [81]. [9] showed that in Cotton *Gossypium hirsutum* L. under salted NaCl stress the increases of SOD activities occurs, as well as decreases of the activities of catalase and ascorbate peroxidase. In the leaves of rice plant, salt stress preferentially enhances the content of H_2O_2 and the activities of SOD, APOX, whereas it decreases catalase activity [82]. These authors reported that NaCl treatment increases the activities of catalase but does not affected the activity of SOD in cucumber plants. The tomato under high salt concentration showed higher among others antioxidant enzyme activities such as SOD, catalase, and ascorbate peroxidase [83]. Similarly, [84] showed that activity of cytosolic CuZn-SOD II, chloroplastic CuZn- SOD II, and mitochondrial and/or peroxisomal Mn-SOD were correlated with increasing concentration of NaCl in pea at the higher NaCl concentrations.

We can consider that heavy metals toxicity is occurred in the ability to bind strongly the oxygen, nitrogen and sulphur ions. This process is related with the free enthalpy of the formation of the product of metals and ligands. By this features heavy metals can inactivate the enzymes by binding to cysteine residues. Heavy metals can also displace one metal with another, which can lead to inhibition or loss the enzyme activities [4], [13]. On the other hand, [85] examined the role of antioxidative enzyme system. They investigated the relation to Cd stress in hyperaccumulator plants of the genus *Alyssum*. In both species superoxide dismutase activity was elevated at high cadmium concentrations, whilst ascorbate peroxidase activity remained unchanged [7], whilst [86] indicated that activity of antioxidant enzymes was estimated as a function of time and concentrations of Pb in roots of lupin. The results of [25] suggested that lead induces oxidative stress in growing rice plants and that SOD and APOX could serve as important components of antioxidative defence mechanism against Pb induced oxidative injury in rice. They observed a Pb dependent increase in the activities of SOD from root tip extract, whilst CAT and APOX activities decreased at higher lead concentrations. Cadmium treatment induced lipooxygenase with simultaneous inhibition of antioxidative enzymes, SOD and CAT [41]. In particular, CAT activity often decreased following exposure to elevated cadmium concentrations [78], [46]. On the other hand, [87] indicated that CAT activities and specific isoenzymes of SOD increased in the leaves and roots of a resistant variety of radish, following exposure to increasing (between 0.25 and 1 mM) concentrations of cadmium. A severe suppression of SOD and CAT, and almost complete loss of APOX activities after 48 h of exposure to 50 μM Cd was observed in pine roots [88]. Cd-induced inhibition of APOX and CAT was also associated with H_2O_2 accumulation and growth retardation in the poplar roots [4]. Also [89] studied the involvement of H_2O_2 and O_2^- in the signaling events that resulted in variation of the transcript levels of CAT, GR and Cu-Zn-SOD in pea plants under Cd stress.

Considering the influence of chemical elements on the defence mechanisms of plants [90] we can conclude the unique importance of Ca^{2+} for stabilization of membranes. High salinity results in increased cytosolic Ca^{2+}, which is transported from the apoplast and the intracellular compartments [91]. This transient increase in the cytosolic Ca^{2+} initiates stress-signal transduction leading to salt adaptation. Adequate levels of calcium are necessary for the membrane to its normal function [31].

Most of the interest in calcium participation in plants has centered on its role in the cytoplasm in controlling the developmental processes [92]. [32] reported the effects of calcium chloride on the sodium chloride-stressed plants of Mung bean *Vigna radiatae* (L.). We could confirm that when $CaCl_2$ was combined with NaCl, $CaCl_2$ altered the overall plant metabolism to ameliorate the deleterious effects of NaCl stress and increased the vegetative growth of plants. [93] indicated that Ca^{2+} prevented Cd-induced increasing of the activity of SOD and restored CAT activity. These results suggested that exogenous application of Ca^{2+} could be advantageous against Cd^{2+} toxicity, and could confer tolerance to heavy metal's stress in plants.

Our results confirm also the disruption of oxidative homeostasis related with environments of glycophytes through higher degree of lipid peroxidation in leaves in the anthropogenic environments and lower level of MDA in roots of plants from wetlands than from remaining environments. However, we found a higher SOD activity in the roots of glycophytes from agriculture areas in comparison with sodium manufactures and wetlands. Simultaneously, the activity of the same enzymes in green parts of plants was lower in wetland areas than in remaining environments (Table 2). All of these results indicated that plants probably can mitigate the oxidative damage initiated by ROS (especially at agriculture) by the complex of defensive antioxidative system [71]. Moreover, [15] indicated that both roots and shoots of plants had malonyldialdehyde (MDA) contents cultivated at the optimal salt concentration (50 mM NaCl) were lower than in the control areas. This was related to enhanced activities of antioxidant enzymes, like superoxide dismutase, catalase and peroxidase, especially in shoots. SOD and CAT activities have been reported to be negatively correlated with the degree of damage to plasmalemma, chloroplasts and mitochondrial membrane systems and positively correlated with stress resistance indices [7], [32]. Simultaneously, we can not exclude other factors as draught or climate conditions, which could generate free radicals or reactive oxygen species (i.e., draught, climate conditions) and thus impact on high activity of antioxidant enzymes in organs in the control environments (C). We should also note that the level of cadmium was higher in leaves of plants from Tuchola Forestry than from remaining environments. Similarly, in roots of plants from these areas, cadmium concentrations did not differ from other environments, but was higher than in the agricultures. Cadmium is probably a factor, which can indirectly upset cell membrane. Simultaneously, we found relationships between biochemical parameters and concentrations of Na, Ca, Cu, Zn and Mn, which stimulated the activity of CAT, and Na-SOD and Ca-SOD and APOX. Instead, we found differences between the content of MDA (higher in shoots of plants from anthropogenic than from sodium manufactures) but the activity of SOD, CAT and APOX did not differ in the organs of Common glasswort Salicornia europaea between two examined environments. It could indicate that Common glasswort, as obligatory halophyte, is adapted to high salted environments (sodium manufactures). Thus mutual and differentiated relationships between chemical elements and antioxidant enzymatic mechanisms as plant responses to the environmental disturbs, are probably another factor influenced the higher level of lipoperoxidation in anthropogenic environments.

We also found that Na, Fe, and Pb are accumulated in higher level in roots of glycophytes in opposite to Ca. Zn, Cu and Mn at all examined environments and they showed more mobility, which depends on the type of environment. Interestingly, the concentrations of Cd and Pb, as toxic elements were also higher in roots of plants at all environments of Kujawy region but not at the control sites, whilst the activity of antioxidant enzymes did not differ significantly between organs of plants in the examined environments, except of SOD and APOX at SM. We found higher level of these enzymes in the leaves than in roots of plants. It was probably related with the concentrations of Ca and also Cu, Zn, and Cd. We could also conclude that Zn and Cu are co-factors of SOD so further analyses are needed. However, the location of chemical elements rather not deflected on the activity of enzymes in glycophytes. On the other hand, the concentrations of Na were higher in shoots than roots of Common glasswort Salicornia europaea, which is connected with adaptation to life in high salinity (e.g., osmoregulation); [13]. It is interesting we found the higher level of MDA and higher lead concentrations in roots, similar to activity of APOX. However, we could not confirm relations among these parameters because we did not find significant correlation. Furthermore, the roots are more markedly affected by saline conditions than leaves, because of being the first part at the plant to encounter soil salinity [94]. [81] confirm that roots of halophyte Sea fennel *Crithmum maritimum* were distinguished from leaves by malondialdehyde concentration, however activity of APOX was higher in leaves, despite the fact that stress factor was only a high salinity. Simultaneously, [95] and [96] stated that root tissues markedly exhibited by higher APOX activity in comparison of shoots of plants under salinity stress.

The mechanisms of salt tolerance are of two main types: those minimizing the entry of salt into the plant, and those minimizing the concentration of salt in the cytoplasm. Halophytes, as naturally salt tolerant plants, have both types of mechanisms. They exclude salt well, but effectively compartmentalize in vacuoles the salt that inevitably gets in. This allows them to growth of long period of time in saline soil. Instead, glycophytes avoid sodium to maintain ionic homeostasis [13], [15]. We can thus conclude basing also on our results showed in Table 5 that higher concentrations of sodium in roots and leaves of Common glasswort *Salicornia europea* than in glycophytes are resulting from the saline determinations of the environment and predominantly from the interactions with sodium. We also conclude that glycophytes are subjected on higher salinity stress, which cause probably higher level of ROS, and lipoperoxidation in the organs (except green parts at AE) than in halophytes. On the other hand, activity of APOX is higher in the roots of halophyte than in glycophytes in both environments in opposite to green parts. We could suspect that this were connected with higher concentrations of Ca, elevated salinity effect [32] and indicated the activity of APOX in the green parts of plants more sensitive on salt damages than roots [97]. Similar activity of CAT is also higher in green parts of our glycophytes than halophyte. It is indicated that glycophytes are well adapted to stressful conditions however there are still many indistinct topics, required to estimation.

The area of mutual relationships: plants-environmental stressors and conducted research in this field has been gaining ground in the recent years. However, in the view of considerable variations in the protective mechanisms against activated oxygen species in different plant species, the further work, especially in natural conditions under many stressors is required to establish the general validity of various regularities and processes in salinity tolerance and also disturbed ionic homeostasis.

The positive and negative correlations between the level of chemical elements and antioxidative responses of plants indicate the complexity of enzymatic antioxidative processes and simultaneously the participation of chemical elements, which does not influence lipid peroxidation but stimulate or injure SOD, CAT or APOX activity. However, the concentrations of Na, Ca, Cu and Fe stimulate lipoperoxidation but can also mobilize antioxidant responses of glycophytes in natural salted Pomeranian region, whilst Cd in the control environment influence MDA content and increasing APOX activity in roots of glycophytes. Simultaneously this toxic metal may impact the activation of many oxidant ways in organs of glycophytes and it depend on the environmental conditions (pH, Ec). Na and Ca concentrations rather stimulate enzymatic antioxidant mechanisms activity, especially in disturbed environments, and they can also stimulate the transition metals which causing increase or decrease of antioxidant enzymatic activity. However, Cd and Pb can defect the activity of APOX and CAT in organs of glycophytes, which is opposite to SOD activity. Similarly, in the organs of Common glasswort *Salicornia europaea*, Fe, Zn, and Cu level can stimulate SOD activity but they can also inhibit CAT activity. These processes depend on the environmental factors (pH, Ec) and types of plant organs and toxic heavy metals Cd and Pb concentration. They defect APOX and CAT activity but influence the increase of SOD. We also found correlation between APOX and calcium concentration (alcalinity) in shoots of Common glasswort and in green parts of glycophytes.

The differences in MDA content and SOD activity either in roots or in green parts of glycophytes indicate that plants can probably mitigate the oxidative damage initiated by reactive oxygen species (ROS) by complexes of defensive enzymatic antioxidative system. Instead we found only differences between MDA content (higher in shoots from anthropogenic than in more salted environments) but the activity of SOD, CAT and APOX didn't differ in organs of Common glasswort *Salicornia europaea*, which indicates that the impact of other factors than examined elements influenced higher level of lipoperoxidation in anthropogenic environments. Antioxidant enzymatic activity didn't differ significantly between plants organs in examined environments except of SOD and APOX in strong salted environment (SM). We found higher level of these enzymes in leaves than in roots of plants studied. It was probably related with Ca and also Cu, Zn, and Cd concentrations. However, the level of lipoperoxidation was significant higher in green parts of glycophytes (anthropogenic, agricultural environments), which is opposite to common glasswort (both environments). Instead in glycophytes avoiding sodium to maintain ionic homeostasis we found higher Na concentrations in roots and leaves of halophytes (Common glasswort *Salicornia europaea*) than in glycophytes. We can conclude that glycophytes subjected to higher salinity stress develop probably higher level of ROS, thus lipoperoxidation in their organs (except of green parts in AE) is higher than in halophytes. On the other hand, APOX activity is higher in roots of halophytes than in glycophytes in both environments in opposite to their green parts. We thus can suspect that this is connected with higher concentrations of Ca (alcalinity) and elevated salinity effect indicated APOX activity in green parts of plants more sensitive on salt damages than roots. Similarly, CAT activity is also higher in green parts of glycophytes than in halophytes. This indicate that glycophytes are better adaptation to stressful conditions, however there are many indistinct questions which require further study.

Author details

Piotr Kamiński, Beata Koim-Puchowska, Monika Wieloch and Karolina Bombolewska
Nicolaus Copernicus University, Collegium Medium in Bydgoszcz, Department of Ecology and Environmental Protection, Bydgoszcz, Poland

Piotr Kamiński
University of Zielona Góra, Faculty of Biological Sciences, Institute of Biotechnology and Environment Protection, Department of Biotechnology, Zielona Góra, Poland

Piotr Puchowski
Government Forestry in Toruń; Zamrzenica Forestry District, Bysław, Poland

Leszek Jerzak
University of Zielona Góra, Faculty of Biological Sciences, Institute of Biotechnology and Environment Protection, Department of Environmental Protection and Biodiversity, Zielona Góra, Poland

Acknowledgement

We thank Professor Brendan P. Kavanagh (Royal College of Surgeons in Ireland, Medical University of Bahrain) for his help with improving English language of the paper.

6. References

[1] Mittler R. 2002. Oxidative stress, antioxidant and stress tolerance. Trends Plant Sci. 7: 405-410.

[2] Bowler C., Fluhr R. 2000. The role of calcium and activated oxygens as signals for controlling cross-tolerance. Trends Plant Sci. 5: 241-246.

[3] Panda S.K., Khan M.H. 2004. Changes in growth and superoxide dismutase activity in *Hydrilla verticillata* L. under abiotic stress. Braz. J. Plant Physiol., 16(2):115-118.

[4] Schutzendubel A., Polle A. 2002. Plant responses to abiotic stresses: heavy-metal induced oxidative stress and protection by mycorrhization. J. Exp. Bot. 53: 1351-1365.

[5] Scandalios J.G. 1993. Oxygen stress and superoxide dismutase. Plant Physiol. 101: 7-12.

[6] Sairam R.K., Srivastava G.C. 2002. Changes in antioxidant activity in sub-cellular fractions of tolerant and susceptible wheat genotypes in response to long term salt stress. Plant Sci. 162: 897-904.

[7] Elkahoui S., Hernandez J.A., Abdelly Ch., Ghrir R., Limam F. 2005. Effects of salt on lipid peroxidation and antioxidant enzyme activities of *Catharanthus roseus* suspension cells. Plant Sci. 168: 607-613.

[8] Dionisio-Sese M.L., Tobita S. 1998. Antioxidant responses of rice seedlings to salinity stress. Plant Sci. 135: 1-9.

[9] Gossett D.R., Millhollon E.P., Lucas M.C. 1994. Antioxidant responses toNaCl stress in salt tolerant and salt-sensitive cultivars of cotton, Crop Sci. 34: 706-714.

[10] Sreenivasulu N., Ramanjulu S., Ramachandra-Kini K., Prakash H.S., Shekar-Shetty H., Savitri H.S., Sudhakar C. 1999. Total peroxidase activity and peroxidase isoforms as modified by salt stress in two cultivars of fox-tail millet with differential salt tolerance. Plant Sci. 141: 1–9.

[11] Loureiro S., Santos C., Pinto G., Costa A., Monteiro M., Nogueira A.J.A., Soares A.M.V.M. 2006. Toxicity assessment of two soils from Jales Mine (Portugal) using plants: growth and biochemical parameters. Arch. Environ. Contam. Toxicol. 50: 182-190.

[12] Bidar G., Pruvot Ch., Garçon G., Verdin A., Shirali P., Douay F. 2009. Seasonal and annual variations of metal uptake, bioaccumulation, and toxicity in *Trifolium repens* and

Lolium perenne growing in a heavy metal-contaminated field. Environ. Sci. Pollut. Res. 16: 42-53.

[13] Przybył K., Woźny A. 2004. Komórka w warunkach stresu środowiskowego. t 2. Wyd. UAM, Poznań, 172 pp.

[14] Mittova V., Tal M., Volokita M., Guy M. 2003. Up-regulation of the leaf mitochondrial and peroxisomal antioxidative systems in response to salt-induced oxidative stress in the wild salt-tolerant tomato species *Lycopersicon pennellii*. Plant Cell Environ. 2: 845-856.

[15] Ben Amor N., Ben Hamed K., Debez A., Grignon C., Abdelly C. 2005. Physiological and antioxidant responses of the perennial halophyte Crithmum maritimum to salinity. Plant Sci. 168, 889-899.

[16] Bowler C., Van Montagu M., Inze D. 1992. Superoxide dismutase and stress tolerance. Ann. Rev. Plant Physiol. Plant Mol. Biol. 43: 83-116.

[17] Baum J.A., Scandalios J.G. 1981. Isolation and characterization of cytosolic and mitochondrial superoxide dismutases of maize. Arch. Biochem. Biophys. 206, 249-261.

[18] Stroiński A. 1999. Some physiological and biochemical aspects of plant resistance to cadmiumeffect. I. Antioxidative system. Acta Physiologiae Plantarum 21 (2): 999:175-188.

[19] Duke M.V., Salin M.L. 1985. Purification and characterization of an iron-containing superoxide   dismutase from eukaryote, *Ginko biloba*. Arch. Biochem. Biophys. 243: 305-314.

[20] Scandalios J.G. 1990. Response of plant antioxidant defence genes to environmental stress. Adv. Genet. 28: 1-41.

[21] Chen G.X., Asada K. 1989. Ascobate peroxidase in tea leaves. Occurrence of two isozymes and their differences in enzymatic and molecular properties. Plant Cell Physiol. 30: 987-998.

[22] Asada K. 1992. Ascorbate peroxidase - a hydrogen peroxide-scavenging enzyme in plants. Physiol. Plant. 85: 235-241.

[23] Barcelo A.R., Ferrer M.A., Florenciano E.G., Munoz R. 1991. The tonoplast localization of two basic isoperoxidases of high pI in Lupinus. Bot. Acta 104: 272-278.

[24] Takahama U., Oniki T. 1992. Regulation of peroxidase-dependent oxidation of phenolics in the apoplast of spinach leaves by ascorbate. Plant Cell Physiol. 33: 379-387.

[25] Verma S., Dubey R.S. 2003. Lead toxicity induces lipid peroxidation and alters the activities of antioxidant enzymes in growing rice plants. Plant Sci. 164: 645-655.

[26] Grassmann J., Hippeli S., Elstner E.F. 2002. Plant's defence and its benefits for animals and medicine: role of phenolics and terpenoids in avoiding oxygen stress. Plant Physiol. Biochem. 40: 471-478.

[27] Hamilton III E.W., Heckathorn S.A. 2001. Mitochondrial adaptations to NaCl stress: Complex I is protected by anti-oxidants and small heat shock proteins, whereas Complex II is protected by proline and betaine. Plant Physiol. 126: 1266-1274.

[28] Mittova V., Guy M., Tal M., Volokita M. 2004. Salinity up-regulates the antioxidative system in root mitochondria and peroxisomes of the wild salt-tolerant tomato species *Lycopersicon pennellii*. J. Exp. Bot. 55: 1105-1113.

[29] Banuls J., Legaz F., Primo-Millo E. 1991. Salinity-calcium interactions on growth and ionic concentrations of citrus plants. Plant Soil, 133: 39-46.

[30] White P.J., Broadley M.R. 2003. Calcium in plants. Ann. Bot. 92: 487-511.

[31] Jaleel C.A., Manivannan P., KishoreKumar A., Sankar B., Panneerselvam R. 2007. Calcium chloride effects on salinity-induced oxidative stress, proline metabolism and indole alkaloids accumulation in *Catharanthus roseus*. C.R. Biol. 330: 674-683.

[32] Manivannan P., Jaleel C.A., Sankar B., Somasundaram R., Murali P.V., Sridharan R., Panneerselvam R. 2007. Salt stress mitigation by calcium chloride in *Vigna radiate* (L.) Wilczek. Acta Biol. Cracov. Ser. Bot. 49, 2: 105-109.

[33] Nagajyoti P.C., Lee K.D., Sreekanth T.V.M. 2010. Heavy metals, occurrence and toxicity for plants: a review. Environ. Chem. Lett. 8: 199-216.

[34] Kabata-Pendias A., Mukherjee A.B. 2007. Trace Elements from Soil to Human. Springer-Verlag, Berlin-Heidelberg-New York, 550 pp.

[35] Bartosz G. 2006. Druga twarz tlenu. Wolne rodniki w przyrodzie. PWN-Pol. Sci. Publ., Warszawa, 447 pp.

[36] Kaznina N.M., Laidinen G.F., Titov A.F., Talanov A.V. 2005. Effect of lead on the photosynthetic apparatus of annual grasses. Izv. Akad. Nauk Ser. Biol. 2: 184-188.

[37] Andra S.S., Datta R., Sarkar D., Makris K.C., Mullens C.P., Sahi S.V., Bach S.B.H. 2010. Synthesis of phytochelatins in vetiver grass upon lead exposure in the presence of phosphorus. Plant Soil. 326: 171-185.

[38] Singh R., Tripathi R.D., Dwivedi S., Kumar A., Trivedi P.K., Chakrabarty D. 2010. Lead bioaccumulation potential of an aquatic macrophyte Najas indica are related to antioxidant system. Bioresource Technol. 101: 3025-3032.

[39] Pal M., Horvath E., Janda T., Paldi E., Szalai G. 2006. Physiological changes and defense mechanisms induced by cadmium stress in maize. *J. Plant Nutr. Soil Sci.* 2006 (169)239-246.

[40] Mishra S., Srivastava S., Tripathi R.D., Dwivedi S., Shukla M.K. 2008. Response of Antioxidant Enzymes in Coontail (*Ceratophyllum demersum* L.) Plants Under Cadmium Stress. Environ. Toxicol. 23: 294-301.

[41] Somashekaraiah B.V., Padmaja K., Prasad A.R.K. 1992. Phytotoxicity of cadmium ions on germinating seedlings of mung bean (*Phaseolus vulgaris*): involvement of lipid peroxides in chlorophyll degradation. Physiol. Plant. 85: 85-89.

[42] Stohs S.J., Bagchi D. 1995. Oxidative mechanisms in the toxicity of metal ions. Free Rad. Biol. Med. 18: 321-336.

[43] Shaw B.P. 1995. Effect of mercury and cadmium on the activities of antioxidative enzymes in the seedlings of *Phaseolus aureus*. Biol. Plant. 37: 587-596.

[44] Gallego S.M., Benavides M.P., Tomaro M.L. 1996. Effect of heavy metal ion excess on sunflower leaves: evidence for involvement of oxidative stress. Plant Sci. 121: 151-159.

[45] Balestrasse K.B., Gardey L., Gallego S.M., Tomaro M.L. 2001. Response of antioxidant defence system in soybean nodules and roots subjected to cadmium stress. Aust. J. Plant Physiol. 28: 497-504.

[46] Fornazier R.F., Ferreira R.R., Vitória A.P., Molina S.M.G., Lea P.J., Azevedo R.A. 2002. Effects of cadmium on antioxidant enzyme activities in sugar cane. Biol. Plant. 45: 91-97.

[47] Cho U., Seo N. 2004. Oxidative stress in Arabidopsis thaliana exposed to cadmium is due to hydrogen peroxide accumulation, Plant Sci. 168: 113-120.

[48] Benavides M.P., Gallego S.M., Tomaro M.L. 2005. Cadmium toxicity in plants. Braz. J. Plant Physiol. 17(1): 21-34.

[49] Szafer W., Kulczyński S., Pawłowski B. 1986. Polish Plants. PWN-Pol. Sci. Publ., Warszawa.

[50] Balnokin Yu.V., Myasoedov N.A., Shamsutdinov Z.Sh., Shamsutdinov N.Z. 2005. Significance of Na^+ and K^+ for sustained hydration of organ tissues in ecologically distinct halophytes of the family Chenopodiaceae. Russ. J. Plant Physiol. 52, 6: 779-787.

[51] Ungar I.A. 1977. The relationship between soil water potential and plant water potential in two inland halophytes under field conditions. Bot. Gazette, 138: 498-501.

[52] Egan T.P., Ungar I.A. 2000. Mortality of the Salt Marsh Species Salicornia europaea and Atriplex prostrata (Chenopodiaceae) in Response to Inundation. Ohio J. Sci. 100, 2: 24-27.

[53] Demirezen D., Aksoy A. 2006. Common hydrophytes as bioindicators of iron and manganese pollutions. Ecol. Indicators, 6: 388-393.

[54] Juszczuk I., Malusà E., Rychter A.M. 2001. Oxidative stress during phosphate deficiency in roots of bean plants (Phaseolus vulgaris L.) J. Plant Physiol. 158: 1299-1305.

[55] Bradford M.M. 1976. A rapid and sensitive method for the quantification of microgram quantities of protein utilizing the principle of protein-dye binding. Anal Biochem. 72 248-254.

[56] Beauchamp C., Fridovich I. 1971. Superoxide dismutase: improved assaysand applicable to acrylamide gels. Anal. Biochem. 44: 276-287.

[57] Aebi H. 1984. Catalase in vitro. Method. Enzymol. 105: 121-126.

[58] Scebba F., Sebastiani L., Vitagliano C. 1998. Changes in activity of antioxidative enzymes in wheat (Triticum aestivum) seedlings under cold acclimation. Physiol. Plant. 104: 747-752.

[59] Nakano Y., Asada K. 1981. Hydrogen peroxide is scavenged by ascorbate-specific peroxidase in spinach chloroplast. Plant Cell Physiol. 22: 867-880.

[60] Okhawa H., Ohishi N., Yagi Y. 1979. Assay of lipid peroxides in animal tissue by thiobarbituric acid reaction. Analyt. Biochem. 95: 351-358.

[61] Stanisz A. 2006. Przystepny kurs statystyki z zastosowaniem STATISTICA PL na przykładach z medycyny. Wyd. Statsoft Polska Sp. z o.o., 531 pp.

[62] Samecka-Cymerman A., Kempers A.J. 2001. Concentrations of heavy metals and plant nutrients in water, sediments and aquatic macrophytes of anthropogenic lakes (former open cut brown coal mines) differing in stage of acidification. Sci. Total Environ. 28: 87-98.

[63] Sundareshwar P.V., Morris J.T., Koepfler E.K., Fornwalt B. 2003. Phosphorous limitation of coastal ecosystem processes. Science, 299: 563-565.

[64] Bargagli R. 1998. Trace Elements in Terrestrial Plants. An Ecophysiological Approach to Biomonitoring and Biorecovery. Springer, Berlin, 324 pp.

[65] Kłosowska K. 2010. Reakcje roślin na stres solny. KOSMOS, Probl. nauk biol. 59: 539-549.

[66] Kopcewicz J., Lewak S. 2002. Fizjologia roślin. PWN-Pol. Sci. Publ., Warszawa, 806 pp.

[67] Wrochna M., Gawrońska H., Gawroński S.W. 2006. Wytwarzanie biomasy i akumulacja jonów Na+, K+, Ca2+, Mg2+, Cl- w warunkach stresu solnego, przez wybrane gatunki roślin ozdobnych. Acta Agrophysica, 134: 775-785.

[68] Kabata-Pendias A., Pendias H. 1984. Trace elements in soils and plants. CRC Press, Boca Raton, 315 pp.

[59] Baker D.E., Senft J.P. 1995. Copper. In: Alloway B.J. (Ed.). Heavy Metals in Soils. 2nd ed. Blackie Acad. Professional, London, pp. 179-205.

[70] Allen S.E. 1989. Chemical Analysis of Ecological Materials. 2nd ed. Blackwell Sci. Publ., Oxford.

[71] Foyer C.H., Noctor G. 2000. Oxygen processing in photosynthesis: regulation and signalling. New Phytol. 146: 359-388.

[72] Reddy A.R., Chaitanya K.V., Vivekanandan M. 2004. Droughtinduced responses of photosynthesis and antioxidant metabolism in higher plants. J. Plant Physiol. 161: 1189-1202.

[73] Kim S.Y., Lim J.H., Park M.R., Kim Y.J., Park T.I., Seo Y.W., Choi K.G., Yun S.J. 2005. Enhanced antioxidant enzymes are associated with reduced hydrogen peroxide in barlay roots under saline stress. J. Biochem. Mol. Biol. 38: 218-224.

[74] Li L., Van Staden J. 1998. Effects of plant growth regulators on the antioxidant system in callus of two maize cultivars subjected to water stress. Plant Growth Regul. 24: 55-66.

[75] Chen K.M.,1, Gong H.J.,, Wang S.M., Zhang C.L., 2007. Antioxidant defense system in *Phragmites communis* Trin. ecotypes. Biologia Plantarum 51 (4): 754-758

[76] Sudhakar C., Lakshmi A., Giridarakumar S. 2001. Changes in the antioxidant enzymes efficacy in two high yielding genotypes of mulberry (*Morus alba* L.) under NaCl salinity. Plant Sci. 161: 613-619.

[77] Dixit V., Pandey V., Shyam R. 2001. Differential oxidative responses to cadmium in roots and   leaves of pea (*Pisum sativum* L cv. Azad). J. Exp. Bot. 52: 1101-1109.

[73] Sandalio L.M., Dalurzo H.C., Gomez M., Romero-Puertas M.C., del Río L.A. 2001. Cadmium- induced changes in the growth and oxidative metabolism of pea plants. J. Exp. Bot. 52: 2115-2126.

[79] Shalata A., Tal M. 1998. The effect of salt stress on lipid peroxidation and antioxidant in the leaf of the cultivated tomato and its wild salt-tolerant relative *Lycopersicon pennellii*. Physiol Plant. 104: 169-174.

[80] Chaparzadeh N., D'Amico M.L., Khavari-Nejad R.A., Izzo R., Navari-Izzo F. 2004. Antioxidatve responses of *Calendula officinalis* under salinity conditions. Plant Physiol. Biochem. 42: 695-701.

[81] Ben Hamed K., Castagna A., Salem E., Ranieri A., Abdelly Ch. 2007. Sea fennel (*Crithmum maritimum* L.) under salinity conditions: a comparison of leaf and root antioxidant responses. Plant Growth Regulation, 53:185-194.

[82] Lechno S., Zamski E., Telor E. 1997. Salt stress-induced responses in cucumber plants. J. Plant Physiol. 150: 206-211.

[83] Rodriguez-Rosales M.P., Kerkeb L., Bueno P., Donaire J.P. 1999. Changes induced by NaCl in lipid content and composition, lipoxygenase, plasma membrane H+ATPase and

antioxidant enzyme activities of tomato (*Lycopersicon esculantum* Mill.) calli. Plant Sci. 143: 143-150.

[84] Hernandez J.A., Campillo A., Jimenez A., Alacon J.J., Sevilla F. 1999. Response of antioxidant systems and leaf water relations to NaCl stress in pea plants. New Phytol 141: 241-251.

[85] Schickler H., Caspi H. 1999. Response of antioxidant enzymes to nickel and cadmium stress in hyperaccumulator plants of the genus Alyssum. Physiol. Plant. 105: 39-44.

[86] Rucińska R., Waplak S., Gwóźdź E. 1999 Free radical formation and activity of antioxidant enzymes in lupin roots exposed to lead. Plant Physiol. Biochem. 37: 187-194

[87] Vitoria A.P., Lea P.J., Azevedo R.A. 2001. Antioxidant enzymes responses to cadmium in radish tissues. Phytochemistry 57: 701-710.

[88] Schutzendubel A., Schwanz P., Teichmann T., Gross K., Langenfeld-Heyser R., Godbold A., Polle A. 2001. Cadmium-induced changes in antioxidative systems, H_2O_2 content and differentiation in pine (*Pinus silvestris*) roots. Plant Physiol. 127: 887-898.

[89] Romero-Puertas M.C., Rodríguez-Serrano M., Corpas F.J., Gomez M., del Rio L.A., Sandalio L.M. 2004. Cadmium-induced subcellular accumulation of O_2^- and H_2O_2 in pea leaves. Plant Cell Environ. 27: 1122-1134.

[90] Demiral T., Turkan I. 2006. Exogenous glycine-betain affects growth and proline accumulation and retards senescence in two rice cultivars under NaCl stress. Environ. Exp. Bot. 56: 72-79.

[91] Knight H., Trewaves A.J., Knight M.R. 1997. Calcium signalling in Arabidopsis thalliana responding to drought and salinity. Plant J. 12: 1067-1078.

[92] Arshi A., Ahmad A., Aref I.M., Iqbal M. 2010. Effect of calcium against salinity-induced inhibition in growth, ion accumulation and proline contents in *Cichorium intybus* L. J. Environ. Biol. 31(6): 939-944.

[93] Di H, Yun-guo L, Yu-e H, , Xiang-jin L, Wei Z (2007) Effects of calcium on chlorophyll and antioxidant enzymes in *Phragmites australis* under cadmium stress. J. Agro-Environ. Sci., 2007-2001.

[94] Di Baccio D., Navari-Izzo F., Izzo R. 2004. Seawater irrigation: antioxidant defence responses in leaves and roots of a sunflower (*Helianthus annuus* L.) ecotype. J. Plant Physiol. 161: 1359-1366.

[95] Meneguzzo S., Navario-Izzo F., Izzo R. 1999. Antioxidative responses of leaves and roots of wheat to increasing NaCl concentrations. J. Plant Physiol. 155: 274-280.

[96] Bandeoglu E., Eyidogan F., Yucel M., Oktem H.A. 2004. Antioxidant responses of shoots and roots of lentil to NaCl salinity stress. Plant Growth Regulation, 42: 69-77.

[97] Tester M., Davenport R. 2003. Na^+ tolerance and Na^+ transport in higher plants. Ann. Bot. 91: 503-527.

Plant-Microbe Relation

Effects of White Root Rot Disease on *Hevea brasiliensis* (Muell. Arg.) – Challenges and Control Approach

Victor Irogue Omorusi

Additional information is available at the end of the chapter

1. Introduction

Monoclonal *Hevea brasiliensis* (Willd. ex Adr. De Juss) Muell. Arg.)is principally valued for its latex content, the latex or Natural Rubber (NR) is very significant in world's industrialization. This importance has been expressly emphasized in the production of elastomers, the use of which is indispensable in space, water, and ship technologies (Jacob 2006). The dependence of world industrialization on NR production is further underscored especially now considering the diminishing reserves of petroleum with increasing environmental hazards.

The rubber tree is subject to a plethora of economically important pathological problems, mainly of fungal origin (the basisdiomycetes) (Igeleke, 1998). In Nigeria, the most serious diseases of rubber seedlings and budded plants in the nursery are leaf diseases (Begho, 1990), while In mature plantation, the most devastating leaf disease is the South American Leaf Blight (SALB), and *Corynespora* Leaf Fall Disease (CLFD) appears to be next to SALB. In field plantations, root diseases pose a serious problem especially in the first few years after planting. In Nigeria, the white root rot disease of rubber is the most serious. It accounts for about 94% of incidences of all root diseases and kills up to five *Hevea* trees/ha (Otoide, 1978). Over a period of time, half of the rubber trees in a plantation are lost to the disease. The infective fungal organism of the white root rot disease is *Rigidoporus lignosus* (klotzsch) Imazeki. The brown root rot disease (*Phellinus noxius*) Corner-Cunn., unlike the white root rot, is most the most serious root disease in *Hevea* plantations in Liberia while *R. lignosus* and Armillaria root rots occur to a lesser extent (Nandris *et al.* 1987). Similarly, in Cote d'Ivoire, *R. lignosus* is the main cause of *Hevea* tree losses with 40-60% of the trees destroyed over a period of up to 21 years (Nandris *et al.* 1987). The white root rot incidence is absent in

India, however, it is serious in Malaysia, Sri Lanka and Congo (Rajalakshmy and Jayarathnam 2000), aside its severe occurrences in Nigeria and Cote d'Ivoire.

Rubber tree exhibits natural resistance to invading root pathogens. Resistance often breaks down due to effect of pathogens that colonize living tissues of the tree to obtain nutrients as a result of the damaging and weakening of the plant with toxins or by preventing the plants defense mechanism (Jayasuriya 2004). A number of certain defense mechanisms in *Hevea* against *R. lignosus* and *P. noxius* have been identified. These include cellular hypertrophy and hyperplasia, cambium activity stimulation, lignifications and suberification of certain cell walls (Jayasuriya 2004., Nicole *et al* 1985).

The process of pre- infection involves pathogen breaking down the host cuticle and cell wall. Plants respond to infection process by producing anti-microbial compounds of low molecular weight (phytoalexins) (Darvill *et al* 1984). *Hevea* plants produce anti-microbial phenolic compounds such as coumarins, flavonoids, triterpenes among others (Jayasuriya 2004) that can partially or completely inhibit microbial infection.

The growth and spread of infective fungal pathogens from existing population have been on the increase with great virulency and inflicting damages even to resistant genotypes. The impact of fungal pathogens results in crop losses. The production of phenylalanine ammonia-lyase (PAL) is implicated as key enzyme in the plant phenyl propanoid pathway to catalyse the synthesis of phenyl lignin and phytoalexin from L-phenylalanine (Jones 1984). The synthesis of these anti-microbial compounds and the subsequent increase in PAL concentration are often useful resistance indicator in the host (Nicholson and Hammerschmidt 1992). Also, oxidases and peroxidases are known to be actively involved in polymerization of phenolic compounds in the lignin formation. In resistance mechanism action of peroxidase is related to initiation of hypersensitive cell collapse (Simons and Rose 1971).

Pathogenesis-related proteins(PR) is yet another defense response of rubber against pathogen infection (Narasimhan *et al* 2000) The PR protein induced by polyacrylic, acetyl salicylic (aspirin) and salicylic acids are known to increase resistance to pathogens (Gianinazzi 1984).

There is variation among high yielding genotypes in disease tolerance level. In this regard, some clones are resistant to virtually most of the diseases but are susceptible to few diseases. However, certain clones exhibit tolerance to few diseases but some others are susceptible to many diseases. There are few clones which tolerate characteristics of pathogens, and as a result of abiotic factors, produce new strains, and these strains can be more aggressive against rubber clones. Mutation is seen to be responsible for variability in pathogens, which involves changing sequence bases in the nuclear DNA, either by way of substitution or addition or deletion of one or many base pairs.

The *Hevea* root fungal parasites- *R. lignosus*, and *P. noxius* are polyporaceae major causes of rubber tree losses in plantation causing the decay of lignified root tissues (Geiger *et al* 1986). *R. lignosus* is partially involved lignin consumption where as *P. noxius* degrade the polysaccharide fraction of but not lignin (Cowling 1961., Kirk 1971), however, findings by

Geiger *et al* (1986) showed that *R. lignosus* and *P. noxius* degrade both the lignin and polysaccharide fractions of the wood. Although *P. noxius* exhibits preferential degradation of polysaccharides, whereas, *R. lignosus* degrades both the lignin and polysaccharides in a relatively balanced manner but with slight preference for lignin.

The *H. brasiliensis* belongs to the family Euphorbiaceae of laticiferous plants. The tree growth can reach a height of over 20 m. Root decaying pathogens, *Rigidoporus lignosis* and *Phellinus noxius* are considered economically important in Africa and Asia due to the fact that they are most dangerous root parasites of the Hevea trees. The R. lignosus is most notorious in its destructive effects on rubber plants.

2. Mechanism of disease infection cycles

The host-parasite interactions involve attacks by *R. lignosus* on the tap root of the *Hevea* tree. The process of disease infection is basically through three stages namely penetration, colonization and degradation. The pathogen penetrates the root system and colonize the tissues. The mycelium of the pathogen, there after degrade the host's cell structures. The root rot pathogen of *R. lignosus* must repeatedly carry out penetration and colonization of their host cell wall. *R. lignosus* carries out its disease infection activities either by enzymatic digestion of the tissues characterized by differentiation of specialized structures (Nicole *et al.* 1986), or by mechanically penetration through colonized natural openings or wounds. Affected host tissues colonized by *R. linosus* through perforation and digestion of cell walls or by penetration through pores and pits of the vascular tissues (Nandris *et al* 1987). Usually, fungal hyphae can be observed as intra and inter interactions of the cell wall. Such observations have revealed some distortions of the cell wall, as well as digestion of the middle lamella and of the cell walls. The implication of these distortions is that enzymatic actions are involved in degradation of cell wall polymers. The root rot path-produce a host range of enzymes which are cell wall degrading enzymes (CWDE) that may correspond to the diverse polymers in plant cell wall of the host and parasite. Infected tissues parasitized are said to contain three enzymes as CM-cellulose, pectinase and Laccase for which their involvement in the pathogenic activity may be presumed (Geiger *et al* 1986). Enzymatic actions in infected tissues are much higher when compared to enzyme activities in healthy tissues. In defense response, the host reacts to parasite infection with increased enzymes from the stimulation of the biosynthesis of enzymes that are present in healthy tissues (Geiger *et al* 1986). Pegg (1977) claimed the enzymes involved take part in host defense responses to the parasite by degrading structural polymers in the mycelia wall. A hypothesis most frequently proposed and verified (Albershiem *et al* 1969) relates to excretion of parasite produced enzymes into the host tissues. The enzymes would be involved in pathogenesis by degrading the polymers in the invaded tissues. Studies of some 3-(1-3) glucanases when synthesized by tomato plantlets, they degrade the β-(1-3) glucans of the invading parasite. *Verticillium alboatrum* (Pegg & Young 1981).

Analysis of root tissues (Geiger *et al* 1986) indicated that some enzymes – (CM – cellulose, pectinase, laccase) are present only in parasite tissues. It is explained that these enzymes are biosynthesized by the parasite and not by the host. However, it is yet to be shown that the fungi are able to perform biosyntheses of those enzymes.

3. Degrading enzymes of cell wall effects in pathogenicity

Involvement of CWDE goes through certain criteria. The synthesis of CWDE is not a proof of involvement in disease. Pathogens with low potentials for the production of CWDE, the ability of cells to reduce the viscosity of a polysaccharide or to grow on a polysaccharide implies synthesis of the relevant polysaccharide, and lack of growth can sometimes reflect inability to metabolize the end product.

Studies of transmission electron microscopy show valuable indications to the participation of CWDE. Microcopic alterations of walls of the infected tissue is displacement of wall fibrils indicating mechanical penetration, whereas extensive wall dissolution signifies freely diffusible extracellular CWDE. Loss of the middle Lamella implies action by pectic enzymes (Baker *et al* 1980) which may also be revealed by lost affinity for the specific stains, ferric hydroxylamine. Alteration or loss of wall microfibrils resulting from cellulose activity is also obvious as reduced birefringence under polarized light.

In wood degradation by enzymes of white root rot fungi the structural elements cellulose, hemicelluloses and lignin synthesized and deposited in the plant cell walls reinforce the mechanical strength and rigidity of the stems of higher plants.

In the host specificity of wood rotters (Tuor *et al* 1995), hard and softwood are distinguished by structural elements building the phenylypropane backbone of the lignin component. Lignin is a three dimensional, optically inactive phenyl propanoid polymer randomly synthesized from coniferyl, p-coumaryl and sinapyl alcohol precursors (Sarkanen and Ludwig 1971). Soft wood on the other hand is referred to as guaiacyl lignin, having over 95% coniferly alcohol (4-hydroxy-3-methoxy-cinnamyl alcohol) units.

4. Pathogenicity of *rigidoporus lignosus* and mechanism infection cycle

Hevea trees are usually killed by root rot pathogens infecting the plants and detection is difficult at the early stages of disease development. Symptoms of root diseases in the above ground level are somewhat similar but differ in the below ground level (Farid *et al* 2009). Generally, the presence of above ground symptoms shows that the trees are now untreatable due to the fast and increased rate of disease infection their make death of plants imminent.

Infected trees show a general foliage discoloration, proceeded sometimes by premature flowering and fruiting. Affected tree branches die back until the whole canopy is destroyed and the tree eventually dies. In Nigeria, the foliage symptoms appear only when the tree is beyond treatment and recovery. The pathogen *R. lignosus,* forms large firm semi fleshy often tiered brackets, on the collar of infected trees in the advanced stage of the disease Normally formations of fructification come up only and after the trees have been dead for a while. Distinctive features of fructification (basidiocarps) (Fig.1), show the upper surface of concentric zones that is brownish-orange with a bright yellow margin when fresh, while the lower surface is reddish-brown.

Figure 1. Basidiocarps of *R.lignosus* on dead *Hevea* tree.

When roots of infected are exposed, profusely branched white rhizomorphs are readily seen. The rhizomorphs are flattened mycelia strands of 1-2 mm thick that grow firmly attached to the surface of infected roots (Fig. 2).

The rhizomorphs grow rapidly and ahead of the rot and extend many meters through the soil freely hindrances from woody substrate. The internal progression of the development of the root rot pathogen is rather an ectotrophic growth characteristic. At infection point, the parasite penetrates the taproot down the soil. Nandris *et al.* (1987) explained for infection to take place, the rhizomorphs are subjected to morphogenetically state into infectious hyphae

Figure 2. Rhizomorph strands on the collars of root of *Hevea* tree.

exhibiting degrading extracellular enzymes capable of wood rotting. The authors further stated that this mechanism is strictly regulated by partial anoxia conditions in the soil. Following root infection, colonization within the taproot progresses towards the collar region and other parts of the root. Newly killed wood is brownish thereafter turns cream and soft. This shows fading of coloration along a gradient from the progression front of the parasite toward the tissues that were colonized before now. The effect of *R. lignosus* causing white root rot extracellular enzymes degradation of lignin in the cell walls of the root system. The basidiocarps seen at the collar region of the tree produce a large number of basidiospores especially during rainy season, but appear less functional in the dissemination of the disease. According to Nandris *et al.*(1987), this role is one of the most controversial points in the biology of *R. lignosus.* The spores are viable, however, John (1965) asserted that there is agreement now that the probability of a spore germinating in situ on a receptive substrate is rather very low. In newly established *Hevea* plantation after clearing of a forest, mycelia of *R. lignosus* do cause infection to take place. However, in the second planting, spores can become inoculum source for infecting the stump surfaces of old rubber trees that are existing between the planting rows. In plate cultures, *R. lignosus* growth on potato dextrose agar (PDA) or malt agar (MA) forms a superficial, extensive white fluffy myceilia (Fig.3).

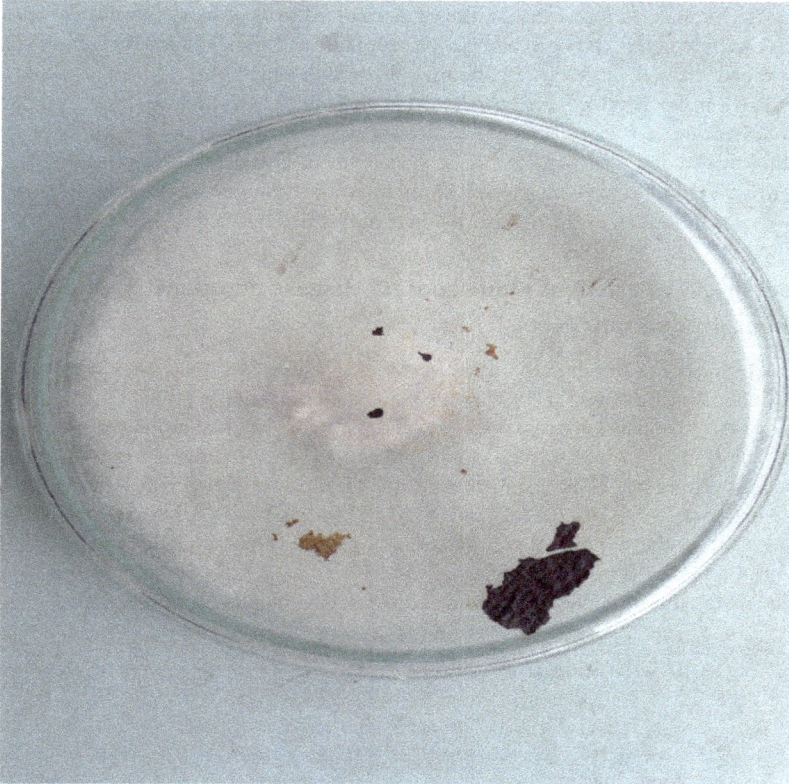

Figure 3. Fluffy growth of *R.* lignosus Disease Symptom Expression

Foliage of infected trees shows general discoloration, often preceded by premature flowering and fruiting. Branches of infected tree die back until the whole canopy is destroyed and the tree eventually dies. In Nigeria, usually the foliage symptoms appear only when the tree is no longer treatable. Exposure of roots around the collar infected trees is carried out by searching the collars with wooden like spades for the presence of rhizomorph filaments. The stick-trapping method is also a useful tool in detecting the growth of *R. Lignosus*. The sticks of *Hevea* wood are poked down the soil around the collar of the tree. The developments of mycelium on the stick are checked after three weeks of insertion into the soil. Another technique is the use of mulch around the collar for three weeks to provide a damp microclimate for the superficial mycelium growth on the collars of the trees. After three weeks the mulch is removed, mycelia growth of *R. lignosus* can be seen.

Generally, two phases characterize the spatial spread of root rot disease in *Hevea* plantations (Nandris *et al* 1985). In the disease cycle of white root rot (Fig. 4), the processes involve infection and colonization by mycelia of the root system of young rubber trees growing from stumps of infected forest tree or otherwise referred to as primary inoculums. The other aspect is progress of the pathogen along roots from infected trees (second inoculums) toward healthy rubber trees around (Fig. 4). The mycelia growth length of R. *Lignosus* is about 2.5 m while it is 0.7 m for P. *noxius* per year (Nandris *et al.* 1987). This clearly shows comparative growth rates of the two root pathogens of *Hevea* that reveals differences in rapidity of spread. According to Nandris *et al.* (1985), disease development and death of *Hevea* trees are most rapid during the first few years after planting. Infection cycle development in diseased trees usually results in sudden death, delayed death or survival of infected trees.

5. Assessment of effect of white root rot disease on quantity and quality of rubber production

The level of incidence of white root rot disease and its effect on *Hevea* latex were evaluated at the Rubber Research Institute of Nigeria, main station, Iyanomo (Table 1). The plantation was previously cropped with cassava, yam, and plantain. The plantation consists of 36 sub plots of one hectare each with nine clones planted out in a completely randomized block designed. Six *Hevea* test clones comprising three local clones – NIG 800, 801, 802, 803, 804, 805, and exotic clones – GT1, PR 107, and RRIM 700 were assessed for the study. Assessment of the severity of the white root rot disease was based on disease index method by Parry (1990) on a scales of 0 = no infection, 1 = light infection, 2 = moderate infection, and 3 = severe infection. Infection was then calculated using the following formula, thus:

$$\text{Disease index(DI)} = \frac{(0 \times a)+(1 \times b)+(2 \times c)+(3 \times d)}{a+b+c+d} \times \frac{\times 100}{x}$$

where,

0, 1, 2, and 3 are infection categories
a, b, c, and d are plants that fall into the infection categories
x is the maximum disease category which is 4

	Hevea clones					
Rep.	NIG 800	NIG 800	NIG 800	GT 1	PR 107	RRIM 707
1	25.10	23.18	28.00	9.34	25.30	22.83
2	31.72	21.10	20.13	10.75	27.47	26.48
3	28.13	29.19	29.90	10.61	33.61	37.11
4	26.53	34.00	32.00	28.14	35.80	22.00
Mean	27.87	26.87	27.51	14.71	30.55	26.86

Table 1. Disease indices of incidence of white root rot of six *Hevea* test clones.

Disease indices recorded showed highest susceptibility score in PR 107 (30.55) and lowest score in GT 1 (14.71). Intermediate scores among other clones were equally high compared with PR 107. The lowest susceptibility in GT 1 indicated that GT 1 showed a significant level of resistance to the white root rot disease.

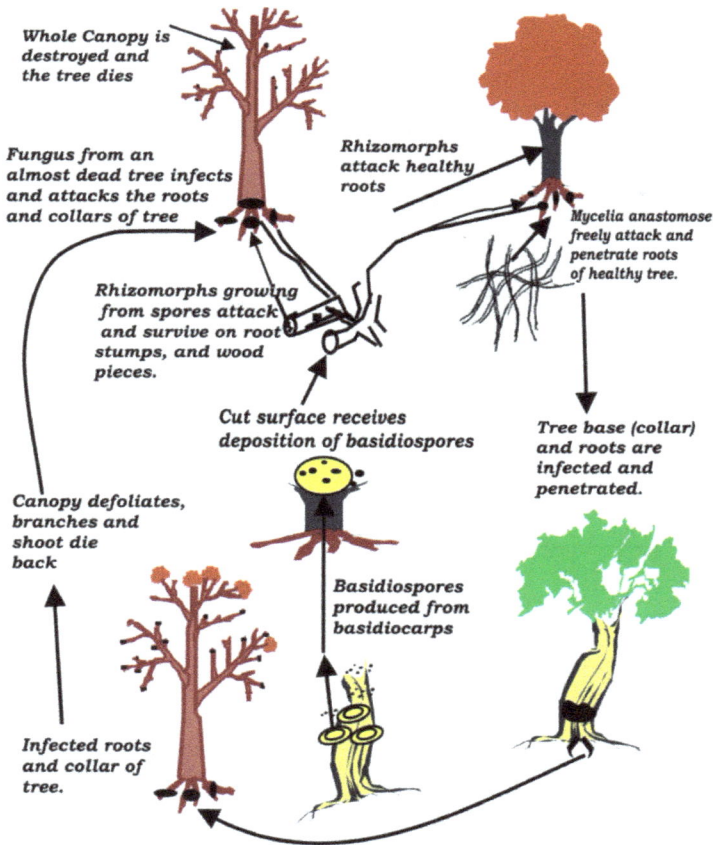

Figure 4. Diagramatic illustration of white root rot disease cycle.

The result of the white root rot incidence in six clones are shown in Table 1.

Based on the susceptibility status recorded, the GT 1 Lowest susceptibility status was assessed for the volume of latex and dry weight of cup lumps from effects at different infection categories of white root rot incidence (Table 2). The 'O' infection category had the highest volume of latex (228.93 cm³), however, showed no significant difference with the rest clones. The highest infectivity category gave the highest dry weight of cup lumps (118.83g), followed by '0' '1' and '2' categories but no significant difference among the clones was obtained. This may be attributed to increase in DRC in response to effect of the white root rot disease.

Infection category	Volume of latex (cm³)	Dry weight of cup lumps (g)
0	228.93	114.00
1	192.44	103.71
2	191.17	94.00
3	224.18	118.88
Cv (%)	26.91	28.24

Table 2. Volume of latex and dry weight of cup lumps under infection categories in GT 1

6. White root rot observations of *hevea* clones

Observations recorded at Rubber Research Institute of Nigeria (RRIN), Iyanomo, on the roots of some *Hevea* clones after ten years of planting are shown in Tables 3 & 4. Results of root rot observations showed that the white root rot accounted for almost all the root rots investigated indicating that there was high incidence of white root rot disease. Inspite of the high incidence of the white root rot, only a small proportion of the trees with root rot infection were actually killed by the disease. From this study, sixteen of the RRIN clones ('C' clones) examined recorded no loss due root disease infection. However, the remaining fourteen 'C' clones and PB 86 had losses due to root rot infection.

Clone	No. Planted Out	No. 9 Survival	No. with White Root Rot	No. of Dead From Root Rot	% Infection	% Survival
RRINSC*1	9	5	3	1	44.44	55.55
3	9	7	5	1	55.55	77.77
8	9	8	6	1	66.66	88.88
11	9	7	5	1	55.55	77.77
21	9	9	6	-	66.66	100.00
27	9	9	7	-	77.77	100.00
28	9	8	6	1	66.66	88.88
29	9	6	4	3	44.44	66.66
42	9	9	5	-	55.55	100.00
43	9	8	4	1	44.44	88.00
47	9	9	8	1	88.88	100.00
51	9	8	8	1	88.88	88.88
52	9	9	7	-	77.77	100.00
54	9	9	6	-	66.66	100.00
55	9	6	6	3	66.66	66.66
58	9	8	6	1	66.66	88.88
75	9	8	6	1	66.66	88.88
76	9	7	4	2	44.44	77.77
82	9	7	5	1	55.55	77.77
83	9	9	6	-	66.66	100.00

Clone	No. Planted Out	No. 9 Survival	No. with White Root Rot	No. of Dead From Root Rot	% Infection	% Survival
85	9	9	8	-	88.88	100.00
86	9	9	8	-	88.88	100.00
87	9	9	6	-	66.66	100.00
97	9	9	6	-		
98	9	5	4	4	44.44	55.55

*RRINSC - Rubber Research Institute of Nigeria 'C' Clones

Table 3. Root Rot Observation after 10years of Planting at RRIN

Clone	No. Planted Out	No. 9 Survival	No. with White Root Rot	No. of Dead From Root Rot	% Infection	% Survival
RRINSC						
104	9	6	1	2	11.11	66.66
105	9	9	7	-	77.77	100.00
106	9	9	5	-	55.55	100.00
114	9	6	5	1	55.55	66.66
118	9	9	4	-	44.44	100.00
PB86	162	121	87	25	53.70	74.69

* Rubber Research Institute of Nigeria 'C' Clones

Table 4. Impact of *Rigidoporus lignosus* on *H. brasiliensis* clones-root rot observations after 10 years of Planting.

7. Fungicide curative approaches

Reduction of root diseases is performed through some control approaches. Proper clearing of land is recommended especially mechanically clearing is the most effective clearing method. In the recent past in Nigeria, treatment of infected roots with root rot diseases involved excavating the soil around the roots of infected trees, and removal of surface rhizomorphs and roots sections penetrated by the pathogen. A long-lasting collar protectant dressing fungicides such as pentachloronitrobenzene (PCNB) is then smeared around the collar, tap root and basal portion of the main laterals of the tree, before replacing the soil. The same treatment is also given to immediate tree neighbors, especially those along the same row. This approach is rather labor intensive and quite difficult to apply on a large scale plantation.

Results of investigation on the effectiveness of calixin fungicide (a.i. tridemorph) at the RRIN mainstation, Iyanomo (Table 5) revealed significant reduction of white root rot disease of *Hevea* trees. Virtually all clones treated with calixin recovered satisfactorily. Of the fourteen test clones treated with calixin, seven of the clones were over 80% cured of the *R.lignosus* infection while five were over 70% cured. However, these percentages of cured plants were highly significantly different at P<0.01. Only a minor proportion of two clones with over 50 and 66% were cured and significantly different at P<0.05.The percentages of cured trees related to the number of clones that survived from the number of dead clones against the number of clones planted out.

Clones	Dead/ No. Planted	No of Survivors	Pecentage of Plants cured
PRINSC 1	2/9	7	77.77
3	1/9	8	88.88
8	1/9	8	88.88
11	1/9	8	88.88
28	1/9	8	88.88
55	3/9	6	66.66
58	2/9	7	77.77
75	1/9	8	88.88
76	2/9	7	77.77
82	1/9	8	88.88
98	4/9	5	55.55
104	2/9	7	77.77
114	2/9	7	77.77
PB 86	30/162	133	82.09

P< 0.01 (Clones with over 70-80% cured)
P<0.05 (Clones with over 50-60% cured)
* RRIN SC – Rubber Research institute of Nigeria 'C'clones

Table 5. Clones with White Root Rot treated with Calixin P< 0.01 (Clones with over 70 -80% cured)

8. Preventive measures against incidence of *r.lignosus*

Foliage discoloration of *Hevea* trees is external indicator of the presence of white root rct incidence when the foliage is watched of three months intervals. Notice of foliage discoloration is often preceded by premature flowering and fruiting. There after branches begin to die back and eventually the whole canopy is destroyed and the tree is dead. This phenomenon above ground described serves as a warning sign of the presence of white root rot disease. Below ground level reveals the presence of rhizomorphs, is usually by inspection of the diseased collar region as well as neighboring trees.

Clearing methods such as uprooting and poisoning of old trees are found to reduce root disease incidence in replanting. The stumps from felled trees should be poisoned and the cut surface painted with creosote. This will prevent the root fungi of food sources.

Planting of legume cover crops helps to reduce root disease since they cause roots and stumps to rot faster (Malaysian Rubber Board 2000). It also asserted that about 30g of powdered sulphur should be amended into the planting hole. Sulphur is known to promote fungal growth antagonistic to incidence of fungal root diseases.

The use of collar protectant to paint the collar part of the tree is said to prevent root disease infection for few years (Malaysian Rubber Board 2000). Suitable fungicidal wound dressing should be applied to any damaged part of the tree. This prevents infective spores from gaining entry into the plant tissues. The digging of isolation trenches of one meter between infected and healthy trees prevents the spread of disease from infected plants to health trees. This method is rather difficult and may not be satisfactorily effective.

9. Treatment measures

On exposure of infected root system, all dead roots are removed, bulked and burnt. Soil particles attached to the roots are shaken off. After exposure, root surface are dried and suitable long lasting collar protect ant fungicide pentachloronitrobenzene (PCNB) is painted over the exposed infected root system, including collar, tap root and basal portion of the main laterals of the tree, before replacing the soil. The same treatment is carried out for the immediate tree neighbors, especially those along the same row. This method of exposing the root system and subsequent application of a collar protect ant fungicide is quite labour intensive exercise and uneconomical. The PCNB use is hazardous, being carcinogenic and has been banned (Rajalakshmy and Jayarathnam 2000). This method has been abandoned in most rubber growing regions of the world.

A less laborious approach is digging a slight furrow around the base of the infected tree and drenching the collar region with two liters of 0.5% calixin (a.i.tridemorph). Infected trees as well as two direct neighbors in the same row are treated every six months. Other suitable fungicides for the treatment of white root rot disease is the use of Bayleton (a.i. triadimefon), Bayfidan (a.i. tridimenol), Anvil (a.i. hexaconazole), Folicur (a.i. tebuconazole), Contaf (a.i. hexaconazole) and Daconil (a.i. chlorothaconil) (Jayaratne et al 2001). The use of these chemical fungicides as described above involving loosening the soil and forming funnel – like furrow around the base of the trunk, and then drench the chemicals into the furrow along the trunk of the tree with about 10-20 ml at every four to six months is highly effective.

The purpose of controlling root diseases is to remove their sources of infection or inocula at the very young stages of the trees in order to prevent the rest trees from being infected as they mature. Due to possible sources of infection it is necessary to put in place adequate management measures. In embarking on preparations for the management, the role of tapers is necessary since tapers know the locations of the diseased trees in the plantations and this however, simplifies task of locating the diseased trees. This explains the need to train tapers on how to recognize root diseases and other *Hevea* disease symptoms. Treating diseased plants is to curtail the spread of the disease since it takes a higher cost of treating older trees and which may be abandoned and isolated by trenches to reduce the spread of root disease to healthy trees.

Author details

Victor Irogue Omorusi
Plant protect Division, Rubber Research Institute of Nigeria, PMB, Iyanomo, Benin City, Nigeria

10. References

Begho, E.R. 1990. Nursery diseases of *Hevea brasiliensis* in Nigeria and their control. 103—106 in : *Proceedings of a Natural Workshop on fruit / tree crop seedling Production* held at NIFOR, Benin City, Aug 14-17, 1990. 160 PP.

Cowling, E.D. 1961. Comparative Biochemistry of the decay of sweet gum sapwood by white rot and brown rot fungi. *U.S. Department of Agriculture, Technical Bulletin* No. 1258. 79p.

Darvill , A.G., and Albersheim, P. 1984. Phytoalexins and their elicitors – A defence against microbial infection in plants. *Annual Review Plant Physiology* 35:243-275.

Geiger, J.P., Rio, B., Nicole, M., and Nandris, D. 1986. Biodegradation of *Hevea Brasiliensis* wood by *Rigidoporus lignosus* and *Phellinus noxius*. *Eur.J . For. Path;* 16:147-159.

Gianinazzi, S.1984. Genetic and molecular aspects of resistance induced by infections or chemicals. In : *Plant –microbe Interactions Molecular and genetic perspectives.* Vol 1. (Eds; Kosuge, T and Nester, E.W), 321-342 PP. Macmillan, New York.

Igeleke, C.L. 1988. Diseases of Rubber (*Hevea brasiliensis)* and their control. Paper presented at the 10th Annual conference of the Horticultural society of Nigeria, 6-8 Nov. 1988

Jacob, C.K 2006. Corynespora leaf disease of *Hevea brasiliensis:* A threat to natural rubber production (9-16pp) in Corynespora leaf disease of *H. brasiliensis:* strategies for management (Ed. C.K. Jacob). Rubber Research institute of India.

Jayasuriya, K.E. 2004. Factors affecting disease tolerance of rubber tree and research needs for developing disease tolerant genotypes for the sustainability of rubber industry. *Bulletin of the Rubber institute of Sri Lanka*, 45 : 1-10.

Jayaratne, R, Wettasinghe, P.C., Siriwardene, D., and Peiris D. 2001. Systemic fungicides as a drench application to control white root disease of rubber. *Journal of the Rubber Research Institute of Srilanka*, 84:1-17.

Jones, D. H. 1984. Phenylanine ammonia – lyase: regulation of its induction, and its role in plant development. *Phytochemistry*, 23:1349-1369.

Kirk, T.K . 1971. Effect of microorganisms on ligin. *Annual Review of Phytopathology*, 9:185-210.

Malaysian Rubber Board. 2000. *Treatment of maladies and injuries and control of pests.* In: Rubber Plantation and Processing technologies. Malaysian Rubber Board, 2000.

Nandris, D., Nicole, M. and Geiger, J.P. 1987. Root Rot Diseases of Rubber Trees. *Plant Disease*, 71 (4) : 298-306.

Narasimhan, K.T and Kothandaraman, R. 2000. Detection of pathogenesis related proteins in *Hevea brasiliensis* infected by *Phytophthora meadii*. *Indian Journal of Natural Rubber*, 13:30-37.

Nicholson, R.L,. and Hammerschmidt, R. 1992. Phenolic compounds and their role in disease resistance. *Annual Review of Pyhtopathology* 30:369-389.

Nicole, M., Geiger, J.P., and Nandris, D. 1985. Variability among Africa populations of *Rigidoporus lignosus and Phellinus noxius, Eur. J. for. Pathol.* 15: 293 -300.

Otoide, V.O. 1978. Further observations on the pre-treatment of forest trees for root disease control in *Hevea* plantings. Paper presented at RRIN Seminar, 1978, 7Pp.

Rajalakshmy, V.K, and Jayarathnam, K.2000. Root diseases and non- microbial maladies. In: *Natural Rubber: Agro-management and crop processing* (Eds. P.J George and C. Kuruvilla Jacob) Rubber Research Institute of India, Kohayam, Kerala, India.

Simons, T.J., and Rose, A.F. 1971. Metabolic changes associated with systemic induced resistance to tobacco mosaic virus in Samsun -NN tobacco. *Phytopathology*, 61:293 -300.

Chapter 5

Plant Defense Enzymes Activated in Bean Plants by Aqueous Extract from *Pycnoporus sanguineus* Fruiting Body

José Renato Stangarlin, Clair Aparecida Viecelli,
Odair José Kuhn, Kátia Regina Freitas Schwan-Estrada,
Lindomar Assi, Roberto Luis Portz and Cristiane Cláudia Meinerz

Additional information is available at the end of the chapter

1. Introduction

The common bean (*Phaseolus vulgaris* L.) can be affected for more than 300 diseases caused by virus, bacteria, fungi, and nematodes. The semibiotroph *Pseudocercospora griseola* (Sacc.) Crous & Braun (sin. *Phaeoisariopsis griseola* (Sacc.) Ferraris), the causal agent of angular leaf spot represents one of the main fungal pathogens of this crop, manifesting on the stem, leaf, and pod [1].

Traditionally, the control of the angular leaf spot has been done with the use of resistant cultivars, seeds free of pathogen and fungicides. The last one, at a short time, has it's advantages, but for a long period of time, can cause problems due the residues accumulation and environmental pollution [2]. Thus, with the objective to find new technologies, ecologically or environmentally safer, for the control of plant diseases, mainly in organic growth, alternative methods for the control of phytopathogens are been development. This kind of alternative methods are been investigated by our 'Biological and Alternative Control of Plant Diseases' research group [3].

The induction of resistance in plants involves the activation of defense latent mechanisms [4] in response to the treatment with elicitor agents, protecting against subsequent infection by pathogens. Among the non-conventional elicitors can be included the extracts of medicinal plants and essential oils [5], homeopathic drugs [6], as well as the extracts obtained from mushrooms [7-9]. Among the basidiomycetes with elicitor proprieties stands out *Pycnoporus sanguineus* (L. ex Fr.) Murr. [10], utilized since the medicine [11,12] to the alternative control of plant diseases [13,14]. These previous works had shown that the biological properties of

P. sanguineus depending of its crude or aqueous extract and not of its individual compounds, like cinnabarin, or extracts obtained from organic solvents.

Previous works had shown the potential of *P. sanguineus* components for controlling plant diseases. Aqueous extracts, obtained from liquid medium-culture filtrate (MCF) [15] and from mycelium (AEM) [16] of *P. sanguineus*, were capable to reduce in 82% and 49% to MCF, and in 93% and 50% to AEM, in greenhouse and field conditions, respectively, the severity of angular leaf spot in bean plants. However, the effect of *P. sanguineus* fruiting bodies in that pathosystem was not investigated. Against plant pathogenic bacteria, fruiting body extracts from *P. sanguineus* were efficient for the control of common bacterial blight in bean, caused by *Xanthomonas axonopodis* pv. *phaseoli* (Smith – Vauterin, Hoste, Kersters & Swings) which can occur either by direct antimicrobial activity and by resistance induction involving the activation of some pathogenesis-related proteins [17].

In another experiment, the *in situ* detection of reactive oxygen species (ROS), mainly hydrogen peroxide (H_2O_2) and superoxide (O_2^-), was searched in bean plants treated with aqueous extracts of the mycelium (AEM) and basidiocarps or fruiting bodies (AEB) of *P. sanguineus* and inoculated after three days with *Colletotrichum lindemuthianum* ((Sacc. & Magn.) Scrib.). It was possible to detect H_2O_2 at 48 hours after inoculation (hai) only to the treatment with basidiocarp extract. The O_2^- was detected mainly to the treatment with mycelium extract at 48 hai. All the treatments showed reaction for H_2O_2 and O_2^- in epidermal and mesophyllic cells at 192 hai, probably due the infection development. These results suggest that *P. sanguineus* extracts promote oxidative burst in bean plants, in early infection process, reducing anthracnose severity [18].

Thus, this work aimed to investigate the potential of *P. sanguineus* for controlling angular leaf spot in common bean, evaluating the *in vitro* antimicrobial activity against *P. griseola* and the induction of resistant enzymes as peroxidase, polyphenoloxidase and β-1,3-glucanase, as well as the influence on physiological mechanisms related to the energy supply, as the protein content and chlorophyll.

2. Materials and methods

Pathogen isolate: *Pseudocercospora griseola* was obtained from bean plants naturally infected, and cultivated in tomato juice (200 mL of tomato juice, 15 g of agar, 4,5 g of $CaCO_3$ and 800 mL of distillated water) for 14 days at 24 °C and dark [19].

Aqueous extract of *Pycnoporus sanguineus* **fruiting body:** This process was carried out as methodologies [7] and [13]. Fruiting bodies or basidiocarps of *P. sanguineus* were collected in western Paraná State, Brazil, and identified according [10]. To obtain the aqueous extract, dehydrated powder from basidiocarps was suspended into distilled water (14 mL g^{-1}) and, after 24 h incubation at 4ºC, the suspension was filtered through a common filter paper (8 g cm^{-2}) and centrifuged at 20,000 g for 25 min. The supernatant obtained, after this procedure, was considered as the crude aqueous extract.

Inhibition of conidia germination: This assay was done in microscopic slide covered with a thin layer of agar-water 1% (700 µL per slide) [20]. Aliquots of 40 µL of aqueous extracts in concentrations 0, 1, 5, 10, 15 and 20%, sterilized in autoclave or filtrated in nitrocellulose membrane (0.45 µm of pore diameter) and aliquots of 40 µL of conidia suspension of *P. griseola* (1×10^4 conidia mL⁻¹) obtained of a culture with 14 days old, were distributed in the surface of the slide, which were incubated in moist chamber under dark at 24 °C [21]. As control treatments were utilized fungicide (azoxystrobin: 40 mg L⁻¹) and acibenzolar-S-methyl (ASM: 75 mg L⁻¹). The percentage of the germination was determined after 24 hours with the addition of 40 µL of lactophenol cotton-blue in each slide to paralyze the conidia germination.

Inhibition of the mycelial growth and sporulation of *P. griseola*: The extracts of basidiocarp of *P. sanguineus* were incorporated in concentrations of 0, 1, 5, 10, 15 and 20%, in tomato juice culture medium. The extracts were sterilized in autoclave and also by filtration in nitrocellulose membrane (0.45 µm of diameter of pore) [22]. As control treatments were utilized fungicide (azoxystrobin: 40 mg L⁻¹) and acibenzolar-S-methyl (ASM: 75 mg L⁻¹). For transferring *P. griseola* to Petri dishes, 100 µL of spore suspension (1×10^4 conidia mL⁻¹) were added to the medium and homogenized with Drygalski loop. The Petri dishes were sealed with plastic film and maintained at 24 °C and dark. Were evaluated the diameter and the number of colonies 14 days after the beginning of the experiment. At the end of the assay of mycelial growth inhibition, it was evaluated the sporulation of the fungus. For this, was prepared a suspension of conidia by the addition of 10 mL of distillated water per plate and determined the number of conidian per mL in Neubauer chamber.

Experiment in greenhouse: Two plants of common bean (cultivar IAPAR 81 – Carioca) were cultivated in plastic pots containing 5 L of a mixture of sterilized soil and sand (proportion 2:1). To resistance induction assay, were used aqueous extract of *P. sanguineus* basidiocarp at concentration of 10% and 20%. As control treatments were used water, fungicide (azoxystrobin - 40 mg L⁻¹) and acibenzolar-S-methyl (75 mg L⁻¹). The extracts and the control treatments were sprayed in the 3rd leaf (vegetative stage V4) (3 mL per leaf).

Field experiment: The experiment consisted in three randomized blocks, with five plots per block. Each plot consisted of three lines of 3 m of length, spaced 0.5 m between them, with 10 plants (cultivar IAPAR 81 – Carioca) per meter. The central line, discounting 0.5 m from the anterior and posterior borders, was considerate as useful area for evaluation. For the assay of resistance induction, were sprayed aqueous extracts of *P. sanguineus* basidiocarp at concentrations of 10% and 20%, and as control treatments were used water, fungicide (azoxystrobin: 40 mg L⁻¹) and the acibenzolar-S-methyl (ASM: 75 mg L⁻¹). The extracts (5 mL per plant) were applied twice, the first one in vegetative stage (V3) and the second in reproductive stage (R3).

Pathogen inoculation: The conidia suspension of *P. griseola* was prepared in water with Tween 20 (one drop 500 mL⁻¹), and the concentration adjusted to 4×10^4 conidian mL⁻¹. The inoculation in the greenhouse was done three days after the application of extracts and control treatments, in the 3rd treated leaf, as well as in the 4th non-treated leaf (vegetative stage V4), to verify a putative systemic resistance induction. After the inoculation, the plants were maintained in humidity chambers and dark at 24 °C during 48 hours and, later, maintained in greenhouse, according to methodology used by [19]. In the field, two inoculations were done, the first in vegetative stage (V3) and the second in reproductive stage (R3), both three days after the application of extracts and control treatments.

Severity evaluation: The severity of the angular leaf spot in the greenhouse was evaluated in the 3rd and 4th leaves at 8, 12, 16, 20 and 24 days after the inoculation, using diagrammatic scale prepared by [23]. In the field, the evaluations started when the first symptoms of disease appeared (seven days after the inoculation), and were obtained five evaluations on the lower middle canopy of the plant. In the second application of extracts and control treatments, the severity was evaluated as the same way that was done in the first application, but only evaluating the upper middle canopy. With the severity data was calculated the area under the disease progress curve (AUDPC) of angular leaf spot as in reference [24].

Biochemical analysis: Leaf disc with 3.46 cm^2 (three disc per sample) were collected at 48, 72, 96, and 120 hours after the inoculation (hai) and also after the symptoms appearance (144 hai). Each collected sample was immediately wrapped in aluminum foil and freeze at -20 °C. Samples were collected from the 3rd treated and inoculated leaf, as well as from the 4th non-treated but inoculated leaf, from the same plant [25].

Obtaining the protein extracts: the samples of leaves were mechanically homogenized in 2 mL of extraction buffer sodium phosphate 0.01 M (pH 6.0), in a porcelain mortar. The homogenate was centrifuged at 6.500 g during 10 min at 4 °C. The supernatant was considerate the enzymatic extract, for later determination of peroxidase, polyphenoloxidase and β-1,3-glucanase activities and protein content [25].

Peroxidase activity: the peroxidase activity was determined at 30 °C, by spectrophotometer at 470 nm during 2.15 min [26]. The peroxidase activity was expressed in absorbance min^{-1} g of fresh $mass^{-1}$.

Polyphenoloxidase activity: the polyphenoloxidase activity was determined according the methodology in reference [27]. The results were expressed in absorbance min^{-1} g of fresh $mass^{-1}$.

β-1,3 glucanase activity: the enzyme activity was evaluated according to [19]. The reaction was determined by colorimetric quantification of glucose released from laminarin, using ρ-hydroxybenzhydrazide. The results were expressed in µg of glucose min^{-1} g of fresh $mass^{-1}$.

The protein content: the total protein content was evaluated as [28]. The concentration of proteins, expressed in equivalent of bovine serum albumin (BSA) in one mL of sample (µg protein mL^{-1}), was determined utilizing standard curve of concentrations of BSA, varying of 0 to 20 µg mL.

The chlorophyll content: for the quantification of chlorophyll was utilized an adapted methodology [29]. The samples of plant tissue (0.1 g) were packed in glass tube with 10 mL of acetone 80%, during 7 days in the dark at 25 °C. After this time was determined the absorbance at 663 nm and 645 nm for chlorophyll a and b, respectively. The concentration of chlorophyll a was obtained by the equation $(0.0127A_{663}) - (0.00269A_{645})$ and of chlorophyll b by the equation $(0.0029A_{645}) - (0.00468A_{663})$. The total chlorophyll content was obtained by adding the results of chlorophyll a and b. The values were expressed in mg g of fresh mass^{-1}.

Statistical analysis: The experiments were arranged in randomized blocks, with five treatments. The analyzes of variance (ANOVA) was done using the statistical program JMP (Statistical Analysis System SAS Institute Inc. USA, 1989 – 2000 version 4.0.0.), and the average compared by the Dunnett's test in level of 5% of probability.

3. Results

There was no significant effect of the concentrations of *P. sanguineus* basidiocarp extract on the spores germination, mycelial growth and sporulation of *P. griseola* (data not shown), indicating the absence of direct antimicrobial activity of these extracts on the pathogen.

However, in the greenhouse and field experiments, the area under disease progress curve (AUDPC) of angular leaf spot showed that the plants treated with *P. sanguineus* extract at 10% and 20% differ from the water-treatment, with reduction of 42% and 54% in the 3rd leaf, respectively. In the 4th leaf was observed reduction of 69% in the AUDPC for the plants treated with basidiocarp extract at 20%, not differing from ASM and fungicide control treatments (Table 1), indicating systemic resistance induction.

In the lower middle canopy, there was no statistical difference in AUDPC for basidiocarp extract when compared to water and ASM control treatments. The difference was observed just to fungicide treatment, which presented the better protection against angular leaf spot. To the upper middle canopy, was verified a reduction of 64% in the AUDPC for the treatment with basidiocarp extract at 20%, better then ASM, which is a commercial resistance inducers product. So, these results indicate the great potential of *P. sanguineus* extracts for the control of *P. griseola* in common bean, with local and systemic effects.

The biochemistry analysis reveled induction of peroxidase activity due the treatment with basidiocarp extract on the 3rd treated leaf, as well as in the 4th non-treated and inoculated leaf, demonstrating the systemic induction effect of *P. sanguineus* (Figure 1). In the 3rd leaf the basidiocarp extract at 10% reduced the activity of the enzyme when compared to the water-treatment and fungicide three days after the inoculation (DAI). To five DAI differed from ASM and fungicide, and to seven DAI from ASM. The basidiocarp at 20% presented higher activity of peroxidase at four DAI, and inhibition to three, five and seven DAI, when compared to ASM, ASM and fungicide and ASM, respectively. The 4th non-treated leaf showed the same pattern of 3rd leaf for peroxidase activity.

The polyphenoloxidase activity was influenced by treatments with extracts of *P. sanguineus*, in the 3rd leaf treated, as well as in the 4th non-treated and inoculated leaf, demonstrating systemic effect (Figure 2). In the treated leaf, the extract of basidiocarp at 10% reduced the enzyme activity when compared to the water and ASM at three DAI. To four DAI the enzyme activity was stimulated when compared to water and fungicide, and to five DAI was smaller than ASM. The basidiocarp at 20% showed increase in activity of polyphenoloxidase in relation to fungicide at three DAI, at four and five DAI, when compared to the three control treatments and at seven DAI in relation to fungicide. The 4th non-treated leaf presented similar effects to the 3rd leaf on the activity of polyphenoloxidase

Treatments	AUDPC			
	Greenhouse		Field	
	3rd Leaf*	4th Leaf*	LMC****	UMC****
Basidiocarp 10%	24.5[1,2,3]	5.3	51.7[3]	54.2[1,3]
Basidiocarp 20%	19.5[1,2,3]	2.7[1]	49.9[3]	32.3[1,2,3]
Water	41.9	8.8	59.4	89.0
ASM**	2.3	5.2	41.9	67.7
Fungicide***	6.7	2.4	6.3	16.1
C.V. (%)	31.4	7.3	47.5	53.9

Averages followed by a bold number differ statistically (Dunnett's test, P≤0.05) of the control treatments water (1), ASM (2) or fungicide (3);
*3rd leaf: treated and inoculated; 4th leaf: non-treated and inoculated from same plant;
**ASM: acibenzolar-S-methyl (75mg L-1);
***Fungicide: azoxystrobin (40 mg L-1);
****LMC and UMC: Lower middle canopy and upper middle canopy of the plant, respectively.

Table 1. Area under disease progress curve (AUDPC) of angular leaf spot in common bean after the application of aqueous extracts of *P. sanguineus* basidiocarp, in greenhouse and field conditions.

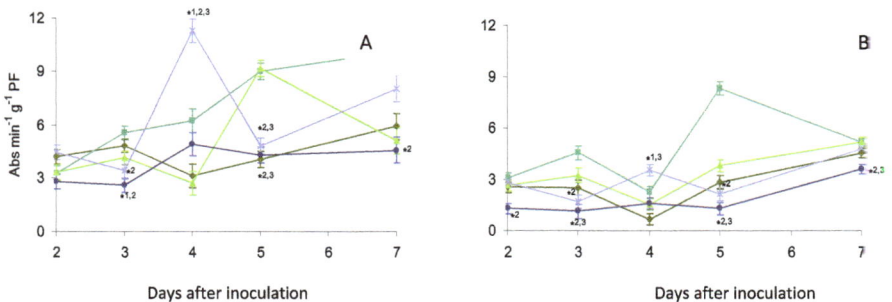

Figure 1. Peroxidase activity in bean plants inoculated with *Pseudocercospora griseola* tree days after the application of water (♦), acibenzolar-S-methyl (ASM 75mg L-1) (■), fungicide (azoxystrobin 40mg L-1) (▲) and the aqueous extracts of basidiocarp of *Pycnoporus sanguineus* at 10% and 20% (● and ✕). A and B represent the 3rd treated and inoculated leaf and 4th non-treated and inoculated leaf, respectively. Bars indicate an average ± standard error. Average followed by * differ statistically (Dunnett's test, P≤0.05) from the control treatments water (1), ASM (2) or fungicide (3). PF: fresh mass.

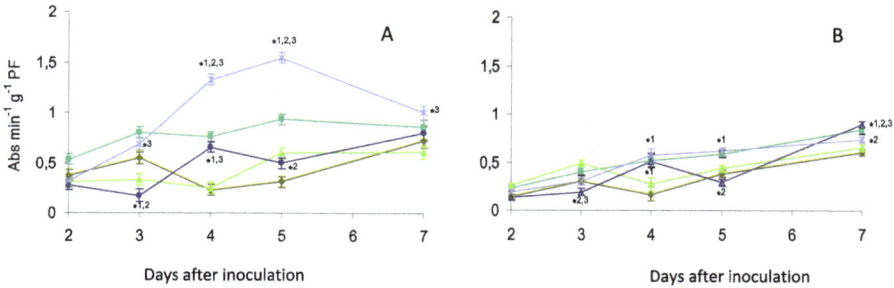

Figure 2. Polyphenoloxidase activity in bean plants inoculated with *Pseudocercospora griseola* tree days after the application of water (♦), acibenzolar-S-methyl (ASM 75mg L^{-1}) (■), fungicide (azoxystrobin 40mg L^{-1}) (▲) and the aqueous extracts of basidiocarp of *Pycnoporus sanguineus* at 10% and 20% (● and ✕). **A** and **B** represent the 3rd treated and inoculated leaf and 4th non-treated and inoculated leaf, respectively. Bars indicate an average ± standard error. Average followed by * differ statistically (Dunnett's test, P≤0.05) from the control treatments water (1), ASM (2) or fungicide (3). PF: fresh mass.

The activity of β-1,3-glucanase was influenced by the treatments with extract of *P. sanguineus*, in the 3rd leaf, as well as in the 4th non-treated and inoculated leaf (Figure 3). In the treated leaf, the extracts of basidiocarp at 10% and 20% reduced the enzyme activity in relation to control treatments at five DAI, and at seven DAI for the extract at 20% when compared to water and fungicide. In the 4th non-treated leaf there was similar effect to the 3rd leaf on the activity of β-1,3-glucanase.

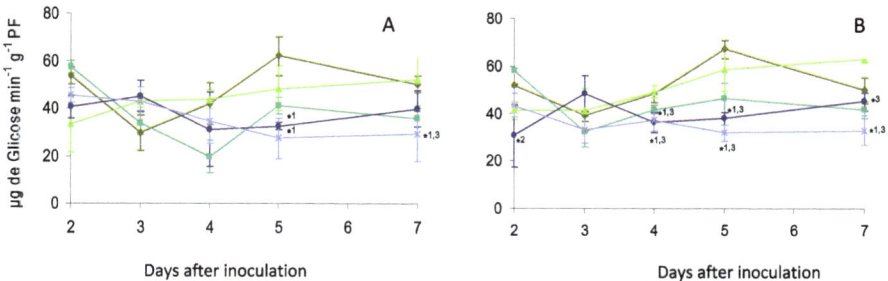

Figure 3. β-1,3-glucanase activity in bean plants inoculated with *Pseudocercospora griseola* tree days after the application of water (♦), acibenzolar-S-methyl (ASM 75mg L^{-1}) (■), fungicide (azoxystrobin 40mg L^{-}) (▲) and the aqueous extracts of basidiocarp of *Pycnoporus sanguineus* at 10% and 20% (● and ✕). **A** and **B** represent the 3rd treated and inoculated leaf and 4th non-treated and inoculated leaf, respectively. Bars indicate an average ± standard error. Average followed by * differ statistically (Dunnett's test, P≤0.05) from the control treatments water (1), ASM (2) or fungicide (3). PF: fresh mass.

The content of protein was significantly altered, both in the leaf treated with *P. sanguineus* extracts, as well as in the non-treated leaf (Figure 4). The effect of basidiocarp extract at 10% was significant at four DAI, stimulating the content of protein in relation to water and to fungicide. At five and seven DAI the effect was superior to water and to ASM. The basidiocarp at 20% was superior to water and to fungicide (three and four DAI) and at five and seven DAI was bigger than the three control treatments. This effect of *P. sanguineus* extracts on the protein content was similar for the 4th non-treated leaf.

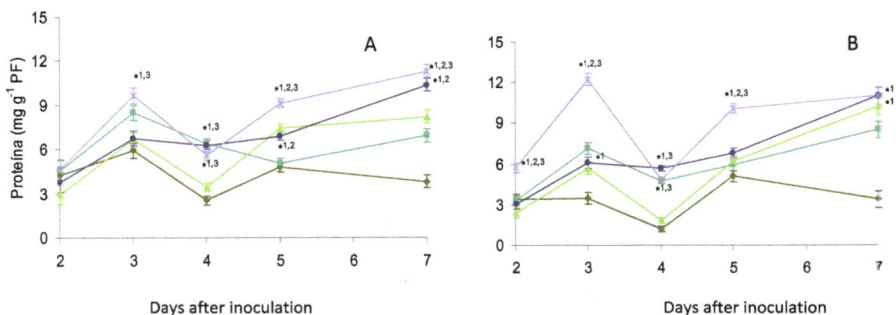

Figure 4. Protein content in bean plants inoculated with *Pseudocercospora griseola* tree days after the application of water (♦), acibenzolar-S-methyl (ASM 75mg L^{-1}) (■), fungicide (azoxystrobin 40mg L^{-1}) (▲) and the aqueous extracts of basidiocarp of *Pycnoporus sanguineus* at 10% and 20% (● and ✕). A and B represent the 3rd treated and inoculated leaf and 4th non-treated and inoculated leaf, respectively. Bars indicate an average ± standard error. Average followed by * differ statistically (Dunnett's test, P≤0.05) from the control treatments water (1), ASM (2) or fungicide (3). PF: fresh mass.

The content of chlorophylls *a*, *b* and total in common bean treated with aqueous extracts of *P. sanguineus* basidiocarp and challenged with *P. griseola* was altered significantly, with increments in the level of pigments (Figure 5). It was verified a similar behavior in the chlorophyll content in the 4th non-treated leaf, emphasizing the systemic effects of *P. sanguineus* on the content of total chlorophyll in common bean.

4. Discussion

The basidiocarp extracts did not present direct antimicrobial activity on *P. griseola*. This is a satisfactory result, since for a product be considerate a resistance inductor, this should not show antimicrobial activity *in vitro* or *in vivo* assays [30].

In greenhouse and field it was observed reduction of AUDPC in plants treated with the extracts of *P. sanguineus*. In [13] the authors obtained similar results with aqueous extracts of the same basidiocarp, reducing the severity of anthracnose in common bean in a systemic way. In reference [31] was demonstrated partial reduction in the severity of anthracnose in cucumber leaf pre-treated with the fruiting bodies extracts of *Lentinula edodes* and *Agaricus blazei*, in a systemic way. The protection effect was dose-dependent and, in a minor degree

time-dependent when it is considered the time interval between induction and pathogen
inoculation.

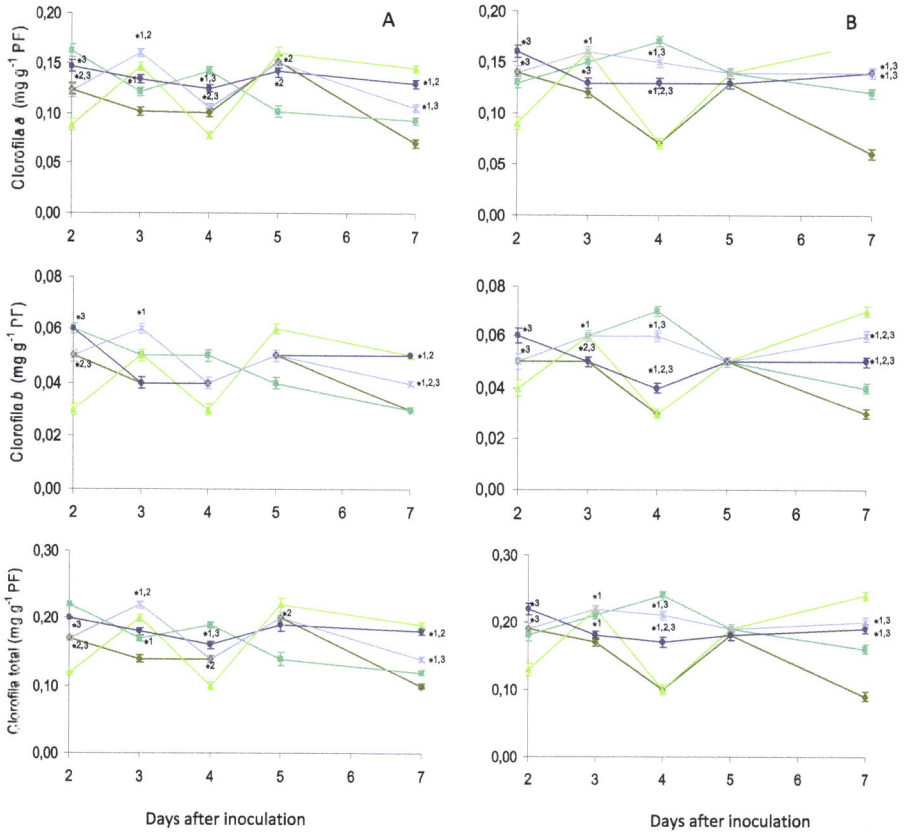

Figure 5. Chlorophylls *a, b* and total content in bean plants inoculated with *Pseudocercospora griseola* tree
days after the application of water (♦), acibenzolar-S-methyl (ASM 75mg L⁻¹) (■), fungicide
(azoxystrobin 40mg L⁻¹) (▲) and the aqueous extracts of basidiocarp of *Pycnoporus sanguineus* at 10%
and 20% (● and ×). **A** and **B** represent the 3rd treated and inoculated leaf and 4th non-treated and
inoculated leaf, respectively. Bars indicate an average ± standard error. Average followed by * differ
statistically (Dunnett's test, P≤0.05) from the control treatments water (1), ASM (2) or fungicide (3).
PF: fresh mass.

The activities of peroxidase, polyphenoloxidase and β-1,3 glucanase, and the content of total
proteins and chlorophylls were altered in plants treated with *P. sanguineus* extract. Changes in
the activities of peroxidase have being frequently correlated to the answer of resistance or
susceptibility in different pathosystems. The peroxidase is responsible for the remove of atoms

of hydrogen of the hydroxyl cinnamic alcohols groups, whose radical polymerize to form the lignin. This polymer, together with cellulose and other polysaccharides occurring in the cell wall of the superior plants, works as a physical barrier to the pathogen penetration [4].

In [13] was verified peroxidase induction, local and systemically, in bean plants treated with aqueous extract of *P. sanguineus* and challenged with *C. lindemuthianum*, agreeing with the results obtained in this work. In reference [14] the authors evaluated the effects of organic extracts of *P. sanguineus* basidiocarp, and verified that the dichloromethane extract for sorghum and soybean, and ethanolic extract for soybean, inhibit the activity of peroxidase, while the hexanic extract promotes the activity for sorghum and soybean. In another work [32], the peroxidase activity in the common bean was influenced, in a time-dependent way, by the number of inducer applications. The inducer ASM promoted increase in the enzyme activity in a more accentuated way and faster than the biotic inducer *Bacillus cereus*. In [33] it was not found significant increments in activity of this enzyme in the common bean treated with *P. sanguineus* extract and inoculated with *C. lindemuthianum*. In [32] was observed that the activity of polyphenoloxidase in common bean was not altered by the treatment with *Bacillus cereus* and ASM, while in [34] was verified induction in the activity of these enzyme in tomato leaves treated with essential oil of *Cymbopogon citratus* and inoculated with *Alternaria solani*.

In the resistance induction, the increment of β-1,3-glucanase is related with the plant defense. This enzyme hydrolyzes β-1,3-glucan, which, together chitin, is the main component of fungal cell wall [35]. In another pathosystem [33] was observed increase in specific activity of β-1,3-glucanase in common bean treated with *P. sanguineus* extract and challenge with *C. lindemuthianum*. In the 1st leaf, treated and inoculated, the activity increased 28%, while in the 2nd leaf, non-treated, but inoculated, the culture filtrate at 5% and the mycelium extract at 10% increased in 331% and 1,057%, respectively, the enzyme activity.

Extracts from other basidiomycetes fruiting bodies have also induced the activity of β-1,3-glucanase. It was verified an increase in this enzyme in passion fruit inoculated with *Xanthomonas campestris* pv. *passiflorae* and treated with extracts of *L. edodes* and *A. blazei* in concentration of 20% and 40% [36]. According [19], bean plants challenged with *P. griseola* did not presented induction in the β-1,3-glucanase activity, however, when challenged with *Uromyces appendiculatus*, was verified induction of this enzyme [25]. This behavior indicates the differential interaction among the elicitors treatment and pathogens challenging, in the activation of defense mechanisms in plants. The plant could invest in the production of compounds that normally would be produced in the presence of the pathogen, however, with greater efficiency when pre-disposed to an elicitor.

In this work, the protein content was significantly altered both in the leaf treated with the *P. sanguineus* extracts as in the non-treated leaf, however, there was a faster response on protein synthesis in the 4th leaf. This result could be related to the age of the leaves in the moment of treatments application and pathogen inoculation, since the 4th leaf has probably more physiological activity than 3rd one, optimizing the protein synthesis and plant resistance response [37].

Protein synthesis could be related with the increase of the demand for substrates, necessary to the production of plant defense mechanisms induced by *P. sanguineus* treatment. Among the proteins, there are the pathogenesis related proteins (PR-proteins) which are induced in plant tissues due to inoculation with pathogens/microorganisms, systemically or local, as well as with treatments with chemical agents [38]. The activation of protein synthesis leads to a phase of plant resistance [37]. In [32] was verified reduction in protein content of bean plants when treated with *Bacillus cereus*, contrary to the treatment with ASM, demonstrating specificity in the physiological response of this host to the elicitor treatment.

The chlorophyll content *a*, *b* and total in bean plants treated with *P. sanguineus* extract and challenged with *P. griseola* was altered significantly, with increase in the levels of these pigments. These results suggest the need to generate energy for the synthesis of compounds involved in plant defense, considering that the chlorophyll molecules *a* and *b* constitute the two pigment systems responsible for the absorption and transference of the radiant energy [39]. The energy generated by the process of photosynthesis can, at a given time, be directed to the production of secondary metabolic compounds, as for example, in case of attack by pathogens [37]. An increase was observed in the levels of chlorophyll in regions infected with *U. appendiculatus*, which occurs both in the common bean cultivar moderately susceptible as in cultivar highly susceptible to pathogen [25].

5. Conclusions

In this work can be concluded that the extract from *P. sanguineus* basidiocarp reduce the severity of the angular leaf spot in common bean, by increasing the activity of defense enzymes peroxidase, polyphenoloxidase and β-1,3-glucanase, local and systemically. Additionally, physiological changes in the content of protein and chlorophylls were verified, probably due the apparatus energy synthesis required for plant defense mechanism involved in the reduction of this disease. In this way, the use of extracts from *P. sanguineus* fruiting bodies for the control of plant diseases in organic growth shows promising.

Author details

José Renato Stangarlin*, Odair José Kuhn, Lindomar Assi and Cristiane Cláudia Meinerz
Western Paraná State University – UNIOESTE, Marechal Cândido Rondon, Paraná, Brazil

Clair Aparecida Viecelli
Assis Gurgacz Foundation – FAG, Cascavel, Paraná, Brazil

Kátia Regina Freitas Schwan-Estrada
Maringá State University – UEM, Maringá, Paraná, Brazil

Roberto Luis Portz
Paraná Federal University – UFPR, Palotina, Paraná, Brazil

* Corresponding Author

Acknowledgement

The authors thank the Araucaria Foundation and the FINEP for the financial support for the realization of this work. JRS and KRFSE thank to CNPq for the scholarship of research productivity.

6. References

[1]   Schwarrtz HF, Steadman JR, Hall R, Forster RL. Compendium of Bean Diseases. St. Paul: APS; 2005.

[2]   Stangarlin JR, Kuhn OJ, Schwan-Estrada KRF. Control of Plant Diseases by Plant Extracts. Revisão Anual de Patologia de Plantas 2008;16: 265-304.

[3]   Stangarlin JR, Kuhn OJ, Assi L, Schwan-Estrada KRF. Control of Plant Diseases using Extracts from Medicinal Plants and Fungi. In: Méndez-Vilas A. (ed.) Science against microbial pathogens: communicating current research and technological advances. Badajoz: Formatex; 2011. p1033-1042.

[4]   Stangarlin JR, Kuhn OJ, Toledo MV, Portz RL, Schwan-Estrada KRF, Pascholati SF. Plant Defense against Pathogens. Scientia Agraria Paranaensis 2011;10(1): 18-46.

[5]   Schwan-Estrada KRF, Stangarlin JR. Extracts and Essential Oils of Medicinal Plants in the Resistance Induction against Plant Pathogens. In: Cavalcanti LS, Di Piero RM, Cia P, Pscholati SF, Resende MLV, Romeiro RS (ed.) Resistance induction in plants against pathogens and insects. Piracicaba: FEALQ; 2005. p125-138.

[6]   Toledo MV, Stangarlin JR, Bonato CM. Homeopathy for the Control of Plant Pathogens. In: Méndez-Vilas A. (ed.) Science against microbial pathogens: communicating current research and technological advances. Badajoz: Formatex; 2011. p1063-1067.

[7]   Di Piero RM, Wulff NA, Pascholati SF. Partial Purification of Elicitors from *Lentinula edodes* Basidiocarps Protecting Cucumber Seedlings against *Colletotrichum lagenarium*. Brazilian Journal of Microbiology 2006;37(2): 169-174.

[8]   Di Piero RM, Novaes QS, Pascholati SF. Effect of *Agaricus brasiliensis* and *Lentinula edodes* Mushrooms on the Infection of Passionflower with Cowpea Aphid-borne Mosaic Virus. Brazilian Archives of Biology and Technology 2010;53: 269-278.

[9]   Fiori-Tutida ACG, Schwan-Estrada KRF, Stangarlin JR, Pascholati SF. Extracts of *Lentinula edodes* and *Agaricus blazei* on *Bipolaris sorokiniana* and *Puccinia recondita* f. sp. *tritici, in vitro*. Summa Phytopathologica 2007;33: 287-289.

[10] Nobles MK, Frew BP. Studies in Wood-inhabiting Hymenomycetes. The Genus *Pycnoporus* Karst. Canadian Journal of Botany 1962;40: 987-1016.

[11] Smânia A, Monache FD, Smânia EF, Gil ML, Benchetrit LC, Cruz FS. Antibacterial Activity of a Substance Produced by the Fungus *Pycnoporus sanguineus*. Journal of Ethnopharmacology 1995;45: 177-81.

[12] Smânia EFA, Smânia A, Loguercio-Leite C, Gil ML. Optimal Parameters for Cinnabarin Synthesis by *Pycnoporus sanguineus*. Journal of Chemical Technology and Biotechnology 1997;70: 57-59.

[13] Assi L. Control of *Colletotrichum lindemuthianum* (Sacc. Et Magn.) Scrib, in common bean (*Phaseolus vulgaris* L.) by *Pycnoporus sanguineus* (L. ex. Fr.) extract. Master thesis. Western Parana State University; 2005.

[14] Peiter-Beninca C, Franzener G, Assi L, Iurkiv L, Eckstein B, Costa VC, Nogueira MA, Stangarlin JR, Schwan-Estrada KRF. Phytoalexin Induction and Peroxidase Activity in Sorghum and Soybean Treated with Basidiocarp Extracts of *Pycnoporus sanguineus*. Arquivos do Instituto Biológico 2008;75: 285-292.

[15] Viecelli CA, Stangarlin JR, Kuhn OJ, Schwan-Estrada KRF. Induction of Resistance in Beans against *Pseudocercospora griseola* by Culture Filtrates of *Pycnoporus sanguineus*. Tropical Plant Pathology 2009;34: 87-96.

[16] Viecelli CA, Stangarlin JR, Kuhn OJ, Schwan-Estrada KRF. Resistance Induction in Bean Plants against Angular Leaf Spot by Extracts from *Pycnoporus sanguineus* Mycelium. Summa Phytopathologica 2010;36: 73-80.

[17] Toillier SL, Iurkiv L, Meienrz CC, Baldo M, Viecelli CA, Kuhn OJ, Schwan-Estrada KRF, Stangarlin JR. Control of Bacterial Blight (*Xanthomonas axonopodis* pv. *phaseoli*) and Biochemical Analyses of Bean Resistance Treated with *Pycnoporus sanguineus* Extracts. Arquivos do Instituto Biológico 2010;77: 99-110.

[18] Baldo M, Stangarlin JR, Franzener G, Assi L, Kuhn OJ, Schwan-Estrada KRF. *In situ* Detection of Reactive Oxygen Species in Bean Treated with *Pycnoporus sanguineus* Extracts and Inoculated with *Colletotrichum lindemuthianum*. Summa Phytopathologica 2011;37(4): 174-179.

[19] Stangarlin JR, Pascholati SF, Labate CA. Effect of *Phaeoisariopsis griseola* on Ribulose-1,5-bisphosphate Carboxylase-oxygenase, Chlorophyllase, β-1,3 glucanase and Chitinase Activities in *Phaseolus vulgaris* cultivars. Fitopatologia Brasileira 2000;25(1): 59-66.

[20] Stangarlin JR, Schwan-Estrada KRF, Cruz MES, Nozaki MH. Medicinal Plants and Alternative Control of Phytopathogens. Biotecnologia Ciência & Desenvolvimento 1999;11: 16-21.

[21] Fiori ACG, Schwan-Estrada KRF, Stangarlin JR, Vida JB, Scapim CA, Cruz MES. Antifungal Activity of Leaf Extracts and Essential Oils of some Medicinal Plants against *Didymella bryoniae*. Journal of Phytopathology 2000;148: 483-487.

[22] Stangarlin JR, Franzener G, Schwan-Estrada KRF, Cruz MES. Estratégias de Seleção e Uso de Extratos de Plantas no Controle Microbiano *in vitro*. In: Scherwinski-Pereira JE (ed.) Contaminações microbianas na cultura de células, tecidos e órgãos de plantas. Brasília: Embrapa; 2010. p293-345.

[23] Godoy CV, Carneiro SMTPG, Iamauti MT, Dalla Pria M, Amorim L, Berger RD, Bergamin Filho A. Diagrammatic Scales for Bean Diseases: Development and Validation. Zeitschrift fur Planzenkrankheiten und Pflanzenschutz 1997;104(4): 336-345.

[24] Shaner G, Finney RE. The Effect of Nitrogen Fertilization in the Expression of Slow Mildewing Resistance in Knox Wheat. Phytopathology 1977;67: 1051-1056.

[25] Stangarlin JR, Pascholati SF. Activities of ribulose-1,5-bisphosphate Carboxylase-oxygenase (rubisco), chlorophyllase, β-1,3 glucanase and Chitinase and Chlorophyll Content in Bean Cultivars (*Phaseolus vulgaris*) infected with *Uromyces appendiculatus*. Summa Phytopathologica 2000;26(1): 34-42.

[26] Lusso MFG, Pascholati SF. Activity and Isoenzymatic Pattern of Soluble Peroxidases in Maize Tissues after Mechanical Injury or Fungal Inoculation. Summa Phytopathologica 1999;25: 244-249.

[27] Duangmal K, Apenten RKO. A Comparative Study of Polyphenoloxidases from Taro (*Colocasia esculenta*) and Potato (*Solanum tuberosum* var. Romano). Food Chemistry 1999;64: 351-359.

[28] Bradford M.M. A Rapid and Sensitive Method for the Quantification of Microgram Quantities of Protein Utilizing the Principle of Protein-dye Binding. Analytical Biochemistry 1976;72: 248-254.

[29] Arnon DI. Copper Enzymes in Isolated Chloroplasts. Polyphenoloxidase in *Beta vulgaris*. Plant Physiology 1949;24: 1-15.

[30] Sticher L, Mauch-Mani B, Métraux J-P. Systemic Acquired Resistance. Annual Review of Phytopathology 1997;35: 235-270.

[31] Di Piero MR, Pascholati SF. Induced Resistance Cucumber Plants against *Colletotrichum lagenarium* by Application of Fruiting Body Extracts from *Lentinula edodes* and *Agaricus blazei*. Summa Phytophathologica 2004;30(2): 243-250.

[32] Kuhn OJ. Resistance Induced in bean Plants (*Phaseolus vulgaris*) by Acibenzolar-S-methyl and *Bacillus cereus*: Physiological and Biochemical Aspects, Growth and Production Parameters. PhD thesis. ESALQ; 2007.

[33] Baldo M. Histological and Biochemical Aspects of the Resistance Induction in Bean Plants and Antifungal Activity by *Pycnoporus sanguineus*. Master thesis. Western Paraná State University; 2008.

[34] Itako, AT, Schwan-Estrada KRF, Tolentino Jr, Stangarlin JR, Cruz MES. Antifungal Activity and Protection of tomato Plants by Extracts of Medicinal Plants. Tropical Plant Pathology 2008;33(3): 241-244.

[35] Cornelissen BJC, Melchers LS. Strategies for Control of Fungal Diseases with Transgenic Plants. Plant Physiology 1993;101: 709-712.

[36] Fiori-Suzuki CCL. Induction of Resistance in Yellow Passion Fruit (*Passiflora edulis* f. *flavicarpa*) by shiitake (*Lentinula edodes*) and *Agaricus blazei*. PhD thesis. Maringa State University; 2008.

[37] Larcher W. Plant Ecophysiology. São Carlos: RiMa; 2000.

[38] Guzzo SD. Pathogen Related Proteins. Revisão Anual de Patologias de Plantas 2003;11: 283-332.

[39] Taiz L, Zeiger E. Plant Physiology. Sunderland: Sinauer; 2006.

Chapter 6

The Role of the Mycorrhizal Symbiosis in Nutrient Uptake of Plants and the Regulatory Mechanisms Underlying These Transport Processes

Heike Bücking, Elliot Liepold and Prashant Ambilwade

Additional information is available at the end of the chapter

1. Introduction

The mycorrhizal symbiosis is arguably the most important symbiosis on earth. Fossil records indicate that arbuscular mycorrhizal interactions evolved 400 to 450 million years ago [1] and that they played a critical role in the colonization of land by plants. Approximately 80 % of all known land plant species form mycorrhizal interactions with ubiquitous soil fungi [2]. The majority of these mycorrhizal interactions is mutually beneficial for both partners and is characterized by a bidirectional exchange of resources across the mycorrhizal interface. The mycorrhizal fungus provides the host plant with nutrients, such as phosphate and nitrogen, and increases the abiotic (drought, salinity, heavy metals) and biotic (root pathogens) stress resistance of the host. In return for their beneficial effect on nutrient uptake, the host plant transfers between 4 and 20% of its photosynthetically fixed carbon to the mycorrhizal fungus [3]. In contrast to mutually beneficial mycorrhizal interactions, some mycoheterotrophic plants (approximately 400 plant species from different plant families, such bryophytes, pteridophytes, and angiosperms) rely on mycorrhizal fungi for their carbon supply. These plants have lost their photosynthetic capabilities and parasitize mycorrhizal fungi that are associated with neighbor autotrophic plants.

Primary focus of this chapter is on mutually beneficial ectotrophic and arbuscular mycorrhizal interactions, because of their high economic and ecological significance and their application potential. Arbuscular mycorrhizal fungi colonize the roots of many agriculturally important food and bioenergy crops and could serve as 'biofertilizers and bioprotectors' in environmentally sustainable agriculture. Ectomycorrhizal fungi on the other hand colonize a smaller number of plant species, but play as symbiotic partners of tree and shrub species a key role in forest ecosystems [4], and could be a critical component in phytoremediation and/or revegetation applications [5, 6].

2. Structural diversity of mycorrhizal interactions

Arbuscular mycorrhizal (AM) and ectomycorrhizal (ECM) associations differ in their structural characteristics and in the plant and fungal species that they involve. In AM roots the fungus penetrates intercellularly and intracellularly into the root cortex, whereas in ECM roots the fungus only penetrates intercellularly into the root cortex. Figure 1 illustrates the main structural differences between AM and ECM associations of angiosperms or gymnosperms, which are discussed in greater detail below.

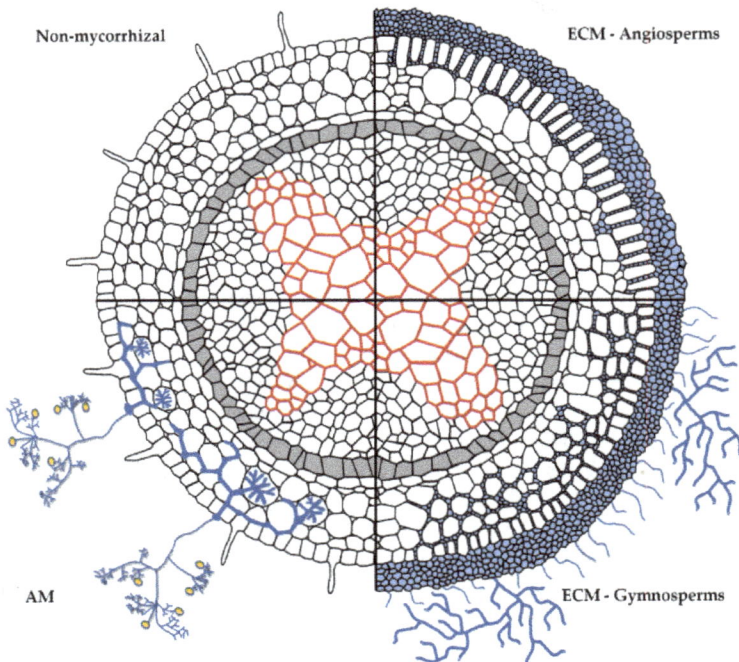

Figure 1. Structural characteristics of arbuscular mycorrhizal (AM) or ectomycorrhizal (ECM) roots of gymnosperms or angiosperms.

2.1. Arbuscular mycorrhizal interactions

Arbuscular mycorrhizas are the most common form of mycorrhizal interactions. They are formed by a wide variety of host plants (approximately 65% of all known land plant species) [2], including many agricultural important crop species, such as soybean, corn, rice, and wheat. All AM fungi have been classified into the separate fungal phylum Glomeromycota [7], which is composed of approximately 150 fungal species [1] with a high genetic and functional diversity within each species. AM fungi are classified into three classes (Archaeosporomycetes, Glomeromycetes, and Paraglomeromycetes), and the five orders: Archaeosporales (e.g. *Geosiphon pyriformes, Archaeospora trappei*), Diversisporales (e.g. *Scutellospora calospora, Acaulospora laevis, Entrophospora infrequens*), Gigasporales (e.g. *Gigaspora margarita, G. rosea*), Glomerales (e.g. *Glomus intraradices, G. mosseae, G. geosporum*) and Paraglomerales (e.g. *Paraglomus occultum, P. laccatum*). This group of fungi is unique due to its age, lifestyle and genetic make-up. AM fungi may have evolved over 1000 million years ago and can be seen as living fossils because they co-exist relatively morphologically unaltered with plants for more than 400 million years [8]. The symbiosis is frequent in all early diverging lineages of the major plant clades. Non-mycorrhizal species or other mycorrhizal types developed in plant lineages of more recent origin. This suggests that this symbiosis is the ancestral form of mycorrhizal interactions and that it played a critical role in the evolution of land plants [1]. In comparison, the symbiosis with nitrogen-fixing Rhizobia bacteria evolved much later (approximately 60 million years ago), and this symbiosis is restricted to only one plant clade.

AM fungi are coenocytic and hyphae and spores contain hundreds of nuclei [9]. The polymorphic nature of these nuclei and the relatively large genome of these fungi has made genome sequencing and annotation of this important group of fungi particularly challenging [8, 10], but recently the first transcriptome of the AM fungus *Glomus intraradices* became available [11]. They are asexual, but an exchange of genetic material between closely related fungi via anastomosis has been observed.

2.1.1. Structural characteristics of arbuscular mycorrhizal roots and fungal life cycle

AM fungi are obligate biotrophs and rely on their autotrophic host to complete their life cycle and to produce the next generation of spores (Figure 2). The spores are able to germinate without the presence of a host, but the spores respond with an increase in hyphal branching and metabolic activity to root exudates [12-14]. Plant roots release for example strigolactones that are able to induce pre-symbiotic growth of AM fungal spores [15].

On the host root surface, AM fungi form a specific appressorium – the hyphopodium. Fungal hyphae emerging from this hyphopodium penetrate into the root through the prepenetration apparatus, which guides the fungal hyphae through the root cells toward the cortex. In the cortex the hyphae enter the apoplast, and grow laterally along the root axis, and penetrate into inner root cortical cells. In 'typical' AM associations of the 'Arum type' enters the fungus the cell by small hyphal branches that by continuous dichotomous branching develop into characteristic highly branched arbuscules (Figure 2, 3d). By contrast, in 'Paris type' mycorrhizas spreads the fungus primarily from cell to cell and develops extensive intracellular hyphal coils that sometimes show an arbuscular like branching [1]. The fungus does not enter the plant symplast and is excluded from the host cytoplasm by the enlarged periarbuscular membrane (PAM) of the host. Some fungi also form vesicles, fungal storage organs in the root apoplast.

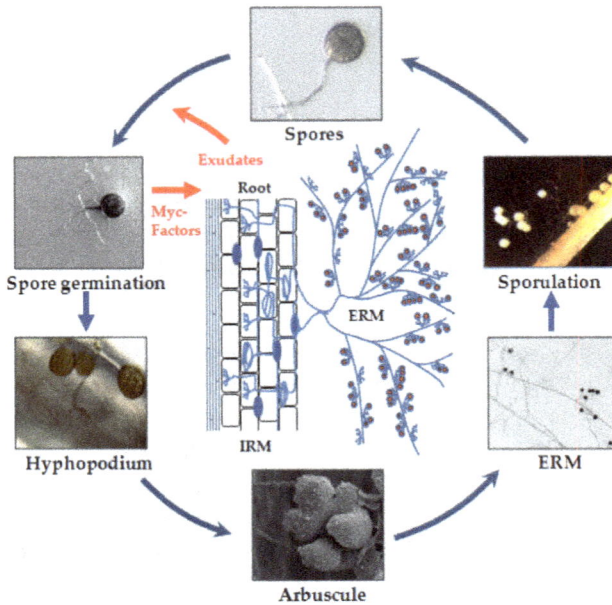

Figure 2. Life cycle of an AM fungus and the different steps during AM development.

Despite its coenocytic nature, the mycelium that is formed within the root, the intraradical mycelium (IRM) differs morphologically and functionally from the extraradical mycelium (ERM), the mycelium that grows into the soil. The ERM absorbs nutrients from the soil and transfers these nutrients to the host root. The IRM on the other hand releases nutrients into the interfacial apoplast and exchanges them against carbon from the host. The fungus uses these carbon resources to maintain and to enlarge the ERM, for cell metabolism (e.g. active uptake processes, nitrogen assimilation), and for the development of spores, which are able to initiate the colonization of a next generation of host plants (Figure 2, 3a).

2.1.2. Colonization of the root with arbuscular mycorrhizal fungi

Similar to Nod factors that play an important role in root nodulation, AM fungi release Myc factors that lead to an expression of plant symbiosis related genes and prepare the root for AM symbiosis. One active Myc factor has been identified as lipochitooligosaccharide [16]. Nod factors are also lipochitooligosacccharides and have a similar composition. It has been suggested that Nod factors developed from Myc factors, and that the functions of Myc and Nod factors overlap [17]. This is also supported by the fact that AM and rhizobial symbiosis share parts of the same signal transduction pathway - the so-called common symbiosis pathway. So far seven genes (SYM genes) of the common symbiosis pathway have been identified that are required for both root symbioses (Table 1).

Gene	Predicted gene function
SYMRK	Leucine-rich receptor–like kinase that plays an essential role for root endosymbioses with Rhizobia bacteria, AM fungi and *Frankia* bacteria, and is involved in the signal transduction to the cytoplasm after the perception of Nod or Myc factors [8, 18].
NUP85/ NUP133	Putative components of the nuclear pore complex that are involved in the transport of macromolecules through the nuclear envelope [8].
CASTOR/ POLLUX	Cation channels in the nuclear envelope that are essential for the perinuclear calcium spiking after the perception of Nod or Myc factors [8, 19].
CCaMK	Calcium and calmodulin-dependent protein kinase with three calcium binding motifs that acts as sensor of the nuclear calcium signatures and is involved in the phosphorylation of *CYCLOPS* [8, 20].
CYCLOPS	Protein with unknown function that acts as phosphorylation target of CCaMK downstream of the nuclear calcium spiking and is presumably the branchpoint of the common SYM pathway [8, 20].

Table 1. Genes of the common symbiosis pathway and their predicted function

One common signaling component is the receptor-like kinase *SymRK* that is involved in the direct or indirect perception of fungal or rhizobial signals and transduces the signal through its intracellular kinase domain to the cytoplasm [17, 18]. The two nucleoporins *NUP85* and *NUP133* act downstream and could be involved in the transport of *CASTOR* and/or *POLLUX* to the inner nuclear envelope. *CASTOR* and *POLLUX* are ion channels that are involved in the oscillation of the calcium concentration in the nucleus and perinuclear cytoplasm (calcium spiking) that can be observed shortly after the plant perceives signals from its root symbionts [8, 19]. The calcium-calmodulin dependent protein kinase *CCAMK* with its calmodulin- and calcium-binding domains is localized in the nucleoplasm and acts likely as the sensor of the calcium signatures that are induced by the perception of Myc or Nod factors. *CCAMK* is known to phosphorylate the last identified SYM gene *CYCLOPS*, which encodes a protein with no sequence similarity to proteins with known function. *CYCLOPS* contains a functional nuclear localization signal and a carboxy-terminal coiled-coil domain and it has been suggested that *CYCLOPS* represents a branch point in the common SYM pathway. Infection threat formation and arbuscular development are *CYCLOPS* dependent, but nodule organogenesis is *CYCLOPS* independent [20].

In contrast to the SYM pathway, little is known about the cellular and molecular re-programming that is required for the intracellular colonization and the development of arbuscules in cortical cells. In *Petunia hybrida pam1* mutants (*penetration and arbuscule morphogenesis 1*) fungal hyphae are able to penetrate into the cells, but the intracellular accommodation of the arbuscules is defective and intracellular hyphae are rapidly degraded [21]. Pam1 is a homologue of the *VAPYRIN* gene in *Medicago truncatula* and *VAPYRIN* mutants show a similar phenotype. The *PAM1* protein shows an affinity to tonospheres, mobile structures that are associated with the tonoplast, and that can also be found in the vicinity of intracellular hyphae in mycorrhizal roots. The first physical contact between both partners leads to a local upregulation of two CAAT-box binding transcription factors (*MtCbf1* and *MtCbf2*) in arbusculated cells of *Medicago truncatula*. Both transcription factors are able to interact with a large range of promoters, and could play a role in the sequential re-programming of root tissues during the establishment of an AM symbiosis [22].

2.1.3. The mycorrhizal interface in arbuscular mycorrhizal associations

Critical for the mutualism in the AM symbiosis is the bidirectional exchange of nutrients across the mycorrhizal interface. The interface between the fungus and the host includes the PAM and the fungal plasma membrane, the fungal cell wall and the periarbuscular space between the fungal cell wall and the PAM. The PAM differs in its protein composition from the plant plasma membrane of non-arbusculated cells and is characterized by mycorrhiza-inducible transporters that facilitate the uptake of nutrients from the mycorrhizal interface. One of these transporters is Pt4, a high affinity phosphate (P) transporter that is only expressed in mycorrhizal roots and that is involved in the acquisition of P delivered by the fungus [23, 24]. A high-affinity ammonium (NH_4^+) transporter (AMT2;2) is also localized in the PAM. This transporter is exclusively expressed in arbusculated cells of mycorrhizal roots, but not in root nodules [25]. In contrast to other high affinity NH_4^+ transporters of plants, AMT2;2 of *Lotus japonicus* (LjAMT2;2) transfers NH_3 instead of NH_4^+ and it has been suggested that the transporter takes up the positively charged NH_4^+ from the mycorrhizal interface and releases uncharged NH_3 into the plant cytoplasm. The detection of mycorrhiza-inducible sulphate transporters in AM roots suggests that also sulphate is transferred from the AM fungus to the host across the mycorrhizal interface [26, 27]. The transport of carbon from the host to the fungus is driven by a monosaccharide transporter in the fungal arbuscular membrane (MST2) [28]. This transporter takes up glucose but also other monosaccharides, such as xylose, what indicates that the fungus can also use cell wall sugars of the plant as alternative carbon source.

2.2. Ectomycorrhizal interactions

There are approximately 7000 to 10000 fungal species and 8000 plant species that form ectomycorrhizal (ECM) associations [29]. The number of plant species is relatively small (approximately 3%), but the group includes plants with high global and economic importance due to the disproportionate large terrestrial land surface that these plants cover, and as main producers of timber. The plant species include wooden perennials, trees or shrubs from cool, temperate boreal or montane forests, but also species from arctic alpine shrub communities [1, 4]. However, most of these plant species are not exclusively colonized by ECM fungi. Many species, such as *Populus* (see Figure 3d), *Salix*, *Betula* and *Fagus* also form AM interactions, and there are indications that the AM symbiosis is the common mycorrhizal form of this taxon [1].

ECM fungi are relatively closely related to saprotrophic fungi and mainly belong to the Basidiomycota (e.g. *Amanita muscaria*, *Hebeloma cylindrosporum*, *Laccaria bicolor*, *Pazillus involutus, Pisolithus tinctorius, Suillus bovinus, Xerocomus badius*), but also include some Ascomycota (e.g. *Cenococcum geophilum, Tuber borchii, Scleroderma hypogaeum*) [1]. The switch from the presumably ancestral saprotrophic to the symbiotic behavior developed convergently in several fungal families during evolution. In contrast to AM fungi, many ECM fungi can be grown in axenic culture without a host, and this has allowed screening of their ability to use different carbon or nutrient sources [30]. ECM fungi have a dual life style and are considered to be facultative saprotophs. In the soil they are highly competitive in nutrient acquisition and secrete a number of hydrolytic enzymes that allow them to degrade litter polymers, and to use organic nutrient sources [4]. At the same time they live within plant roots as symbionts and this requires a set of adaptation mechanisms to avoid plant parasitism. ECM fungi have for example lost their ability to degrade plant cell wall polysaccharides (cellulose, pectins, and pectates), and this restricts their penetration into the root to the intercellular spaces [31].

2.2.1. Colonization of the root with ectomycorrhizal fungi

Typical for ECM roots are changes in the root morphology, such as the dichotomous branching of lateral roots, e.g. in pines (Figure 3b), the production of a large number of root meristems and as a result an extensive root branching (Figure 3c), the inhibition of root hair formation, and the enlargement of cortical cells. Many of these morphological effects can be observed prior to colonization and can be interpreted as a preparation of the plant to increase root symbiosis.

Prior to the establishment of a functional ECM root and similar to the processes during AM development, there is an exchange of signals and cross-talk between both partners. The fungal tryptophan betaine hypaphorine has been shown to trigger reduced root hair elongation and swelling of the root hair tip and a stimulation of short root formation [32]. ECM fungi also produce phytohormones, including auxins, cytokinins, abscisic acid and ethylene, and it has been shown that the changes in the root morphology are caused by an overproduction of auxin in ECM fungal hyphae and changes in the endogenous hormone levels in the roots. The effect of ECM fungi on lateral root formation is independent from the plant's ability to form ECM associations. The ECM fungus *Laccaria bicolor* can induce lateral root formation also in *Arabidopsis thaliana*, a non-mycorrhizal plant, and the effect is correlated to an accumulation of auxin in the root apices [33]. The auxin accumulation in the root tips and/or other fungal signals could stimulate basipetal auxin transport and lateral root primordia formation by an induction of plant genes involved in auxin transport and signaling.

The fungal partner responds to root exudate components, such as rutin and zeatin, with a stimulation in hyphal growth and branching and growth towards the root and an accumulation of hypaphorine [32, 34]. In response to host signals, ECM fungi also release effector proteins into the rhizosphere, such as the *MYCORRHIZAL INDUCED SMALL SECRETED PROTEIN 7* (*MiSSP7*) of *Laccaria bicolor*. This fungal protein targets the plant nucleus after its uptake, and alters plant gene expression [35]. *MiSSP7* has been shown to be crucial for the establishment of the ECM symbiosis and resembles effector proteins of pathogenic fungi, and bacteria with similar function. A transcriptional response of the host can be observed within hours after an initial contact between both partners has been established. Plant genes encoding proteins involved in stress and defense response, as well as genes involved in signal transduction and communication, and water uptake are upregulated in response to the presence of an ECM fungus in the rhizosphere [36].

2.2.2. Structural characteristics of ectomycorrhizal roots

An established ECM symbiosis is characterized by three structural components: the hyphal sheath or mantle, the Hartig net (in later passages of this text sometimes also referred to as intraradical mycelium or IRM), and the extraradical mycelium. The hyphal sheath or mantle encloses the root completely. The structural composition of the mantle is very diverse and can range from relatively thin, loosely arranged assemblages of hyphae to very thick, multilayered and pseudoparenchymatous mantles (Figure 1, 3bcf). The surface of the mantle can be compact and smooth (Figure 3c) or rough with numerous emerging hyphae and hyphal strands or rhizomorphs (Figure 3b). The fungal sheath is involved in nutrient storage and controls the nutrient transfer to the host. The fungal mantle can represent a significant apoplastic barrier [37, 38] and thereby creates a closed interfacial apoplast, in which the conditions can be controlled by both partners.

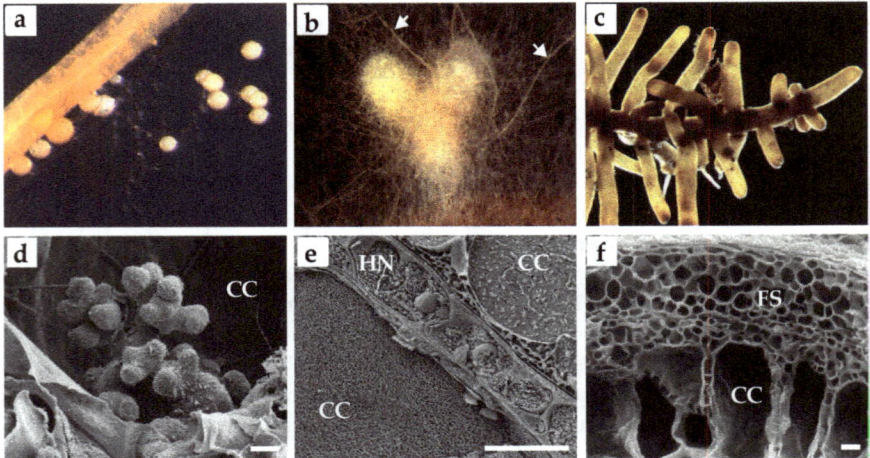

Figure 3. Morphological characteristics of AM (a, d) and ECM (b, c, e, f) roots. Images of the outer root morphology (a-c) and scanning electron microscopical images of fungal structures within the root (d-f). (a) AM root of *Daucus carota* colonized with *Glomus intraradices* with fungal spores and ERM; (b) dichotomous ECM pine root colonized by *Suillus bovinus* with rhizomorphs (arrows); (c) ECM root of beech colonized by an unidentified fungus with extensive root branching; (d) arbuscule of an AM fungus within the cortical cell (CC) of an ECM root of *Populus tremuloides*; (e) Hartig net (HN) region and mycorrhizal interface in an ECM root of *Populus tremuloides*; (f) ECM root with epidermal Hartig net and radially elongated epidermis cells and fungal sheath (FS).

The Hartig net plays the key role in the nutrient transfer between both partners. The Hartig net is formed by hyphae that penetrate into the root cortex intercellularly (Figure 1, 3e). The penetration depth of the Hartig net differs between angiosperms and gymnosperms. Most angiosperms develop an epidermal Hartig net and confine the penetration of the Hartig net to the outer epidermis, which is often radially elongated (Figure 1, 3f). By contrast, the Hartig net in gymnosperms normally encloses several layers of cortical cells and sometimes extends up to the endodermis (Figure 1)[1].

The extraradical mycelium (ERM) of the fungus acts as an extension of the root system and it has been estimated that the ERM of the fungus *Pisolithus tinctorius* can represent 99% of the nutrient-absorbing surface length of pine roots [39]. The ERM of ECM fungi can account for 32% of the total microbial biomass and 700-900 kg ha^{-1} in forest soils [40, 41]. The ERM can have a relatively simple organization with individual hyphae with similar structure that grow into the soil (mainly in ascomycetes) or can be differentiated into singular hyphae and rhizomorphs (Figure 3b). Rhizomorphs are aggregates of hyphae which grow in parallel and whose organization level can range from simple assemblages of undifferentiated and loosely woven hyphae to complex aggregations of hyphae with structural and functional differentiations [42].

2.2.3. The mycorrhizal interface in ectomycorrhizal associations

Transport studies suggest that in ECM associations nutrients are exchanged simultaneously across the same interface [43]. The interface includes the plasma membranes and cell walls of both partners and the interfacial matrix between both partners. The plant transfers photosynthates as sucrose from source to sink organs and ECM roots act as strong carbon sinks in mycorrhizal root systems. It is generally accepted that in contrast to phytopathogenic fungi or ericoid mycorrhizal fungi, AM and ECM fungi are not able to use sucrose as a carbon source, and that they take up simpler sugars, such as glucose or fructose, from the mycorrhizal interface. The presence of invertase genes in fungal genomes is correlated with the nutritional mode and in contrast to other plant-associated fungi, such as pathogens, or endophytes, there are no indications that AM or ECM fungi possess invertase genes [44] or have invertase activity [30]. Consequently, mycorrhizal fungi rely on the invertase activity of the host in the interfacial apoplast for sucrose hydrolysis. Sucrose hydrolysis makes the hexoses glucose and fructose available for the fungus and it has been suggested that glucose is mainly taken up by hyphae of the Hartig net and fructose mainly by hyphae of inner mantle layers [45]. Compared to the ERM, fungal hexose transporters are up-regulated in ECM roots, indicating that the fungus in symbiosis takes up carbon primarily from the mycorrhizal interface [46].

The high affinity NH$_4^+$ importer *AmAMT2* of *Amanita muscaria* is up-regulated in the ERM, but down-regulated in Hartig net and the fungal sheath [47]. The high expression of this transporter in the ERM suggests a high capability of the ERM for NH$_4^+$ uptake. The low expression level in the Hartig net on the other hand indicates that NH$_4^+$ can serve as a potential nitrogen source that is delivered by the mycorrhizal fungus to the host. A low expression level of this NH_4^+ importer in the Hartig net would reduce the re-absorption of NH_4^+ by the fungus from the interfacial apoplast and increase the net transport of NH_4^+ to the host. The potential transport of NH_4^+ across the ECM interface is also supported by the presence and upregulation of plant high affinity NH_4^+ importers in ECM roots [48].

3. Discussion

3.1. Nutrient uptake pathways in arbuscular mycorrhizal or ectomycorrhizal roots

Mycorrhizal plants can take up nutrients from the soil via two pathways: the 'plant pathway' that involves the direct uptake of nutrients from the soil by the root epidermis and its root hairs or the 'mycorrhizal pathway' that involves the uptake of nutrients via the ERM of the fungus and the transport to the Hartig net in ECM interactions or to the IRM in AM interactions, and the uptake by the plant from the interfacial apoplast. The uptake of nutrients from the soil via the plant pathway, however, is often limited by the low mobility of nutrients in the soil. The mobility of for example phosphate (P) is so low that its uptake leads rapidly to the development of depletion zones around the roots and limits the further P uptake via the plant pathway to the low rate of diffusion [49].

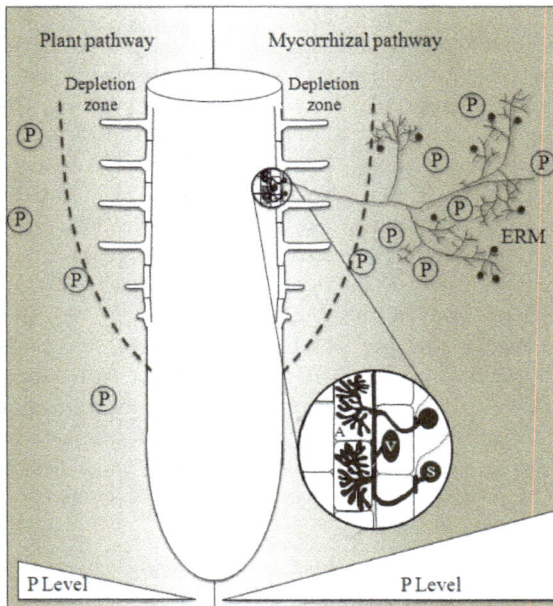

Figure 4. P uptake of the plant via the plant pathway or mycorrhizal pathway. Abbreviations: Extraradical mycelium of the fungus (ERM), vesicles (V) and spores (S) of the arbuscular mycorrhizal fungus.

AM and ECM roots differ in their structural characteristics and this has implications for the nutrient uptake pathways in AM or ECM plants (Figure 1). AM roots do not form a fungal sheath and can theoretically use both pathways for nutrient uptake. It has previously been suggested that in the AM symbiosis both uptake pathways act additively. This led to the assumption that the uptake via the mycorrhizal pathway can be neglected when for example the P availability in the soil is high and mycorrhizal plants not always show a positive growth response. This view, however, is now being questioned [50-52], and it has been argued that the mycorrhizal pathway can dominate the total P uptake and that the true contribution of the mycorrhizal pathway to total P uptake can be "hidden" [52]. It has been demonstrated that even in non-responsive wheat plants, 50% of the plant's P can be taken up via the mycorrhizal pathway [53]. This indicates that mycorrhizal plants change their nutrient acquisition strategy and that even under high P availabilities in the soil, the mycorrhizal fungus can still contribute significantly to the P uptake of the plant. Plant P transporters that are involved in the uptake via the plant pathway are down-regulated in response to the AM symbiosis [54, 55], while mycorrhiza-specific transporters that are involved in the P uptake from the mycorrhizal interface are induced [23, 24, 56].

However, the contribution of the plant or mycorrhizal pathway to total P uptake also depends on the plant and fungal species. *Glomus intraradices* has been shown to suppress the expression of plant P transporters of the plant pathway the most, whereas *G. mosseae* had the least effect [55]. In tomato, almost 100% of the plant's P was taken up by *G. intraradices* via the mycorrhizal pathway, but the contribution of *Gigaspora rosea* to total P uptake was much lower [57]. *Glomus caledonium* always showed a high P uptake and transfer independent on the plant species, but *Glomus invermaium* only transferred significant amounts of P to the host plant flax [58]. This indicates that the contribution of the mycorrhizal pathway to nutrient acquisition also depends on fungal specific effects on the activity of the plant pathway and on the efficiency with which both partners interact and exchange nutrients across the mycorrhizal interface.

By contrast, in ECM tree species such as *Pinus*, the majority of the root surface is composed of roots that do not contribute to nutrient acquisition, such as the condensed tannin zone or the cork zone. Root zones that are active in nutrient acquisition such as the non-mycorrhizal white or ECM roots represent only 2% or 16% of the total root length, respectively [59]. ECM roots are characterized by a more or less dense fungal sheath that surrounds the mycorrhizal root completely. If the fungal mantle is impermeable to nutrient ions, the underlying root tissue would be isolated from the soil solution, and these roots would exclusively rely on the mycorrhizal pathway for nutrient uptake. Whether the fungal sheath represents an apoplastic barrier depends on the ECM fungal species and the structure and properties of the mantle. Some fungi for example have been shown to express and to release hydrophobins during ECM development [60]. Hydrophobins are small hydrophobic proteins that are involved in the adhesion of hyphae to surfaces, but can also increase the water repellency of the fungal sheath, and thereby make the sheath impermeable for water and nutrient ions [61, 62]. The fungal mantle of *Hebeloma cylindrosporum* for example is impermeable to sulfate [63]. The fungal sheath of *Pisolithus tinctorius* and of *Suillus bovinus* is not completely impermeable to nutrient ions, but it seems likely that under normal soil conditions when plant and fungus compete for nutrients, the passive movement of nutrients through the fungal sheath of ECM roots is restricted and that a significant part of the nutrient uptake of these ECM roots is under fungal control [37, 63]. Under consideration that only 2% of the root surface of pines is non-mycorrhizal, and that the ERM of an ECM fungus can represent up to 99% of the nutrient-absorbing surface length of pine roots [39], ECM tree species such as pines are considered to be highly dependent on their fungal symbionts [64, 65], and it can be assumed that the mycorrhizal pathway plays in ECM root systems an even more significant role for nutrient uptake than in AM root systems (Table 2).

Characteristic	ECM symbiosis	AM symbiosis
Fungal life style	Facultative saprotroph	Obligate biotroph
Structural components	Mantle, Hartig net, and ERM with or without rhizomorphs	Arbuscules or intercellular hyphal coils, ERM, vesicles in some types
Penetration	Exclusively intercellularly	Intercellularly and intracellularly
Nutrient uptake pathway	ECM roots represent a significant proportion of the nutrient absorbing surface and nutrient uptake predominately via the mycorrhizal pathway	Theoretically plant and mycorrhizal pathway, but mycorrhizal pathway can dominate nutrient uptake in mycorrhizal roots
Contribution to plant nutrition	Particularly important for N nutrition, but also significant contributions to P nutrition	Particularly important for P nutrition; contributions to N nutrition still under debate
Fungal nutrient resources	Efficient uptake of inorganic and organic nutrient resources	Uptake predominately of inorganic nutrient resources, utilization of organic nutrient resources considered to be small

Table 2. Comparison of nutrient uptake mechanisms in ECM or AM interactions.

4. Mycorrhizal interactions and phosphate nutrition of plants

4.1. Uptake of phosphate from the soil

The positive effect of mycorrhizal fungi on phosphate (P) nutrition is long known and has been attributed to:

- The exploration of large soil volumes by the ERM in which orthophosphate (P_i) is scavenged and delivered to plant cortical cells, bypassing the plant pathway for P uptake [66, 67] (Figure 4);
- The small hyphal diameter that allows the fungus to penetrate into small soil cores in search for P, and higher P influx rates per surface unit [66, 68];
- The capability of mycorrhizal fungi to store P in form of polyphosphates, which allows the fungus to keep the internal P_i concentration relatively low, and allows an efficient transfer of P from the ERM to the IRM [69]; and

- The production and secretion of acid phosphatases and organic acids that facilitate the release of P from organic complexes [70, 71].

Similar to plants, fungi have two uptake systems for P: (1) a high affinity system that works against an electrochemical potential gradient, which takes up P_i from the soil via proton co-transport [72, 73], and (2) a low affinity system which facilitates the diffusion of P_i across the fungal plasma membrane [74]. AM and ECM fungi express high affinity P transporters in the ERM that are involved in the P uptake from the soil [75, 76] (Figure 5). The expression of these transporters is regulated in response to the externally available P concentration, and to the P demand of the fungus. Under P_i starvation the transcript levels generally increase. Interestingly, in the ERM of the ECM fungus *Hebeloma cylindrosporum* two P transporters are expressed, one transporter is up-regulated under low (*HcPT1*), and one transporter is up-regulated under high P supply conditions (*HcPT2*) [76]. The simultaneous expression of two fungal P transporters that respond differently to the P level in the soil, could enable the ERM of the fungus to take up P efficiently also from locally varying P concentrations in the soil (e.g. from nutrient hot spots or from the root rhizosphere with its low P_i concentrations).

4.2. Fungal phosphate metabolism

Orthophosphate (P_i) that is absorbed by the ERM can (a) replenish the cytoplasmic, metabolically active P_i pool; (b) be incorporated into phospholipids, RNA-, DNA- and protein-phosphates; (c) or can be transferred into a storage pool of short- or long-chained polyphosphates (polyP) [77] (Figure 5). Inorganic polyP are linear polymers in which P_i residues are linked by energy-rich phospho-anhydride bonds. Two types of polyP can be distinguished in mycorrhizal fungi: short chain polyP with a length of up to 20 P_i residues and long chained polyP with more than 20 P_i residues. The average length of short chained polyP in AM fungi has been estimated as 11-20 P_i [78, 79], and of long chained polyP as 190 to 300 P_i residues [80, 81]. Mycorrhizal fungi can rapidly store a significant proportion (more than 60%) of their cellular P as polyP [69, 82, 83]. In the mycorrhizal symbiosis, polyP are involved in:

a. **P homeostasis in the hyphae and maintenance of low intracellular P_i levels.** Low P_i levels in fungal hyphae increase the efficiency with which P can be absorbed and reduce the osmotic stress at high internal P concentrations [79, 84];

b. **Long-distance transport from the ERM to the IRM.** Based on the high flux rate of P through the hyphae of mycorrhizal fungi[1] [85, 86], it has been suggested that P is transferred mainly as polyP from the ERM to the root [87, 88]. The chain lengths of polyP in the ERM are longer than in the IRM, suggesting that polyP are primarily formed in the ERM and re-mobilized in the IRM [89]. In ECM and AM fungi a motile tubular vacuole system has been identified [90-92], that allows the polyP transport through the hyphae separately from the cytoplasmic compartment and enables the fungus to fine-tune its local cytoplasmic P_i concentration.

---

[1] between 2×10^{-10} and 2×10^{-9} mol P cm^{-2} s^{-1} have been described for AM fungi

c. **Regulation of P transport.** It has been suggested that mycorrhizal fungi control the intracellular P_i concentration in the IRM and the P flux to the host by regulating the formation and/or turnover rates of polyP in the IRM or in the Hartig net region [93-95]. Long-chain polyP are mainly involved in long term storage of P, whereas short-chain polyP are correlated to the P transport in the symbiosis [93].

d. **Cation homeostasis.** PolyP are polyanions and their negative charge is balanced by cations. The cations K^+ and Mg^{2+} are mainly involved in charge balance [82, 83, 96], but polyP can also serve as trap for toxic cations such as heavy metals [82]. The basic amino acid Arg^+ can also be involved in the charge balance of polyP and it has been suggested that in ECM fungi polyP can also store significant amounts of N [82, 97].

5. Mycorrhizal interactions and nitrogen nutrition

5.1. Uptake of nitrogen from the soil

Many ecosystems in which the nitrogen (N) availability in the soil is low and the supply with N often limits plant growth are dominated by ECM plant communities. ECM fungi can take up inorganic N sources very efficiently from soils [98, 99], but their capability to utilize organic N sources, and to make these sources available for the host plant, is generally seen as an important factor in the N nutrition of ECM plant species [1]. Many ECM fungi can for example mobilize and utilize amino acids and amides, such as glutamine, glutamate and alanine, which can represent a significant N pool, particularly in acid-organic soils [1]. Some amino acids can be taken up intact, and can directly be incorporated into assimilation pathways and can thereby also represent a significant carbon pool for ECM fungi [100].

By contrast, the contribution of AM fungi to total N uptake of plants is still under debate. The mobility of the inorganic N sources nitrate (NO_3^-) and ammonium (NH_4^+) in the soil is relatively high, and it has been suggested that the improved N status of AM plants is only the consequence of an improved P nutrition [101]. However, there are numerous reports in which a significant transport of N by AM fungi to their host has been demonstrated. In AM root organ cultures, 21% of the total N in the roots were taken up by the ERM [102]; and in similar experiments even higher proportions were observed [103]. In maize, 75% of the N in the leaves were taken up by the ERM of an AM fungus [104]. It becomes increasingly clear, that the mycorrhizal pathway can play a role in the N nutrition of AM plants, even if the percentage contribution to total N nutrition of the host plant can vary considerably [50]. Compared to ECM fungi, the capability of AM fungi to utilize organic N sources is considered to be relatively low, but some AM fungi are also able to use organic N sources. AM fungi can take up exogenously supplied amino acids [13, 103, 105] and are able to mobilize N from organic nutrient patches and to transfer these N sources to their host [106].

The ERM of AM and ECM fungi can take up the inorganic N sources ammonium (NH_4^+) or nitrate (NO_3^-) from the soil [99, 103, 105] (Figure 5). NH_4^+ has often been described as the preferred N source of mycorrhizal fungi, because its uptake is energetically more efficient than the uptake of NO_3^- [13, 102, 105]. However, ECM fungi differ in their ability to absorb NO_3^- from the soil, and some ECM fungi have been shown to produce a greater biomass when supplied with NO_3^- compared to NH_4^+ [107]. However, a supply of NH_4^+ leads to a down-regulation of a NO_3^- transporter and a nitrate reductase of *Hebeloma cylindrosporum*, what suggests that also ECM fungi generally prefer NH_4^+ over NO_3^- [1, 108, 109].

Several high affinity NH_4^+ transporters from AM and ECM fungi have been identified. The expression of *AMT1* and *AMT2*, two NH_4^+ transporters of the ECM fungus *H. cylindrosporum*, is regulated by the exogenous NH_4^+ supply. The expression of both transporters is up-regulated under low NH_4^+ supply conditions, but down-regulated in response to an exogenous supply of NH_4^+. It has been suggested that the intracellular level of glutamine is responsible for the repression under high supply conditions. In addition to *AMT1* and *AMT2*, a low affinity NH_4^+ transporter (*AMT3*) is expressed under non-limiting NH_4^+ supply conditions, which enables the fungus to maintain a basal level of N uptake and assimilation also at high exogenous supply conditions [110]. *GintAMT1*, an NH_4^+ transporter of the AM fungus *Glomus intraradices* seems to be mainly involved in the uptake of NH_4^+ by the ERM under low NH_4^+ availabilities [111]. An exogenous supply of NO_3^- stimulates the expression of a fungal NO_3^- transporter in the ERM of *G. intraradices* [112]. Similar to the N repression observed for the NH_4^+ transporters in ECM fungi, the expression of this transporter is repressed by an increase in the internal levels of NH_4^+ or a downstream metabolite, such as glutamine [113].

5.2. Fungal nitrogen metabolism

After its uptake from the soil, NO_3^- is converted into NH_4^+ via nitrate and nitrite reductases in AM [103, 114] and ECM fungi [109]. In AM fungi and most ECM fungi, NH_4^+ is assimilated via the glutamine synthase/glutamine oxoglutarate aminotransferase (GS/GOGAT) pathway into amino acids [103, 112].

In AM fungi, the amino acids glutamine and arginine (Arg) are major sinks for absorbed N and both amino acids become highly labeled after a supply of $^{15}NO_3^-$ or $^{15}NH_4^+$ to the ERM [103]. AM fungi assimilate N into Arg via the anabolic arm of the urea cycle and its key enzymes: carbamoyl phosphate synthase, argininosuccinate synthase, and argininosuccinate lyase[2] [112, 115]. The transcript levels in the ERM of these genes respond within hours to an exogenous supply of NO_3^- [112], what supports the view that Arg is rapidly assimilated in the fungal ERM of the AM symbiosis. Arg has a low C/N ratio and is positively charged and it has been suggested that polyP play a key role for the transport of N from the ERM to the IRM in the AM symbiosis [103, 115, 116] (Figure 5). Also for ECM fungi a potential role of Arg in N translocation has been discussed.

---

[2] carbamoyl phosphate synthase (CPS, production of carbamoyl phosphate from CO_2, ATP and NH_3); argininosuccinate synthase (ASS, synthesis of argininosuccinate from citrulline and aspartate); and argininosuccinate lyase (AL, conversion of L-argininosuccinate to Arg and fumarate).

However, ^{15}N labeling experiments suggest, that N can also be transferred as glutamine from the ERM to the IRM in ECM systems [1].

For AM associations, it has been suggested that N is transferred in inorganic form across the mycorrhizal interface. Current models on N transport in the AM symbiosis propose the breakdown of Arg in the IRM via the catabolic arm of the urea cycle into inorganic N, which is subsequently transferred across the mycorrhizal interface to the host [112, 113]. Several genes that are involved in Arg breakdown are highly expressed in the IRM, and are up-regulated in response to a NO_3^- supply to the ERM [112]. The catabolic enzymes arginase (*CAR1*) and urease (*URE*)[3] are involved in Arg breakdown, and the activities of both enzymes increase in AM roots after a supply of inorganic N sources to the ERM [113, 116]. A supply of NO_3^- to the ERM also results in increasing transcript levels of *CAR1* and *URE* in the IRM and it has been suggested that this may be due to an increase in the internal Arg level in the IRM [112] after a supply of NO_3^- to the ERM.

For ECM associations, it has traditionally been believed that amino acids are transferred across the mycorrhizal interface and that N transport from the fungus to the host is in organic form [1]. If organic N is transferred across the mycorrhizal interface, the carbon skeletons of amino acids would contribute to a significant re-flux of carbon from the fungus to the host. The uptake of amino acids from the interface would also require the presence of efficient plant uptake systems for organic N from the mycorrhizal interface, which have not yet been identified. ECM plants on the other hand have been shown to express high affinity NH_4^+ transporters in the ECM interface [48], what suggests that N could also be transferred as inorganic NH_4^+ across the mycorrhizal interface. This view is also been supported by the observed down-regulation of fungal NH_4^+ transporters in fungal sheath and Hartig net, what would reduce the fungal re-absorption of NH_4^+ from the interface [47]. Whether ECM fungi utilize, similar to AM fungi, the catabolic arm of the urea cycle to release NH_4^+ in the Hartig net, still needs to be investigated (Figure 5). However, ECM fungi are also able to hydrolyze urea via urease [117]. In ECM roots two fungal aquaporins are highly expressed, that belong to the group of Fps-like aquaglyceroporins [118]. When expressed in yeast, both aquaporins increase the permeability of the membrane for NH_4^+ but not for urea. Based on this observation, the authors concluded that an aquaporine mediated leakage of urea does not play a significant role for N transport in the ECM symbiosis and that NH_4^+ is released into the mycorrhizal interface.

5.3. Regulation of nutrient transport across the mycorrhizal interface

The molecular mechanisms that are involved in the regulation of P and N transfer across the mycorrhizal interface are still unknown [50]. Models of nutrient transfer across the mycorrhizal interface generally involve an efflux of P and N from the fungal symplast through the fungal plasma membrane into the interfacial apoplast and the active absorption across the plasma membrane by the host plant (Figure 5).

---

[3] arginase (*CAR1*, Arg breakdown to ornithine and urea); urease (*URE*, hydrolysis of urea to CO_2 and ammonia)

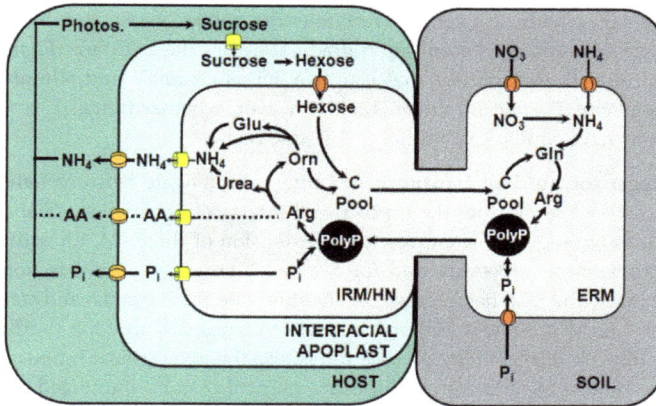

Figure 5. Transport processes in arbuscular and ectomycorrhizal interactions. The model shows the nutrient uptake by the fungal ERM through P_i, NO_3^- or NH_4^+ transporters (red), N assimilation into Arg via the anabolic arm of the urea cycle (only in AM fungi shown) and the conversion of P_i into polyP in the ERM, transport of polyP from the ERM to the IRM, polyP hydrolysis and release of Arg and P_i in the IRM or HN, Arg breakdown to NH_4^+ via the catabolic arm of the urea cycle (only in AM fungi shown), facilitated P_i, NH_4^+, and potential amino acid (AA, only in ECM postulated) efflux through the fungal plasma membrane (yellow) into the interfacial apoplast, plant uptake of nutrients from the mycorrhizal interface through mycorrhiza-inducible P_i or NH_4^+ transporters, stimulation in photosynthesis by improved nutrient supply and facilitated efflux of sucrose through the plant plasma membrane into the interfacial apoplast, sucrose hydrolysis in the interfacial apoplast via an apoplastic plant invertase, and uptake of hexoses by the mycorrhizal fungus through fungal monosaccharide transporters.

The net loss of nutrients from free living fungi is normally regarded as slow, and membrane transport processes are generally favouring fungal re-absorption [77]. Therefore it has been suggested, that in the interface, conditions might exist, that promote the efflux of nutrients from the fungus or reduce the level of competing fungal uptake systems. The following conditions could contribute to a stimulation of nutrient transport across the mycorrhizal interface:

Development and maintenance of a concentration gradient: A concentration gradient across the mycorrhizal interface with high concentrations in the IRM and low concentrations in the interfacial apoplast and in cortical cells would maximize the efflux of nutrients through the fungal plasma membrane, and reduce fungal re-absorption. High P concentrations for example within the hyphae of the Hartig net or in the IRM increase the P transfer to the host [83, 119] and reduce fungal re-absorption [75, 120, 121]. There are also indications that the differential expression of plant and fungal uptake transporters in the mycorrhizal interface plays a role in the development of a strong concentration gradient across the mycorrhizal interface. High affinity P and N transporters of ECM and AM fungi are highly expressed in the ERM, but down-regulated in the IRM [47, 122]. This favours the active absorption of nutrients by the ERM, but reduces the re-absorption of nutrients by the IRM. By contrast, mycorrhiza-inducible high affinity plant P_i [23, 24, 56, 123] and NH_4^+ transporters [25, 48, 124] are localized in plant plasma membranes in the interface region.

The high expression of these transporters facilitates the uptake of resources by the plant and the development of a strong concentration gradient across the interface (Figure 5). Plants express low affinity P_i transporters that can also act as channels and stimulate P_i efflux under low exogenous P_i concentrations [125]. However, whether fungal P_i transporters in the interface may have similar capabilities is still unknown.

Carbon as trigger for nutrient transport: AM fungi are obligate biotrophs and also ECM fungi absorb carbon mainly from the mycorrhizal interface in symbiosis. The carbon from the host provides the required resources for an extension of the ERM, for active uptake or other energy consuming processes, and for the development of new infection units. The supply of carbon by the host has been shown to stimulate the P uptake and transfer in AM [83, 94, 126] and ECM symbiosis [95] and it has been suggested that the P_i efflux from the IRM could be directly linked to the glucose uptake by the mycorrhizal fungus [127]. The P_i efflux from the IRM of the AM fungus *Gigaspora margarita* can be stimulated by an external supply of glucose, and its subsequent phosphorylation is coupled to a breakdown of polyP [128]. Also ECM fungi show an increased P_i efflux after a supply with sucrose under pure culture conditions [121]. There is currently no molecular evidence for a direct linkage between P_i efflux and carbon supply, but it has been shown that the expression of *MST2* (a monosaccharide transporter of *Glomus sp.*) and *PT4* (the mycorrhiza-inducible P transporter in the PAM) are tightly linked [28].

The carbon supply of the host also stimulates N uptake and transport in the AM symbiosis and triggers changes in fungal gene expression [113]. An increase in the carbon availability stimulates the expression of several genes involved in N assimilation and Arg biosynthesis[4] in the ERM, but reduces the transcript levels of a fungal urease. This increases the levels of Arg in the ERM and stimulates the export of Arg to the IRM (Figure 5). In the IRM, higher carbon availability induces fungal arginase and urease activity, but reduces the transcript levels of genes involved in Arg biosynthesis. The low expression of these enzymes will prevent that NH_4^+, which is released in the IRM as a result of the increased arginase and urease activities, is re-converted into Arg. This will lead to an increase in the internal NH_4^+ level in the IRM and will facilitate N release into the mycorrhizal interface [113]. An effect of the photosynthetic activity of the host on N uptake and transport has also been observed in ECM birch seedlings [129].

PolyP hydrolysis: AM and ECM fungi store significant amounts of P as polyP and it is generally assumed that polyP play an important role in the transport of P and N from the ERM to the IRM [116] (Figure 5). An active hydrolysis of polyP in the IRM would release P_i and Arg (for subsequent breakdown into NH_4^+ by the catabolic arm of the urea cycle) into the fungal cytoplasm and would facilitate the efflux into the mycorrhizal interface. AM and ECM fungi regulate the nutrient transport to the host by the accumulation or remobilization of polyP and it has been shown that the carbon supply of the host plant can trigger polyP hydrolysis [83, 94, 95, 113]. It has been proposed that polyP synthesis and breakdown could

---

[4] N assimilation: glutamine synthetase, glutamate synthase; Arg biosynthesis: carbamoyl-phosphate synthase, glutamine chain, argininosuccinate synthase, argininosuccinate lyase

be regulated by the activity of H^+-ATPases in the fungal tonoplast [130] or by a differential regulation of the involved enzymes.

Effects on membrane permeability. The accumulation of particular ions (e.g. H^+, K^+, Na^+, Ca^{2+}) in the interfacial apoplast could lead to a depolarization of the transmembrane electric potential and to an opening of ion channels [131, 132]. Monovalent cations, such as Na^+ and K^+, have been shown to stimulate P_i efflux from ECM fungal hyphae in pure culture experiments and could also have an effect on fungal re-absorption [121]. Lyso-phosphatidylcholine (LPC) acts in AM roots as a lipophilic signal, that induces the expression of mycorrhiza-inducible P transporters and there are indications for an extracellular localization and production of LPC. LPC leads to a rapid extracellular alkalinization of tomato cells in suspension-cultures [133] and could have similar effects also on the fungal membrane potential at the mycorrhizal interface. AM and ECM fungi in symbiosis have been shown to express aquaporins [118, 134]. Aquaporins are integral membrane channels that facilitate the concentration gradient-driven water and/or solute transport through the plasma membrane. In root nodules, aquaporins have been shown to be involved in the NH_4^+ flux through the symbiosome membrane that encloses the N fixing bacteroids [135]. In ECM roots, they could facilitate the transport of NH_4^+ through the fungal plasma membrane into the interfacial apoplast and in combination with a reduced re-absorption increase the net transport of NH_4^+ to the host [118]. Aquaporins could also stimulate the efflux of other nutrients, such as phosphate, through the fungal plasma membrane into the mycorrhizal interface [136].

6. Conclusions

It has been hypothesized that mycorrhizal growth responses follow a mutualism – parasitism continuum [137] and that the outcome of the symbiosis primarily depends on cost (carbon) to benefit (nutrient gain) ratios. When the nutrient availability in the soil is high, growth depressions in AM plants have been observed [e.g. 138, 139], and it was generally assumed that the negative growth responses are the result of the high carbon costs of the symbiosis for the host plant that are not counterbalanced by a net gain in P. Alternatively, it has been suggested that negative growth responses in AM interactions could also be the result of a reduced P uptake via the plant pathway which is not compensated for by an increase in P uptake via the mycorrhizal pathway, leading to an overall reduction in total P uptake and P deficiency for the plant [52]. Similarly, it has been proposed that for ECM plants carbon is an excess rather than a costly resource and that the outcome of the symbiosis for the host is primarily dependent on the nutrient acquisition efficiency of the ECM fungus [140].

Carbon to P exchange processes in the AM symbiosis are driven by biological market dynamics and both partners reciprocally reward beneficial partners with more resources [94]. AM fungi differ in their efficiency with which they suppress the plant nutrient uptake pathway [55]. AM fungi are completely dependent on their host plant for their carbon supply, and it is interesting to speculate that the suppression of the plant pathway could be more a fungal-driven than a host-motivated response. A strong suppression of the plant pathway will shift the ratio between both uptake pathways towards the mycorrhizal pathway and will result in a higher mycorrhizal dependency of the host. A high mycorrhizal dependency increases the carbon allocation to the root system [141], and this will make more carbon available for the fungus, which in return has been shown to trigger P and N transport in the AM symbiosis [83, 94, 113, 126]. This is also consistent with the observation that the N transport to the host was reduced when the fungus had access to an additional carbon source [113], and the mycorrhizal dependency of the fungus was reduced. This indicates that the fungus is more in control of the symbiosis than previously been suggested and that mycorrhizal fungi can gain an advantage in the symbiosis by adjusting their nutrient transfer to the host.

The question arises, whether and how the host plant is able to control the symbiosis. Arbuscules in the AM symbiosis undergo in the host cell a cycle of growth, degradation, senescence and recurrent growth, and the typical life span of arbuscules is only 8.5 days [142]. The life span of arbuscules has been shown to depend on their ability to deliver nutrients to the host and is regulated by the host plant demand. A high supply of the plant with P, leads in roots of *Petunia hybrida* to malformed arbuscules with a low branching pattern. P acts systemically and even a relatively small increase in the P level in the shoot, has a large effect on the AM colonization [143]. This effect was not the result of transcriptional changes in SYM genes, but was correlated to a decrease in the expression of the mycorrhiza-inducible P transporter (Pt4) in the PAM. Pt4 expression plays a critical role for the AM symbiosis and when MtPt4 (Pt4 of *Medicago truncatula*) is not expressed, arbuscules are prematurely degraded and the fungus is unable to proliferate within the host [144]. Interestingly, in MtPt4 mutants under N deficiency the degeneration of arbuscules is suppressed [145], what suggests that plant and fungus can change their resource exchange dynamics according to nutrient demand and supply conditions. The host driven regular turnover of arbuscules seems to provide the AM plant with an instrument to remove and 'to penalize' inefficient AM fungal symbionts [8], and to respond with transformations in its intracellular colonization to short term changes in the exogenous nutrient supply conditions.

Author details

Heike Bücking*, Elliot Liepold and Prashant Ambilwade
South Dakota State University, Biology and Microbiology Department, Brookings, USA

Acknowledgement

Our research was financially supported by the National Science Foundation (IOS awards 0943338 and 1051397), the DOE and the Sun Grant Initiative, and the South Dakota Wheat Commission. We would also like to acknowledge the contributions of Carl R. Fellbaum and Jerry A. Mensah to some of the images in this text.

* Corresponding Author

7. References

[1]   Smith SE, Read DJ. Mycorrhizal Symbiosis. Third edition. Amsterdam: Academic Press;
      2008.

[2]   Wang B, Qiu YL. Phylogenetic Distribution and Evolution of Mycorrhizas in Land
      Plants. Mycorrhiza 2006;16(5): 299-363.

[3]   Wright DP, Read DJ, Scholes JD. Mycorrhizal Sink Strength Influences Whole Plant
      Carbon Balance of *Trifolium repens* L. Plant, Cell and Environment 1998;21: 881-91.

[4]   Finlay RD. Ecological Aspects of Mycorrhizal Symbiosis: With Special Emphasis on the
      Functional Diversity of Interactions Involving the Extraradial Mycelium. Journal of
      Experimental Botany 2008;59(5): 1115-26.

[5]   Bücking H. Ectomycoremediation: An Eco-friendly Technique for the Remediation of
      Polluted Sites. In: Rai M, Varma A (eds.) Diversity and Biotechnology of
      Ectomycorrhizae. Soil Biology. Berlin, Heidelberg: Springer; 2011. p209-29.

[6]   Giri B, Giang PH, Kumari R, Prasad R, Sachdev M, Garg AP, Oelmüller R, Varma A.
      Mycorrhizosphere: Strategies and Functions. Soil Biology 2005;3: 213-52.

[7]   Schüssler A, Schwarzott D, Walker C. A New Fungal Phylum, the Glomeromycota:
      Phylogeny and Evolution. Mycological Research 2001;105: 1413-21.

[8]   Parniske M. Arbuscular Mycorrhiza: The Mother of Plant Root Endosymbioses. Nature
      Reviews Microbiology 2008;6: 763-75.

[9]   Hosny M, Gianinazzi-Pearson V, Dulieu H. Nuclear DNA Content of 11 Fungal Species
      in Glomales. Genome 1998;41(3): 422-8.

[10] Martin F, Gianinazzi-Pearson V, Hijri M, Lammers PJ, Requena N, Sanders IR, Shachar-
      Hill Y, Shapiro H, Tuskan G, Young JPW. The Long Hard Road to a Completed *Glomus
      intraradices* Genome. New Phytologist 2008;180: 747-50.

[11] Tisserant E, Kohler A, Dozolme-Seddas P, Balestrini R, Benabdellah K, Colard A, Croll
      D, Da Silva C, Gomez SK, Koul R, Ferrol N, Fiorilli V, Formey D, Franken P, Helber N,
      Hijri M, Lanfranco L, Lindquist L, Liu Y, Malbreil M, Morin E, Poulain J, Shapiro H, van
      Tuinen D, Waschke, A, Azcón-Aguilar C, Bécard G, Bonfante P, Harrison MJ, Küster H,
      Lammers, PJ, Paszkowski U, Requena N, Rensing SA, Roux C, Sanders IR, Shachar-Hill
      Y, Tuskan G, Young JPW, Gianinazzi-Pearson V, Martin F. The Transcriptome of the
      Arbuscular Mycorrhizal Fungus *Glomus intraradices* (DAOM 197198) Reveals Functional
      Tradeoffs in an Obligate Symbiont. New Phytologist 2011;193: 755-69.

[12] Bücking H, Abubaker J, Govindarajulu M, Tala M, Pfeffer PE, Nagahashi G, Lammers
      P, Shachar-Hill Y. Root Exudates Stimulate the Uptake and Metabolism of Organic
      Carbon in Germinating Spores of *Glomus intraradices*. New Phytologist 2008;180: 684-95.

[13] Gachomo E, Allen JW, Pfeffer PE, Govindarajulu M, Douds DD, Jin HR, Nagahashi G,
      Lammers PJ, Shachar-Hill Y, Bücking H. Germinating Spores of *Glomus intraradices* can
      use Internal and Exogenous Nitrogen Sources for *de novo* Biosynthesis of Amino Acids.
      New Phytologist 2009;184(2): 399-411.

[14] Tamasloukht M, Séjalon-Delmas N, Kluever A, Jauneau A, Roux C, Bécard G, Franken
      P. Root Factors Induce Mitochondrial Related Gene Expression and Fungal Respiration

During Developmental Switch from Asymbiosis to Presymbiosis in the Arbuscular Mycorrhizal Fungus *Gigaspora rosea*. Plant Physiology 2003;131: 1468-78.

[15] Akiyama K, Hayashi H. Strigolactones: Chemical Signals for Fungal Symbionts and Parasitic Weeds in Plant Roots. Annals of Botany 2006;97: 925-31.

[16] Maillet F, Poinsot V, André O, Puech-Pagès V, Haouy A, Gueunier M, Cromer L, Giraudet D, Formey D, Niebel A, Martinez EA, Dirguez H, Bécard G, Dénarié J. Fungal Lipochitooligosaccharide Symbiotic Signals in Arbuscular Mycorrhiza. Nature 2011;469: 58-63.

[17] Bonfante P, Requena N. Dating in the Dark: How Roots Respond to Fungal Signals to Establish Arbuscular Mycorrhizal Symbiosis. Current Opinion in Plant Biology 2011;14: 451-7.

[18] Gherbi H, Markmann K, Svistoonoff S, Estevan J, Autran D, Giczey G, Auguy F, Péret B, Laplaze L, Franche C, Parniske M, Bogusz D. SymRK Defines a Common Genetic Basis for Plant Root Endosymbioses with Arbuscular Mycorrhizal Fungi, Rhizobia, and Frankia Bacteria. Proceedings of the National Academy of Sciences, U.S.A. 2008;105: 4928-32.

[19] Charpentier M, Bredemeier R, Wanner G, Takeda N, Schleiff E, Parniske M. *Lotus japonicus* CASTOR and POLLUX are Ion Channels Essential for Perinuclear Calcium Spiking in Legume Root Endosymbiosis. Plant Cell 2008;20: 3467-79.

[20] Kistner C, Winzer T, Pitzschke A, Mulder L, Sato S, Kaneko T, Tabata S, Sandal N, Stougaard J, Webb KJ, Szczyglowski K, Parniske M. Seven *Lotus japonicus* Genes Required for Transcriptional Reprogramming of the Root during Fungal and Bacterial Symbiosis. Plant Cell 2005;17(8): 2217-29.

[21] Feddermann N, Muni RR, Zeier T, Stuurman J, Ercolin F, Schorderet M, Reinhardt D. The PAM1 Gene of *Petunia*, Required for Intracellular Accommodation and Morphogenesis of Arbuscular Mycorrhizal Fungi, Encodes a Homologue of VAPYRIN. Plant Journal 2010;64(3): 470-81.

[22] Hogekamp C, Arndt D, Pereira PA, Becker JD, Hohnjec N, Küster H. Laser Microdissection Unravels Cell-type-specific Transcription in Arbuscular Mycorrhizal Roots, Including CAAT-Box Transcription Factor Gene Expression Correlating with Fungal Contact and Spread. Plant Physiology 2011;157(4): 2023-43.

[23] Harrison MJ, Dewbre GR, Liu J. A Phosphate Transporter from *Medicago truncatu a* Involved in the Acquisition of Phosphate Released by Arbuscular Mycorrhizal Fungi. The Plant Cell 2002;14: 2413-29.

[24] Xu GH, Chague V, Melamed-Bessudo C, Kapulnik Y, Jain A, Raghothama KG, Levy AA, Silber A. Functional Characterization of LePT4: a Phosphate Transporter in Tomato with Mycorrhiza-Enhanced Expression. Journal of Experimental Botany 2007;58(10): 2491-501.

[25] Guether M, Neuhauser B, Balestrini R, Dynowski M, Ludewig U, Bonfante P. A Mycorrhizal-Specific Ammonium Transporter from *Lotus japonicus* Acquires Nitrogen Released by Arbuscular Mycorrhizal Fungi. Plant Physiology 2009;150(1): 73-83.

[26] Casieri L, Gallardo K, Wipf D. Transcriptional Response of *Medicago truncatula* Sulphate Transporters to Arbuscular Mycorrhizal Symbiosis with and without Sulphur Stress. Planta 2012; DOI: 10.1007/s00425-012-1645-7.

[27] Allen JW, Shachar-Hill Y. Sulfur Transfer Through an Arbuscular Mycorrhiza. Plant Physiology 2009 Jan;149(1): 549-60.

[28] Helber N, Wippel K, Sauer N, Schaarschmidt S, Hause B, Requena N. A Versatile Monosaccharide Transporter that Operates in the Arbuscular Mycorrhizal Fungus *Glomus sp.* is Crucial for the Symbiotic Relationship with Plants. Plant Cell 2011;23: 3812-23.

[29] Taylor AFS, Alexander I. The Ectomycorrhizal Symbiosis: Life in the Real World. Mycologist 2005;19: 102-12.

[30] Salzer P, Hager A. Sucrose Utilization of the Ectomycorrhizal Fungi *Amanita muscaria* and *Hebeloma crustuliniforme* Depends on the Cell Wall-Bound Invertase Activity of their Host *Picea abies*. Botanica Acta 1996;104: 439-45.

[31] Martin F, Aerts A, Ahren D, Brun A, Danchin EGJ, Duchaussoy F, Gibon J, Kohler A, Lindquist E, Pereda V, Salamov A, Shapiro HJ, Wuyts J, Blaudez D, Buee M, Brokstein P, Canback B, Cohen D, Courty PE, Coutinho PM, Delaruelle C, Detter JC, Deveau A, DiFazio S, Duplessis S, Fraissinet-Tachet L, Lucic E, Frey-Klett P, Fourrey C, Feussner I, Gay G, Grimwood J, Hoegger PJ, Jain P, Kilaru S, Labbe J, Lin YC, Legue V, Le Tacon F, Marmeisse R, Melayah D, Montanini B, Muratet M, Nehls U, Niculita-Hirzel H, Oudot-Le Secq MP, Peter M, Quesneville H, Rajashekar B, Reich M, Rouhier N, Schmutz J, Yin T, Chalot M, Henrissat B, Kues U, Lucas S, Van de Peer Y, Podila GK, Polle A, Pukkila PJ, Richardson PM, Rouze P, Sanders IR, Stajich JE, Tunlid A, Tuskan G, Grigoriev IV. The Genome of *Laccaria bicolor* Provides Insights into Mycorrhizal Symbiosis. Nature 2008;452(7183): 88-U7.

[32] Ditengou FA, Lapeyrie F. Hypaphorine from the Ectomycorrhizal Fungus *Pisolithus tinctorius* Counteracts Activities of Indole-3-Acetic Acid and Ethylene but not Synthetic Auxins in Eucalypt Seedlings. Molecular Plant-Microbe Interactions 2000;13(2): 151-8.

[33] Felten J, Kohler A, Morin E, Bhalerao RP, Palme K, Martin F, Ditengou F, Legué V. The Ectomycorrhizal Fungus *Laccaria bicolor* Stimulates Lateral Root Formation in Poplar and *Arabidopsis* through Auxin Transport and Signaling. Plant Physiology 2009;151: 1991-2005.

[34] Martin F, Duplessis S, Ditengou F, Lagrange H, Voiblet C, Lapeyrie F. Developmental Cross Talking in the Ectomycorrhizal Symbiosis: Signals and Communication Genes. New Phytologist 2001;151: 145-54.

[35] Plett JM, Kemppainen M, Kale SD, Kohler A, Legue V, Brun A, Tyler BM, Pardo AG, Martin F. A Secreted Effector Protein of *Laccaria bicolor* is Required for Symbiosis Development. Current Biology 2011;21(14): 1197-203.

[36] Sebastiana M, Figueiredo A, Acioli B, L. S, Pessoa R, Balde A, Pais MS. Identification of Plant Genes Involved on the Initial Contact between Ectomycorrhizal Symbionts

(*Castanea sativa* - European Chestnut and *Pisolithus tinctorius*). European Journal of Soil Biology 2009;45(3): 275-82.

[37] Bücking H, Kuhn AJ, Schröder WH, Heyser W. The Fungal Sheath of Ectomycorrhizal Pine Roots: An Apoplastic Barrier for the Entry of Calcium, Magnesium, and Potassium into the Root Cortex? Journal of Experimental Botany 2002;53: 1659-69.

[38] Ashford AE, Peterson CA, Carpenter JL, Cairney JWG, Allaway WG. Structure and Permeability of the Fungal Sheath in the *Pisonia* Mycorrhiza. Protoplasma 1988;147: 149-61.

[39] Rousseau JVD, Reid CPP, English RJ. Relationship Between Biomass of the Mycorrhizal Fungus *Pisolithus tinctorius* and Phosphorus Uptake in Loblolly Pine Seedlings. Soil Biology and Biochemistry 1992;24(2): 183-4.

[40] Högberg MN, Högberg P. Extramatrical Ectomycorrhizal Mycelium Contributes One-Third of Microbial Biomass and Produces, Together with Associated Roots, Half the Dissolved Organic Carbon in a Forest Soil. New Phytologist 2002;154: 791-5.

[41] Wallander H, Nilsson LO, Hagerberg D, Baath E. Estimation of the Biomass and Seasonal Growth of External Mycelium of Ectomycorrhizal Fungi in the Field. New Phytologist 2001;151: 753-60.

[42] Agerer R. Exploration Types of Ectomycorrhizae. A Proposal to Classify Ectomycorrhizal Mycelial Systems According to their Patterns of Differentiation and Putative Ecological Importance. Mycorrhiza 2001;11: 107-14.

[43] Bücking H, Heyser W. Microautoradiographic Localization of Phosphate and Carbohydrates in Mycorrhizal Roots of *Populus tremula* x *Populus alba* and the Implications for Transfer Processes in Ectomycorrhizal Associations. Tree Physiology 2001;21(2): 101-7.

[44] Parrent JL, James TY, Vasaitis R, Taylor AFS. Friend or Foe? Evolutionary History of Glycoside Hydrolase Family 32 Genes Encoding for Sucrolytic Activity in Fungi and its Implications for Plant-Fungal Symbioses. BMC Evolutionary Biology 2009;9: 148.

[45] Nehls U, Mikolajewski S, Magel E, Hampp R. Carbohydrate Metabolism in Ectomycorrhizas: Gene Expression, Monosaccharide Transport and Metabolic Control. New Phytologist 2001;150: 533-41.

[46] Lopez MF, Dietz S, Grunze N, Bloschies J, Weiss M, Nehls U. The Sugar Porter Gene Family of *Laccaria bicolor*: Function in Ectomycorrhizal Symbiosis and Soil-Growing Hyphae. New Phytologist 2008;180(2): 365-78.

[47] Willmann A, Weiss M, Nehls U. Ectomycorrhiza-Mediated Repression of the High-Affinity Ammonium Importer Gene AmAMT2 in *Amanita muscaria*. Current Genetics 2007;51(2): 71-8.

[48] Selle A, Willmann M, Grunze N, Gessler A, Weiss M, Nehls U. The High-Affinity Poplar Ammonium Importer PttAMT1.2 and its Role in Ectomycorrhizal Symbiosis New Phytologist 2005;168(3): 697-706.

[49] Schachtman DP, Reid RJ, Ayling SM. Phosphorus Uptake by Plants: From Soil to Cell. Plant Physiology 1998;116(2): 447-53.

[50] Smith SE, Smith FA. Roles of Arbuscular Mycorrhizas in Plant Nutrition and Growth: New Paradigms from Cellular to Ecosystem Scales. Annual Review Plant Biology 2011;62: 227-50.

[51] Smith FA, Grace EJ, Smith SE. More than a Carbon Economy: Nutrient Trade and Ecological Sustainability in Facultative Arbuscular Mycorrhizal Symbioses. New Phytologist 2009;182: 347-58.

[52] Smith SE, Jakobsen I, Grønlund M, Smith FA. Roles of Arbuscular Mycorrhizas in Plant Phosphorus Nutrition: Interactions Between Pathways of Phosphorus Uptake in Arbuscular Mycorrhizal Roots have Important Implications for Understanding and Manipulating Plant Phosphorus Acquisition. Plant Physiology 2011;156: 1050-7.

[53] Li HY, Smith SE, Holloway RE, Zhu YG, Smith FA. Arbuscular Mycorrhizal Fungi Contribute to Phosphorus Uptake by Wheat Grown in a Phosphorus-Fixing Soil even in the Absence of Positive Growth Responses. New Phytologist 2006;172: 536-43.

[54] Chiou TJ, Liu H, Harrison MJ. The Spatial Expression Patterns of a Phosphate Transporter (MtPT1) from *Medicago truncatula* Indicate a Role in Phosphate Transport at the Root/Soil Interface. The Plant Journal 2001;25(3): 281-93.

[55] Grunwald U, Guo WB, Fischer K, Isayenkov S, Ludwig-Müller J, Hause B, Yan XL, Kuster H, Franken P. Overlapping Expression Patterns and Differential Transcript Levels of Phosphate Transporter Genes in Arbuscular Mycorrhizal, Pi-Fertilised and Phytohormone-Treated *Medicago truncatula* Roots. Planta 2009;229(5): 1023-34.

[56] Paszkowski U, Kroken U, Roux C, Briggs SP. Rice Phosphate Transporters Include an Evolutionarily Divergent Gene Specifically Activated in Arbuscular Mycorrhizal Symbiosis. Proceedings of the National Academy of Sciences, U.S.A. 2002;99(20): 13324-9.

[57] Smith SE, Smith FA, Jakobsen I. Mycorrhizal Fungi can Dominate Phosphate Supply to Plants Irrespective of Growth Responses. Plant Physiology 2003;133(1): 16-20.

[58] Ravnskov S, Jakobsen I. Functional Compatibility in Arbuscular Mycorrhizas Measured as Hyphal P Transport to the Plant. New Phytologist 1995;129: 611-8.

[59] Taylor JH, Peterson CA. Morphometric Analysis of *Pinus banksiana* Lamb. Root Anatomy During a 3-Month Field Study. Trees 2000;14: 239-47.

[60] Coelho ID, de Queiroz MV, Costa MD, Kasuya MCM, de Araujo EF. Identification of Differentially Expressed Genes of the Fungus *Hydnangium sp* during the Pre-Symbiotic Phase of the Ectomycorrhizal Association with *Eucalyptus grandis*. Mycorrhiza 2010;20(8): 531-40.

[61] Unestam T. Water Repellency, Mat Formation, and Leaf-Stimulated Growth of some Ectomycorrhizal Fungi. Mycorrhiza 1991;1: 13-20.

[62] Unestam T, Sun YP. Extramatrical Structures of Hydrophobic and Hydrophilic Ectomycorrhizal Fungi. Mycorrhiza 1995;5: 301-11.

[63] Taylor JH, Peterson CA. Ectomycorrhizal Impacts on Nutrient Uptake Pathways in Woody Roots. New Forests 2005;30(2-3): 203-14.

[64] Ouahmane L, Revel JC, Hafidi M, Thioulouse J, Prin Y, Galiana A, Dreyfus B, Duponnois R. Responses of *Pinus halepensis* Growth, Soil Microbial Catabolic Functions and Phosphate-Solubilizing Bacteria after Rock Phosphate Amendment and Ectomycorrhizal Inoculation. Plant and Soil 2009;320(1-2): 169-79.

[65] Brundrett MC. Tansley Review No. 134: Coevolution of Roots and Mycorrhizas of Land Plants. New Phytologist 2002;154: 275-304.

[66] Jakobsen I, Abbott LK, Robson AD. External Hyphae of Vesicular-Arbuscular Mycorrhizal Fungi Associated with *Trifolium subterraneum* L.. I. Spread of Hyphae and Phosphorus Inflow into Roots. New Phytologist 1992;120: 371-80.

[67] Jakobsen I, Gazey C, Abbott IK. Phosphate Transport by Communities of Arbuscular Mycorrhizal Fungi in Intact Soil Cores. New Phytologist 2001;149(1): 95-103.

[68] Li HY, Smith FA, Dickson S, Holloway RE, Smith SE. Plant Growth Depressions in Arbuscular Mycorrhizal Symbioses: Not Just Caused by Carbon Drain? New Phytologist 2008;178(4): 852-62.

[69] Hijikata N, Murase M, Tani C, Ohtomo R, Osaki M, Ezawa T. Polyphosphate has a Central Role in the Rapid and Massive Accumulation of Phosphorus in Extraradical Mycelium of an Arbuscular Mycorrhizal Fungus. New Phytologist 2010;186(2): 285-9.

[70] Ezawa T, Hayatsu M, Saito M. A New Hypothesis on the Strategy for Acquisition of Phosphorus in Arbuscular Mycorrhiza: Up-Regulation of Secreted Acid Phosphatase Gene in the Host Plant. Molecular Plant-Microbe Interactions 2005;18(10): 1046-53.

[71] Alvarez M, Huygens D, Diaz LM, Villanueva CA, Heyser W, Boeckx P. The Spatial Distribution of Acid Phosphatase Activity in Ectomycorrhizal Tissues Depends on Soil Fertility and Morphotype, and Relates to Host Plant Phosphorus Uptake. Plant, Cell and Environment 2011;35(1): 126-35.

[72] Bucher M. Functional Biology of Plant Phosphate Uptake at Root and Mycorrhiza Interfaces. New Phytologist 2007;173(1): 11-26.

[73] Javot H, Pumplin N, Harrison MJ. Phosphate in the Arbuscular Mycorrhizal Symbiosis: Transport Properties and Regulatory Roles. Plant, Cell and Environment 2007;30: 310-22.

[74] Thomson BD, Clarkson DT, Brain P. Kinetics of Phosphorus Uptake by the Germ-Tubes of the Vesicular–Arbuscular Mycorrhizal Fungus, *Gigaspora margarita*. New Phytologist 1990;116(4): 647-53.

[75] Maldonado-Mendoza IE, Dewbre GR, Harrison MJ. A Phosphate Transporter Gene from the Extra-Radical Mycelium of an Arbuscular Mycorrhizal Fungus *Glomus intraradices* is Regulated in Response to Phosphate in the Environment. Molecular Plant-Microbe Interactions 2001;14(10): 1140-8.

[76] Tatry MV, Kassis EE, Lambilliotte R, Corratge C, van Aarle I, Amenc LK, Alary R, Zimmermann S, Sentenac H, Plassard C. Two Differentially Regulated Phosphate Transporters from the Symbiotic Fungus *Hebeloma cylindrosporum* and Phosphorus Acquisition by Ectomycorrhizal *Pinus pinaster*. Plant Journal 2009;57(6): 1092-102.

[77] Beever RE, Burns DJW. Phosphorus Uptake, Storage and Utilization by Fungi. Advances in Botanical Research 1980;8: 127-219.

[78] Shachar-Hill Y, Pfeffer PE, Douds D, Osman SF, Doner LW, Ratcliffe RG. Partioning of Intermediary Carbon Metabolism in Vesicular-Arbuscular Mycorrhizal Leek. Plant Physiology 1995;108: 7-15.

[79] Viereck N, Hansen PE, Jakobsen I. Phosphate Pool Dynamics in the Arbuscular Mycorrhizal Fungus *Glomus intraradices* Studied by in *vivo* ^{31}P NMR Spectroscopy. New Phytologist 2004; 162(3): 783-94.

[80] Ezawa T, Cavagnaro T, Smith SE, Smith FA, Ohtomo R. Rapid Accumulation of Polyphosphate in Extraradical Hyphae of an Arbuscular Mycorrhizal Fungus as Revealed by Histochemistry and a Polyphosphate Kinase/Luciferase System. New Phytologist 2003;161: 387-92.

[81] Ezawa T, Kuwahara S, Sakamoto K, Yoshida T, Saito M. Specific Inhibitor and Substrate Specificity of Alkaline Phosphatase Expressed in the Symbiotic Phase of the Arbuscular Mycorrhizal Fungus, *Glomus etunicatum*. Mycologia 1999;91(4): 636-41.

[82] Bücking H, Heyser W. Elemental Composition and Function of Polyphosphates in Ectomycorrhizal Fungi - an X-ray Microanalytical Study. Mycological Research 1999;103: 31-9.

[83] Bücking H, Shachar-Hill Y. Phosphate Uptake, Transport and Transfer by the Arbuscular Mycorrhizal Fungus *Glomus intraradices* is Stimulated by Increased Carbohydrate Availability. New Phytologist 2005;165(3): 899-912.

[84] Ohtomo R, Sekiguchi Y, Mimura T, Saito M, Ezawa T. Quantification of Polyphosphate: Different Sensitivities to Short-Chain Polyphosphate using Enzymatic and Colorimetric Methods as Revealed by Ion Chromatography. Analytical Biochemistry 2004;328: 139-46.

[85] Cooper KM, Tinker PB. Translocation and Transfer of Nutrients in Vesicular-Arbuscular Mycorrhizas. II. Uptake and Translocation of Phosphorus, Zinc and Sulphur. New Phytologist 1978;81: 43-52.

[86] Cooper KM, Tinker PB. Translocation and Transfer of Nutrients in Vesicular-Arbuscular Mycorrhizas. IV: Effect of Environmental Variables on Movement of Phophorus. New Phytologist 1981;88: 327-9.

[87] Callow JA, Capaccio LCM, Parish G, Tinker PB. Detection and Estimation of Polyphosphate in Vesicular-Arbuscular Mycorrhizas. New Phytologist 1978;80: 125-34.

[88] Cox G, Moran KJ, Sanders F, Nockolds C, Tinker PB. Translocation and Transfer of Nutrients in Vesicular-Arbuscular Mycorrhizas. III. Polyphosphate Granules and Phosphorus Translocation. New Phytologist 1980;84: 649-59.

[89] Solaiman MZ, Ezawa T, Kojima T, Saito M. Polyphosphates in Intraradical and Extraradical Hyphae of an Arbuscular Mycorrhizal Fungus, *Gigaspora margarita*. Applied and Environmental Microbiology 1999;65(12): 5604-6.

[90] Uetake Y, Kojima T, Ezawa T, Saito M. Extensive Tubular Vacuole System in an Arbuscular Mycorrhizal Fungus, *Gigaspora margarita*. New Phytologist 2002;154: 761-8.

[91] Olsson PA, van Aarle IM, Allaway WG, Ashford AE, Rouhier H. Phosphorus Effects on Metabolic Processes in Monoxenic Arbuscular Mycorrhiza Cultures. Plant Physiology 2002;130: 1162-71.

[92] Ashford AE, Allaway WG. The Role of the Motile Tubular Vacuole System in Mycorrhizal Fungi. Plant and Soil 2002;244: 177-87.

[93] Takanishi I, Ohtomo R, Hayatsu M, Saito M. Short-Chain Polyphosphate in Arbuscular Mycorrhizal Roots Colonized by Glomus spp.: A Possible Phosphate Pool for Host Plants. Soil Biology and Biochemistry 2009;41: 1571-3.

[94] Kiers ET, Duhamel M, Beesetty Y, Mensah JA, Franken O, Verbruggen E, Fellbaum CR, Kowalchuk GA, Hart MM, Bago A, Palmer TM, West SA, Vandenkoornhuyse P, Jansa J, Bücking H. Reciprocal Rewards Stabilize Cooperation in the Mycorrhizal Symbiosis. Science 2011;333(6044): 880-2.

[95] Bücking H, Heyser W. Uptake and Transfer of Nutrients in Ectomycorrhizal Associations: Interactions between Photosynthesis and Phosphate Nutrition. Mycorrhiza 2003;13: 59-68.

[96] Ryan MH, McCully ME, Huang CX. Location and Quantification of Phosphorus and other Elements in Fully Hydrated, Soil Grown Arbuscular Mycorrhizas: A Cryo-Analytical Scanning Electron Microscopy Study. New Phytologist 2003;160: 429-41.

[97] Bücking H, Beckmann S, Heyser W, Kottke I. Elemental Contents in Vacuolar Granules of Ectomycorrhizal Fungi Measured by EELS and EDXS. A Comparison of Different Methods and Preparation Techniques. Micron 1998;29(1): 53-61.

[98] Brandes B, Godbold DL, Kuhn AJ, Jentschke G. Nitrogen and Phosphorus Acquisition by the Mycelium of the Ectomycorrhizal Fungus Paxillus involutus and its Effect on Host Nutrition. New Phytologist 1998;140: 735-43.

[99] Finlay RD, Ek H, Odham G, Söderström B. Mycelial Uptake, Translocation and Assimilation of Nitrogen from ^{15}N-labelled Ammonium by Pinus sylvestris Plants Infected with four Different Ectomycorrhizal Fungi. New Phytologist 1988;110: 59-66.

[100] Chalot M, Brun A, Finlay RD, Söderström B. Metabolism of C-14 Glutamate and C-14 Glutamine by the Ectomycorrhizal Fungus Paxillus involutus. Microbiology-(UK) 1994;140: 1641-9.

[101] Reynolds HL, Hartley AE, Vogelsang KM, Bever JD, Schultz PA. Arbuscular Mycorrhizal Fungi do not Enhance Nitrogen Acquisition and Growth of Old-Field Perennials Under Low Nitrogen Supply in Glasshouse Culture. New Phytologist 2005;167: 869-80.

[102] Toussaint JP, St-Arnaud M, Charest C. Nitrogen Transfer and Assimilation between the Arbuscular Mycorrhizal Fungus Glomus intraradices Schenck & Smith and Ri T-DNA Roots of Daucus carota L. in an Vitro Compartmented System. Canadian Journal of Microbiology 2004;50: 251-60.

[103] Jin H, Pfeffer PE, Douds DD, Piotrowski E, Lammers PJ, Shachar-Hill Y. The Uptake, Metabolism, Transport and Transfer of Nitrogen in an Arbuscular Mycorrhizal Symbiosis. New Phytologist 2005;168(3): 687-96.

[104] Tanaka Y, Yano K. Nitrogen Delivery to Maize via Mycorrhizal Hyphae Depends on the form of N Supplied. Plant, Cell and Environment 2005;28: 1247-54.

[105] Hawkins HJ, Johansen A, George E. Uptake and Transport of Organic and Inorganic Nitrogen by Arbuscular Mycorrhizal Fungi. Plant and Soil 2000;226: 275-85.

[106] Leigh J, Hodge A, Fitter AH. Arbuscular Mycorrhizal Fungi can Transfer Substantial Amounts of Nitrogen to their Host Plant from Organic Material. New Phytologist 2009;181: 199-207.

[107] Plassard C, Barry D, Eltrop L, Mousain D. Nitrate Uptake in Martime Pine (*Pinus-pinaster*) and the Ectomycorrhizal Gungus *Hebeloma-cylindrosporum* - Effect of Ectomycorrhizal Symbiosis. Canadian Journal of Botany 1994;72(2): 189-97.

[108] Rékangalt D, Pépin R, Verner MC, Debaud JC, Marmeisse R, Fraissinet-Tachet L. Expression of the Nitrate Transporter nrt2 Gene from the Symbiotic Basidiomycete *Hebeloma cylindrosporum* is Affected by Host Plant and Carbon Sources. Mycorrhiza 2009;19(3): 143-8.

[109] Jargeat P, Rékangalt D, Verner MC, Gay G, Debaud JC, Marmeisse R, Fraissinet-Tachet L. Characterisation and Expression Analysis of a Nitrate Transporter and Nitrite Reductase Genes, two Members of a Gene Cluster for Nitrate Assimilation from the Symbiotic Basidiomycete *Hebeloma cylindrosporum*. Current Genetics 2003;43(3): 199-205.

[110] Javelle A, Morel M, Rodriguez-Pastrana BR, Botton B, Andr, B, Marini AM, Brun A, Chalot M. Molecular Characterization, Function and Regulation of Ammonium Transporters (Amt) and Ammonium-Metabolizing Enzymes (GS, NADP-GDH) in the Ectomycorrhizal Fungus *Hebeloma cylindrosporum*. Molecular Microbiology 2003;47(2): 411-30.

[111] Lopez-Pedrosa A, Gonzalez-Guerrero M, Valderas A, Azcon-Aguilar C, Ferrol N. GintAMT1 Encodes a Functional High-Affinity Ammonium Transporter That is Expressed in the Extraradical Mycelium of *Glomus intraradices*. Fungal Genetics and Biology 2006;43: 102-10.

[112] Tian C, Kasiborski B, Koul R, Lammers PJ, Bücking H, Shachar-Hill Y. Regulation of the Nitrogen Transfer Pathway in the Arbuscular Mycorrhizal Symbiosis: Gene Characterization and the Coordination of Expression with Nitrogen Flux. Plant Physiology 2010;153: 1175-87.

[113] Fellbaum CR, Gachomo EW, Beesetty Y, Choudhari S, Strahan GD, Pfeffer PE, Kiers ET, Bucking H. Carbon Availability Triggers Fungal Nitrogen Uptake and Transport in Arbuscular Mycorrhizal Symbiosis. Proceedings of the National Academy of Sciences, U.S.A. 2012;109(7): 2666-71.

[114] Kaldorf M, Schmelzer E, Bothe H. Expression of Maize and Fungal Nitrate Reductase Genes in Arbuscular Mycorrhiza. Molecular Plant-Microbe Interactions 1998;11(6): 439-48.

[115] Govindarajulu M, Pfeffer PE, Jin HR, Abubaker J, Douds DD, Allen JW, Bücking H, Lammers PJ, Shachar-Hill Y. Nitrogen Transfer in the Arbuscular Mycorrhizal Symbiosis. Nature 2005;435: 819-23.

[116] Cruz C, Egsgaard H, Trujillo C, Ambus P, Requena N, Martins-Loucao MA, Jakobsen I. Enzymatic Evidence for the Key Role of Arginine in Nitrogen Translocation by Arbuscular Mycorrhizal Fungi. Plant Physiology 2007;144: 782-92.

[117] Morel M, Jacob C, Fitz M, Wipf D, Chalot M, Brun A. Characterization and Regulation of PiDur3, a Permease Involved in the Acquisition of Urea by the Ectomycorrhizal Fungus *Paxillus involutus*. Fungal Genetics and Biology 2008;45(6): 912-21.

[118] Dietz S, von Bulow J, Beitz E, Nehls U. The Aquaporin Gene Family of the Ectomycorrhizal Fungus *Laccaria bicolor*: Lessons for Symbiotic Functions. New Phytologist 2011;190(4): 927-40.

[119] Bücking H, Heyser W. Subcellular Compartmentation of Elements in Non-Mycorrhizal and Mycorrhizal Roots of *Pinus sylvestris*: an X-ray Microanalytical Study. I. The Distribution of Phosphate. New Phytologist 2000;145(2): 311-20.

[120] Cairney JWG, Smith SE. Efflux of Phosphate from the Ectomycorrhizal Basidiomycete *Pisolithus tinctorius*: General Characteristics and the Influence of Intracellular Phosphorus Concentration. Mycological Research 1993;97: 1261-6.

[121] Bücking H. Phosphate Absorption and Efflux of Three Ectomycorrhizal Fungi as Affected by External Phosphate, Cation and Carbohydrate Concentrations. Mycological Research 2004;108: 599-609.

[122] Harrison MJ, van Buuren ML. A Phosphate Transporter from the Mycorrhizal Fungus *Glomus versiforme*. Nature 1995;378: 627-9.

[123] Rausch C, Daram P, Brunner S, Jansa J, Laloi M, Leggewie G, Amrhein N, Bucher M. A Phosphate Transporter Expressed in Arbuscule-Containing Cells in Potato. Nature 2002;414: 462-6.

[124] Kobae Y, Tamura Y, Takai S, Banba M, Hata S. Localized Expression of Arbuscular Mycorrhiza-Inducible Ammonium Transporters in Soybean. Plant Cell Physiology 2010;51(9): 1411-5.

[125] Preuss CP, Huang CY, Gilliham M, Tyerman SD. Channel-like Characteristics of the Low-Affinity Barley Phosphate Transporter PHT1;6 when Expressed in *Xenopus* Oocytes. Plant Physiology 2010;152(3): 1431-41.

[126] Hammer EC, Pallon J, Wallander H, Olsson PA. Tit for Tat? A Mycorrhizal Fungus Accumulates Phosphorus Under Low Plant Carbon Availability. FEMS Microbiology Ecology 2011;76: 236-44.

[127] Woolhouse HW. Membrane Structure and Transport Problems Considered in Relation to Phosphorus and Carbohydrate Movements and the Regulation of Endotrophic Mycorrhizal Associations. In: Sanders FE, Mosse P, Tinker PB. (eds.). Endomycorrhizas. London: Academic Press; 1975. p.209-39.

[128] Solaiman MZ, Saito M. Phosphate Efflux from Intraradical Hyphae of *Gigaspora margarita* in Vitro and its Implication for Phosphorus Translocation. New Phytologist 2001;151: 525-33.

[129] Kytöviita MM. Role of Nutrient Level and Defoliation on Symbiotic Function: Experimental Evidence by Tracing 14C/15N Exchange in Mycorrhizal Birch Seedlings. Mycorrhiza 2005;15: 65-70.

[130] Ezawa T, Smitha SC, Smithb FA. Enzyme Activity Involved in Glucose Phosphorylation in two Arbuscular Mycorrhizal Fungi: Indication that PolyP is not the Main Phosphagen. Soil Biology and Biochemistry 2001;33(9): 1279-81.

[131] Saito M. Symbiotic Exchange of Nutrients in Arbuscular Mycorrhizas : Transport and Transfer of Phosphorus. In: Kapulnik Y, Douds DD, Jr.. (eds.). Arbuscular Mycorrhizas: Physiology and Function. Netherlands: Kluwer Academic Publishers; 2000. p85-106.

[132] Smith SE, Gianinazzi-Pearson V, Koide R, Cairney JWG. Nutrient Transport in Mycorrhizas: Structure, Physiology and Consequences for Efficiency of the Symbiosis. Plant and Soil 1994;159: 103-13.

[133] Drissner D, Kunze G, Callewaert N, Gehrig P, Tamasloukht M, Boller T, Felix G, Amrhein N, Bucher M. Lyso-Phosphatidylcholine is a Signal in the Arbuscular Mycorrhizal Symbiosis. Science 2007;318: 265-8.

[134] Aroca R, Bago A, Sutka M, Paz JA, Cano C, Amodeo G, Ruiz-Lozano, JM. Expression Analysis of the First Arbuscular Mycorrhizal Fungi Aquaporin Described Reveals Concerted Gene Expression Between Salt-Stressed and Nonstressed Mycelium. Molecular Plant-Microbe Interactions 2009;22(9): 1169-78.

[135] Hwang JH, Ellingson SR, Roberts DM. Ammonia Permeability of the Soybean Nodulin 26 Channel. FEBS Letters 2010;584(20): 4339-43.

[136] Tyerman SD, Niemietz CM, Bramley H. Plant Aquaporins: Multifunctional Water and Solute Channels with Expanding Roles. Plant Cell and Environment 2002;25(2): 173-94.

[137] Johnson NC, Graham JH, Smith FA. Functioning of Mycorrhizal Associations Along the Mutualism-Parasitism Continuum. New Phytologist 1997;135: 575-85.

[138] Peng S, Eissenstat DM, Graham JH, Williams K, Hodge NC. Growth Depression in Mycorrhizal Citrus at High-Phosphorus Supply. Plant Physiology 1993;101: 1063-71.

[139] Colpaert Jv, Van Laere A, Van Assche JA. Carbon and Nitrogen Allocation in Ectomycorrhizal and Non-mMcorrhizal *Pinus sylvestris* L. Seedlings. Tree Physiology 1996;16: 787-93.

[140] Corrêa A, Gurevitch J, Martins-Loução MA, Cruz C. C Allocation to the Fungus is not a Cost to the Plant in Ectomycorrhizae. Oikos 2012;121: 449-63.

[141] Nielsen KL, Bouma TJ, Lynch JP, Eissenstat DM. Effects of Phosphorus Availability and Vesicular-Arbuscular Mycorrhizas on the Carbon Budget of Common Bean (*Phaseolus vulgaris*). New Phytologist 1998;139(4): 647-56.

[142] Alexander T, Toth R, Meier R, Weber HC. Dynamics of Arbuscule Development and Degeneration in Onion, Bean and Tomato with Reference to Vesicular Arbuscular Mycorrhizae in Grasses. Canadian Journal of Botany 1989;67: 2505-13.

[143] Breullin F, Schramm J, Hajrezaei M, Favre P, Druege U, Hause B, Bucher M, Kretzschmar T, Bossolini E, Kuhlermeier C, Martinoia E, Franken P, Scholz U, Reinhardt D. Phosphate Systematically Inhibits Development of Arbuscular Mycorrhiza

in *Petunia hybrida* and Represses Genes Involved in Mycorrhizal Functioning. Plant Journal 2010;64: 1002-17.

[144] Javot H, Penmetsa RV, Terzaghi N, Cook DR, Harrison MJ. A *Medicago truncatula* Phosphate Transporter Indispensable for the Arbuscular Mycorrhizal Symbiosis. Proceedings of the National Academy of Sciences, U.S.A. 2007;104: 1720-5.

[145] Javot H, Penmetsa RV, Breuillin F, Bhattarai KK, Noar RD, Gomez SK, Zhang Q, Cook DR, Harrison MJ. *Medicago truncatula* mtpt4 Mutants Reveal a Role for Nitrogen in the Regulation of Arbuscule Degeneration in Arbuscular Mycorrhizal Symbiosis. Plant Journal 2011;68(6): 954-65.

Plant Biotechnology

Chapter 7

Small Non-Coding RNAs in Plant Immunity

Camilo López, Boris Szurek and Álvaro L. Perez-Quintero

Additional information is available at the end of the chapter

1. Introduction

1.1. The "zigzag model" of plant-pathogen interactions

During millions of years of co-evolution, plants have established sophisticated genetic mechanisms to protect their integrity against invading pathogens. Pathogens in turn have coped with such barriers to gain access to nutrients and proliferate inside the plant. The "zigzag model" illustrates in a simple way the different layers of innate immunity during interactions of plants with their pathogen [1]. This model describes two main immunity responses, the first one relies on plants' ability to recognize so-called microbial-associated molecular patterns (MAMPs), which are highly conserved structures and molecules in all kinds of pathogenic and nonpathogenic microorganisms. This response is known as MAMP-triggered immunity (MTI) and is efficient against non-adapted or non-host pathogens [2]. The best-studied MAMPs are the flagelline peptide, the elongation factor Tu protein (EF-Tu), chitin whis is a major component of fungal cell walls and lipopolysacharides (LPS). MAMPs perception depends on plant pathogen recognition receptors (PRRs), with FLS2 and EFR recognizing flagelline and EF-Tu, respectively. These two PRRs share a similar structural architecture formed by extracellular Leucine Rich Repeats (LRR) and a cytoplasmic kinase domain. CERK1, on the other hand, which is the Arabidopsis PRR involved in the recognition of chitine, contains three extracellular LysM domains and a cytoplasmic kinase domain. In response to MTI, pathogens developed strategies to overcome it by sending effector proteins inside plant cells. These effector proteins abolish MTI by either suppressing early recognition or interfering with down-stream signaling events [3]. A second layer of plant immunity known as effector-triggered immunity (ETI) relies on a more sophisticated mechanism to detect pathogens, based on the specific recognition of particular effector proteins by Resistance (R) proteins. In this case the effector proteins are named Avr factors. The Avr-R proteins interaction can be direct or mediated by another protein, referred to as pathogenicity target. In the later case the R protein guards the pathogenicity target and detects its modification caused by the effector protein [4]. The largest group of R proteins includes a Nucleotide Binding Site (NBS) and a LRR domain. This group can be subdivided in two subclasses based on the presence or not of a Toll-Interleukin Related (TIR) domain in the N terminus. Upon the perception of pathogen's molecules by PRRs or R proteins, a signaling cascade including MAP kinases is activated, leading to a reprogramming in host's gene expression along with the activation of genes with antimicrobial function (PR, pathogenesis related) [2, 5]. In the last years a great effort in research has focused on understanding how gene expression is modified in response to pathogens, revealing some crucial factors for the interaction such as transcription factors, DNA regulatory elements and non-coding small RNAs.

2. An overview of beneficial interactions

Plant-microbe interactions are not always disadvantageous to plants. During millions of years of co-evolution, plants established symbiotic interactions with bacteria and fungi. The best-studied models illustrating this kind of interaction are the rhizobial and mycorrhizal symbiosis, which involve a particular group of bacteria and fungi, respectively. The establishment of symbiosis requires a concerted molecular dialogue involving the correct recognition and the activation/repression of specific signaling pathways [6]. In the rhizobial symbiosis, *Rhizobia* form an intimate relationship with leguminous plants. Plants provide carbon and energy to the bacteria, that in exchange fix atmospheric nitrogen of interest for plants [7]. Compatible *Rhizobium* species perceive plant-secreted flavonoids and induce the expression of bacterial *nod* genes that are essential for the development of nodules in plant hosts. The Nod factors are recognized by specific plant receptors carrying an extracellular LysM domain and an intracellular kinase domain. Upon perception several cytoplasmic events occur at the root epidermal cells, including membrane depolarization, calcium spiking and activation of a calmodulin-dependent kinase signaling [7]. These processes create a favorable cellular environment leading to the establishment of an infection thread branch through which *Rhizobia* penetrate into the host. Once in the cytoplasm, bacteria group and form bacteroids where nitrogen fixation occurs. In contrast to this highly specific interaction between legumes and *Rhizobia*, mycorrhizal fungi establish symbiosis with almost all terrestrial plant species. In this case, fungi provide nutrients from the soil to the plant, particularly P, and in exchange plants feed the fungus with their photosynthetic products [8]. The nutrient transfer occurs in the arbuscules, which are specialized structures formed in cortical root cells. Arbuscular mycorrhizae (AM) depend on the activation of a symbiosis signaling (Sym) pathway, which shares some elements with rhizobial symbiosis [6]. In the case of fungal AM, the perception relies on the recognition of diffusible Myc factors, leading to the reprogramming of the basic metabolism of plant cells and hyphens.

However, these types of beneficial interactions are not always successful. Plants indeed tolerate the invasion of these microorganisms only under nitrogen or nutrient-deficient conditions. In consequence, a sophisticated perception mechanism should exist in plants in order to simultaneously estimate nutrient deficiency and distinguish between beneficial microbes and pathogens. To achieve symbiosis, a fine regulation of the plant immune responses is therefore required for accepting or not candidate microorganisms [6]. In the last years we have learned a lot about how non coding RNAs are a crucial players in regulating such responses as well others.

3. Non-coding small RNAs

In eukaryotes, the endogenous regulation of gene expression is mostly dependent on the control of the RNA polymerase II by accessory proteins including activators, repressors and the mediator complex. Then, small non-coding RNAs (sncRNA) were discovered and found to be new key elements of gene expression regulation [9, 10]. sncRNAs are short molecules of typically 18 to 30 nt, involved in gene expression control, defense against other parasitic nucleic acids, epigenetic modification and heterochromatin regulation.

The best-studied sncRNAs are microRNAs (miRNAs) and small interference RNAs (siRNAs). miRNAs derive from nuclear genes. A gene coding for a miRNA (*MIRNA*) is first transcribed by the RNA polymerase II to a primary miRNA (pri-miRNA), the size of which ranges from 100 nt to several kilobases (kb). A Dicer-like (DCL) protein DCL1 in Arabidopsis along with HYPONASTIC LEAVES1 (HYL1), process the pri-miRNA into a 70 to 400 nt long precursor miRNA (pre-miRNA). This pre-miRNA forms a characteristic hairpin-like structure. A subsequent processing step involving DCL slices the pre-miRNA to form a miRNA:miRNA* duplex (21-22 nt). The duplex is then methylated by HEN1 and exported from the nucleus to the cytoplasm where it will join an AGO protein to form the silencing complex (RISC). Only the mature miRNA strand which is usually the one with less stable 5′-end pairing, is retained in the complex, while the passenger (miRNA*) strand is degraded. The miRNA* degradation process remain unknown, although some family of exoribonucleases encoded by the *SMALL RNA DEGRADING NUCLEASE* (*SDN*) genes degrades mature miRNAs which could also be involved in the miRNA* degradation [11]. The miRNA retained in the RISC complex will then guide the silencing of complementary mRNAs (targets) [12-13].

In contrast, siRNAs originate from transgenes, viruses, transposons or other RNAs that form perfectly complementary double-stranded RNA precursors (dsRNAs). In particular, virus-derived siRNAs also known as virus induced RNAs (vsiRNAs), have been extensively studied in plants. From a siRNA precursor, multiple siRNAs are generated and the silencing signal can be further amplified upon the generation of secondary siRNAs subsequently processed by RNA dependent RNA polymerases (RDRs), SILENCING DEFECTIVE3 (SDE3), NRPD1a and NRPD1b (largest subunits of Pol IVa and Pol IVb isoforms of RNA polymerase IV, respectively) [14-16].

Recently, various new types of ncRNAs have been described. Among them are the trans-acting siRNAs (ta-siRNAs) (21-22nt) which combine both the siRNA and miRNA pathways since they originate from a nuclear *TAS* gene which is transcribed into a mRNA and cleaved by a miRNA. The cleaved product is converted into dsRNA by the DEFECTIVE IN RNA-DIRECTED DNA METHYLATION/SUPPRESSOR OF GENE SILENCING (RDR6/SGS3) processing complex, and leading to specific siRNAs called ta-siRNAs. These mature 21-nt long siRNAs, similar to miRNAs, are able to initiate the cleavage of homologous cellular transcripts, thus acting in trans. Additionally the siRNA signal from ta-siRNAs can also be amplified upon the generation of secondary siRNAs [14, 17-19]. Also reported are natural antisense transcripts-derived siRNAs (nat-siRNAs) (21-24 nt) which are cis-acting siRNAs derived from naturally occurring overlapping regions of sense and antisense transcripts [20]. The long siRNAs (lsiRNAs) (30-40 nt) are DCL1 and AGO7 dependent in their biogenesis and act by decapping or by 5'-3' degradation of target mRNAs [21]. Whilst other types of sncRNA exist, we will focus on the most common ones and report on their role(s) in plant immunity. A comparison of the types of sncRNAs discussed in this chapter can be seen in table 1.

	siRNA	miRNA	ta-siRNA	nat-siRNAs
Derived from	Invasive nucleic acids (virus, transgenes)	non-coding regions	non-coding regions	antisense genes
Transcribed by	Depends of origin	RNA pol II	RNA pol II	RNA pol II
Processed by	DCL, RDR, SDE, NRPD	DCL1, HYL1, HEN1	RDR6/SGS3, DCL, miRNAs	DCL1, HYL1
Targets transcripts in	cis	trans	trans	cis
Binds to	AGO1, AGO2	AGO1	AGO1,AGO7	AGO1?

Table 1. Comparison of important features of common types of scnRNAs found in plants.

4. The "zigzag model" in plant-virus interactions

The seminal work achieved on plant-virus interactions studies led to the discovery of post-transcriptional gene silencing (PTGS) as a genuine plant defense mechanism against virus. Most of the plant viruses are positive single stranded RNAs (ssRNAs). To colonize and multiply into new plant cells, virus have to replicate several thousands of times. During this process of replication and infection, RNA viruses produce double-stranded RNAs (dsRNAs). A DCL protein, usually DCL4 in Arabidopsis, recognizes these dsRNAs and cleaves them producing vsiRNAs. The vsiRNAs are next incorporated into the RNA-Induced Silencing Complex (RISC) where only one of the two RNA strands is retained. This RNA strand is complementary to the viral RNA and exploited to target the RNA viral molecule and degrade it. Some vsiRNAs serve as template and substrate of an RNA-dependent RNA polymerase (RdRP), thereby amplifying the signal and producing more vsiRNAs [16, 22, 23]. These vsiRNAs move through the plasmodesmata of the cell-cell assuring a systemic anti-viral defense response [24, 25]. As a general plant defense mechanism against all viruses, this first branch of resistance can be considered as analogous to MAMP-triggered immunity. In this case, dsRNAs are considered as MAMPs and DCLs that recognize the dsRNAs are viewed as a sort of PRR.

In line with the "zigzag model", virus evolved strategies to overcome this first layer of immunity. As a matter of fact, viruses carry silencing suppressors (SS) are able to act at different levels of the silencing pathway [26]. Considering these suppressors as bona fide pathogen effector proteins, this scenario is reminiscent of effector-triggered susceptibility or ETS. SS proteins were previously considered as pathogenicity factors with an important function in the development of symptoms on plant hosts during infection [26]. A single SS can exert suppressor activity at different steps of the silencing pathways. One of the best-studied and more versatile SS is HC-Pro which is able to block silencing by either interfering with DCL proteins [27] and/or sequestering 21-nt siRNA duplexes [28]. The TCV P38 coat protein [29] and CMV 2b protein [30] affect the processing of dsRNA through the inactivation of DCL proteins. The P21 and P19 proteins respectively produced by the Beet yellows virus (BYV; Closterovirus, Closteroviridae) and the Tomato bushy stunt virus (TBSV, Tombusvirus, Tombusviridae), exert their function by interacting with miRNA duplexes and hairpin derived siRNAs [28, 31, 32]. Additionally, these SS interfere with the small RNAs stability by blocking the activity of the methyltransferase protein HEN1 [33, 34]. The Beet western yellows virus (BWYV; Polerovirus, Luteoviridae) P0 protein interacts via an F-box domain with AGO1 which results in its degradation, illustrating again how important components of the RNA silencing machinery are targeted in order to affect silencing [35].

To overcome virus-deployed strategies suppressing silencing, plants evolved R proteins recognizing specific viral proteins to trigger an immune response, which can be considered as a sort of effector-triggered immunity. This R-protein dependent ETI depends on the recognition of so called Avr proteins, which can virtually be encoded by any viral coding-gene. Examples of viral Avr proteins include the coat, helicase, replicase and movement proteins [36]. More than 15 anti-viral R proteins and belonging to the large class of NBS-LRRs have been characterized, including R proteins N [37], Y-1 [38] and RT4-4 [39] respectively isolated from *Nicotiana tabacum, Solanum tuberosum* and *Phaseolus vulgaris* and confering resistance to TMV, PVY and CMV. On the other hand, the Rx [40], HRT [41] and RCY1 [42] proteins respectively isolated from *Solanum andigena* and Arabidopsis, belong to the CC-NBS-LRR sub-class. The immunity triggered by these proteins is considered as monogenic and dominant, and manifested by an hypersensitive response (HR) [43]. Interestingly in most cases resistance of plants against virus segregates as a recessive trait, and is expressed as a cellular immunity. Remarkably, all recessive resistance genes isolated so far encode translation initiation factors [36]. As mentioned before, once the recognition of the virus is established, a re-programming in host gene expression takes place in plant host cells [44-47]

5. sncRNAs and viruses: new frontiers of defense

Recent studies suggest that sncRNAs are involved in global gene expression changes during plant-virus interaction. It has been proposed that the expression of plant miRNAs targeting plant transcripts is altered in response to virus recognition with the aim of affecting viral replication and spreading. Indeed various plant miRNAs are known to be up- or down-regulated following viral infection [48-52]. For example miR1885 is induced in response to infection of *Brassica rapa* by Turnip mosaic virus and is known to target a TIR-NBS-LRR (TNL) disease-resistance gene [53]. miR164 is also induced upon viral infection and its induction is due to hormone-dependent specific transcriptional activation [54]. However, the effect of this differential regulation in the outcome of the interaction is not well established, as it may also result from the silencing suppression activity of the virus. This is also the case of miR168, which is induced upon infection by various viruses in different plants [49, 55]. It was shown recently that miR168 accumulation upon infection with Cymbidum ringspot virus (CymRSV) is due to the action of the p19 SS [56]. The specific targeting of miR168 by p19 may be crucial for viral infection given that this miRNA is involved in a regulatory loop with AGO1, which forms the RISC complex and in thus involved in various silencing processes [56].

It has also been suggested that miRNAs directly target viral RNAs, as it occurs in animals [57, 58]. Indeed several studies demonstrated that artificial miRNAs (amiRNAs) targeting key components of the viral replication machinery can efficiently impair viral growth upon infection [59-65]. These efforts have revealed that such resistance is cell-autonomous, inheritable, more efficient than siRNA-mediated strategies and successful in blocking viral replication and movement [59, 66]. Furthermore transgenic plants expressing dimeric or polycistronic amiRNAs directed against different viruses result in a wider spectrum of viral resistance [60, 61, 63].

The hypothesis of plant microRNAs naturally evolving to target viral genomes has long being discussed, taking in account the potential disadvantages of the miRNA pathway over the siRNA one. The miRNA pathway is not an adaptive response since the evolution of viral genomes would be fast enough to surpass the evolution of miRNAs rendering them ineffective in a very short term. However some miRNA families may be adapted to target viral genomes, as suggested by bioinformatic analysis [67, 68].

Viruses also encode miRNAs (or similar ncRNAs) directed against the plant genomes (or even their own genomes), that will use the host miRNA machinery to be processed and execute their silencing effect. This mechanism has been described only in animal infecting viruses [69]. So far, the only mechanisms found resembling viral miRNAs in plant viruses refer to sRNAs encoded by the Cauliflower mosaic virus (CaMV) that are partially complementary to regions of the Arabidopsis genome [70] and viral sRNAs that bind the RNAi machinery to divert the silencing machinery from viral promoter and coding regions [71].

A general model of the way sncRNAs mediate the interaction between plants and viruses (as well as with bacteria discussed below) can be seen in Figure 1.

6. The need for auxin: responses to bacteria, fungi and symbiotic microbes

Auxin is a relatively well-known plant hormone mainly implicated in growth which acts, under particular conditions, as a repressor of salicylic acid (SA). SA is a hormone involved in the activation of plant defenses in response to biotrophic pathogens [72, 73]. It is therefore not surprising that plants, in response to microbes, have evolved sophisticated mechanism for fine-tuning of SA-mediated responses.

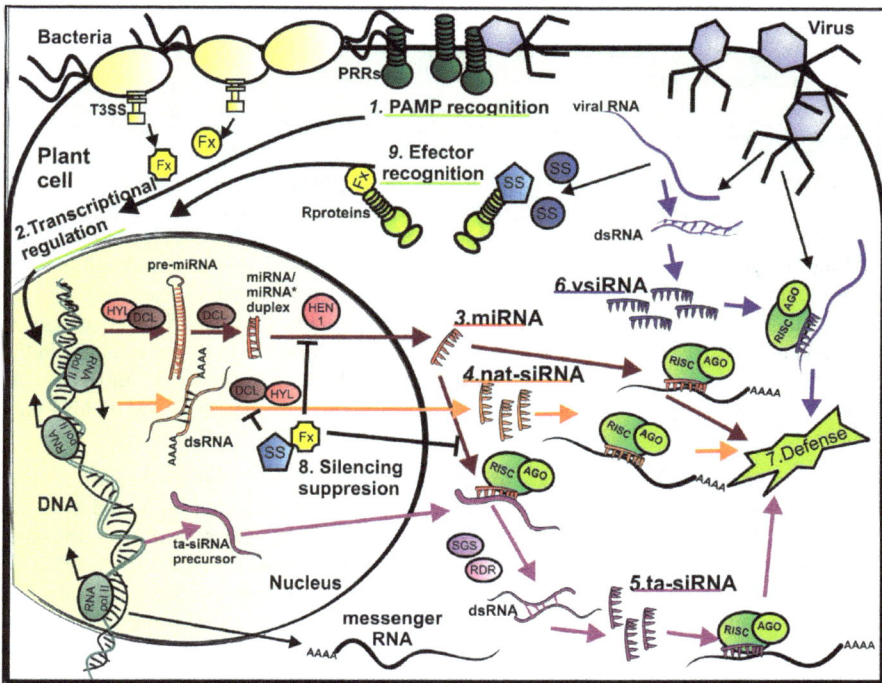

Figure 1. General model of sncRNAs role in plant-pathogen interaction. *1*.Upon recognition of PAMPs from virus and bacteria the *2*. transcription of loci coding for sncRNAs is regulated so the biogenesis pathways of either *4*. miRNAs *5*. nat-siRNAs *6*. ta-siRNAs is activated with the aims of *7*. targeting genes that when silenced would trigger defense responses (like Auxin response factors). Transcription of sncRNAs could be stopped by the plant in response to pathogens so that positive regulators of immunity (like NBS-LRR proteins) can escape miRNA regulation (not shown). *8*. Additionally, in the case of virus, vsiRNAs can be produced during viral replication and they would target viral RNAs thus producing defense. *9*. Virus and bacteria can counterattack by using effectors and silencing suppressors to disrupt silencing and in response *10*. plants could recognize these effectors via resistance proteins. Fx= bacterial effector, SS= silencing supresor, dsRNA= double-stranded RNA, PRRs= pathogen recognition receptors, R proteins = resistance protein, RISC= RNA-induced silencing complex

The best-studied miRNA induced upon bacterial infection is miR393. By comparing the gene expression profile of wild type and transgenic plants expressing several viral SS, it was elegantly demonstrated that upon treatment with flagellin (flg22), some transcripts were more abundant in transgenic plants. Among them was found a transcript coding for the F-box auxin receptor TIR1. By RACE (rapid amplification of cDNA ends) the authors demonstrated that this particular mRNA is targeted and cleaved by miR393. The perception of flagellin by plants induced the expression of miR393 which correlates with a clear reduction of the TIR1 protein content. This led in turn to the stabilization of Aux/IAA proteins which repress auxin signaling by heterodimerization with Auxin Response Factors

(ARF). Flagellin perception leads to a repression of auxin signaling and consequently restricted the growth *of Pseudomonas syringae* pv. tomato (*Pst*). This study provided for the first time a link between auxin response, miRNAs and MTI. In summary plants repress the auxin signaling pathway in response to bacterial hit, favoring the defenses activated by SA compromising vegetative growth [74].

miR393 was shown to be induced in Arabidopsis plants inoculated with *Pst* DC3000 strain mutated in *hrcC* [75]. This strain is defective for type III secretion, unable to deliver effector proteins into the host plant cell and consequently triggers MTI. Employing a small-RNA profiling analysis, Zhang *et al.* [76] investigated the differential expression of miRNAs in plants challenged with *Pst* DC3000 *hrcC* mutant, a virulent strain of the same species carrying an empty vector and avirulent *Pst* DC3000 containing the *avrRpt2* effector [76]. Curiously, miR393 was repressed at 6 hours post-infection (hpi) and induced at 14 hpi in the three treatments. However Northern-blot experiments show an induction in all treatments and at both time points. To explain this discrepancy, the authors suggest that miR393 may regulate auxin signaling at an early stage of bacterial infection [76].

The complex interplay between auxin and miRNAs goes beyond miR393. Several reports have shown that different ARFs and auxin receptors coding genes are regulated by other miRNAs such as miR160, miR166 and miR167, not only in response to phytopathogenic bacteria and fungi but also during beneficial interactions involving *Rhizobium* or AM [72, 76-80]. MiR167 and miR160 which target ARFs genes, are induced upon infection with different *Pst* DC3000 strains [76, 77]. MiR160 was also found to be induced in response to flg22 and bound to AGO1. Transgenic plants over-expressing miR160 show enhanced callose deposition and higher resistance to DC3000 indicating a role for miR160 as positive regulator of plant pathogen response [78].

miR393 is highly conserved and was also detected in cassava plants challenged with *Xanthomonas axonopodis* pv. manihotis, which is the causal agent of Cassava Bacterial Blight (CBB) [81]. Interestingly the expression of miR160 and miR393 is reduced during the infection of Arabidopsis with *Agrobacterium tumefaciens*, thereby increasing auxin signaling [82].

During the symbiosis occurring between soybean and *Bradirhizobium japonicum*, miR160, which targets the auxin repressor ARF17, is down-regulated, suggesting an increase in free auxin during this interaction. In contrast, miR393 was found to be induced, which is in opposition to miR160 effect, as miR393 regulates the auxin receptor TIR1 and consequently inhibits auxin signaling [83]. In AM symbiosis, a strong connection between miRNAs and auxin has also been unveiled. During the interaction of *M. truncatula* plants with *Glomus intraradices*, it was reported that miR160c and miR167 are induced in mycorrhizal roots while miR160 was predominantly localized in the phloem [84]. These authors reported also miR5229a/b to be the most induced miRNA. By in situ hybridization it was demonstrated that it was exclusively expressed in arbuscule-containing cells of the root cortex, albeit with different signal intensities indicating a specific function during different stages of the arbuscule development. The predicted target of miR5229a/b is a transcript encoding for a heme peroxidase playing different roles in the regulation of ROS production, cell wall biosynthesis but also auxin and ethylene metabolism. This provides another example of the relationship between miRNAs and auxin at a more indirect level [84]. Among 33 pathogen-responsive miRNAs detected during the interaction of *Populus beijingensis* with the fungus *Dothiorella gregaria*, the induction of miR393 was observed with a peak of expression at 7 days after inoculation but their levels were reduced at 14 and 21 days [85]. On the other hand, the same study showed the repression of miR160 as was previously reported in pine infected with the fungus *Cronartium quercumm* f. sp *fusiforme* [86]. In wheat interaction with *Erysiphe graminis* *f. sp. tritici*, the expression pattern of miR393 was less expressed in the susceptible cultivar Jingdong8 as compared to the near-isogenic resistant line Jingdong8-Pm30 [87]. These results illustrate the fact that in some cases, conserved miRNAs may play similar functions in different pathosystems. However in other cases, the expression profile and in consequence the function of miRNAs can be specific even if their sequences and targets are conserved. Altogether, these results highlight how crucial is auxin balance in plant-microbe interactions.

7. MTI and silencing: beyond miR393 and auxin

Fahlgreen *et al.* [77] reported on miRNAs repressed upon infection with a strain of *Pst* DC3000 inactivated in *hrcC* which is a major component of the TTSS. One of the identified miRNAs is miR825, which is predicted to target transcripts encoding a Remorin, a transcription factor of the zinc-finger homeobox family and a frataxin-related protein. These targets are known to act as positive regulators of plant defense, it is therefore expected that the miRNAs controlling them are repressed.

A first connection between miRNAs-mediated silencing and MTI emerged from the study of Arabidopsis AGO1 mutant lines, found to be compromised in MTI [78]. These plants are characterized by a reduction in seedling growth inhibition, callose deposition, expression of MTI-markers genes and the activation of MAP kinases and ROS production upon treatment with flagellin. Also, the growth of TTSS-mutant strains was increased in these plants. These results demonstrate that AGO1, and indirectly the silencing pathway, are key elements of MTI. These phenotypes are not observed in AGO7 mutant plants, indicating that only AGO1 activity is associated with MTI. Interestingly, the involvement of miRNAs in regulating plant immunity is not restricted to MTI. It may also be associated to ETI, as exemplified by cases of plant resistance responses against fungal pathogens, where several components of the silencing machinery were showed to play a role [88]. More precisely, Arabidopsis *sgs2-1*, *sgs1-1* and *sgs-3* mutants which are defective in siRNA production, are more susceptible to various strains of *Verticillum dahliae* but not to other pathogens such as *Botrytis cinerea*, *Alternaria brassicicola* and *Plectosphaerella cucumerina*. Intriguingly, while *ago7-2*, *dcl4-2*, *nrpd1a-3* and *rdr2-4* mutants are more susceptible to *V. dahlia*, *ago1-25*, *ago1-27*, *hen1-6* and *hst-1* mutants display enhanced resistance to this pathogen. Finally, *dcl-2*, *sde3-4* and *sde3-5* mutants were as susceptible as the Arabidopsis Col-0 wild type line [88].

8. Towards ETS

Accordingly to the "zigzag model", adapted pathogens overcome MTI to infect particular host plants. Based on the study of Arabidopsis mutants affected in genes involved in the silencing pathway like *dcl* and *hen1*, it could be demonstrated that non-pathogenic bacteria like *P. fluorescens* and *E. coli* were able to grow in these plants and not in the wild type. In addition, an increase in growth was observed upon inoculation of silencing-defective mutant lines with a TTSS-mutant *Pst* DC3000 strain and the non-host pathogen *Pseudomonas phaseolicola* [89]. These ground-breaking observations suggested a pivotal role of silencing in triggering MTI responses. In consequence, adapted pathogens should have acquired effector proteins capable of suppressing silencing-associated MTI pathways. As a matter of fact, the virulent strain *Pst* DC3000 was reported to repress the expression of miR393 which is normally induced upon flagellin recognition as well as other MTI responses, whereas a TTSS-mutant strain did not [89]. Since virulent *Pst* DC3000 wild type strain harbors an intact TTSS, it was concluded that miR393 repression results of the action of injected effectors. Upon *Agrobacterium*-mediated transient expression of particular T3 effectors into Arabidopsis leaves, the expression of miR393 primary transcripts was monitored. This assay successfully demonstrated that T3Es AvrPto and HOPT-1 block the miRNA pathway by targeting the activity of DCL1 and AGO, respectively [89]. Interestingly, it was also demonstrated that pre-inoculation of virus containing SS led to the development of disease-like symptoms and favored multiplication of non-pathogenic and non-host bacteria inoculated subsequently. The authors suggest this as a molecular base explaining the synergistic interactions eventually observed between some viral and bacterial phytopathogens in the field [89].

9. The arms race goes on: miRNA's role in ETI

As mentioned previously, the specific recognition of effectors by R proteins triggers ETI, which involves gene expression reprogramming. In a survey aimed at determining the role of siRNAs in gene expression during R-protein-mediated responses it was found that a 22 nt nat-siRNA was induced specifically by *Pst* DC3000 containing the avirulence gene *avrRpt2* [90]. This nat-siRNA named nat-siRNAATGB2 is produced due to an overlapping region between the *At4g35860* and *At4g35850* transcripts. *At4g35860* encodes a Rab2-like small GTP-binding *(ATGB2)* while *At4g35850* encodes a PPR (pentatricopeptide repeats) protein-like gene *(PPRL)*. The sequence of the nat-siRNAATGB2 is complementary to the 3′ UTR region of the antisense gene *PPRL*. In fact it was demonstrated a correlation between the induced expression of nat-siRNAATGB2 and a repression of *PPRL* after infection with *Pst* carrying *avrRpt2* [90]. The induction of nat-siRNAATGB2 is dependent of the presence of *RPS2* and *NDR1*, two genes required for the induction of the *avrRpt2*-mediated response. The biogenesis of nat-siRNAATGB2 depends on the DCL1-HYL1 complex, which is stabilized upon HEN1-mediated methylation and amplified by RDR6 and SGS3. In concordance, plants mutated in these genes do not show a reduction of the *PPRL* expression. On the other hand, the overexpression of *PPRL* produced a delayed HR and enhanced growth of *Pst* DC3000 carrying *avrRpt2*, indicating that PPRL is a negative regulator of plant defense responses [90]. A novel class of sncRNAs induced in response to *Pst* DC3000 strain carrying *avrRpt2* was also identified [21]. In this case the sncRNAs are long siRNA (lsiRNA) of 30 to 40 nt. Among these lsiRNAs is AtlsiRNA-1 which is generated from a NAT pair between the genes *SRRLK* and *AtRAP*. Interestingly, the biogenesis of AtlsiRNA-1 requires DCL1, DCL4, AGO7, HYL1, HEN1, HST1, RDR6 and Pol IV. The target of AtlsiRNA-1 is the gene *AtRAP*, which encodes for a RAP-domain protein with a role in plant resistance to pathogens. AtlsiRNA-1 does not cleave its target mRNA as most siRNAs usually do, but it guides their degradation through decapping and XRN4-mediated 5'-to-3' decay. A knockout mutation in *AtRAP* increases the resistance of Arabidopsis against virulent and avirulent *Pst* strains. In addition, overexpression of *AtRAP* leads to an increase in bacterial growth [91]. Thus, AtRAP and PPRL can be considered negative regulators of plant immunity.

10. *R* genes, my favorite targets: miRNAs

Since the expression of *R* genes is constitutive in most cases, it should be expected that plants have developed mechanisms regulating their activity and restrain the activation of plant immune responses under pathogen-free conditions. Although an elegant mechanism of regulation of NB-LRR proteins by conformational changes depending of the presence of the effector and hydrolysis of ATP has been described [92], controlling the activity of negative regulators mediated by sncRNAs emerges as an additional strategy to control plant immune responses.

Some reports have indeed demonstrated a direct regulation of *R* genes by siRNAs. In a pioneering study, Yi and Richards [93] detected endogenous siRNAs at the *RPP5* locus with antisense transcription activity. In this locus were identified seven *R* genes of the TNL class interspersed with three related sequences and two other non *R* genes. The genes *RPP4* (Recognition of *Peronospora parasitica*, now referred to as *Hyaloperonospora arabidopsis*) and *SNC1* (suppressor of *npr1-1*, constitutive 1) present in this cluster confer resistance to fungal and bacterial pathogens, respectively and are coordinately regulated by transcription control. It was shown that a production of antisense transcripts generates siRNAs to regulate the mRNA level of these genes. In fact, in *dcl4* and *ago1* mutant Arabidopsis plants the expression of *SNC1* mRNA was elevated suggesting a role of siRNAs involved in its regulation [93]. A similar observation was reported in the symbiotic interaction of *M. truncatula* with *Shinorhizobium meliloti* where genome-wide analysis of small RNAs revealed a relatively high proportion of 21-nt sRNAs corresponding to NBS-LRR genes [94].

Another example of regulation of *R* genes mediated by sncRNAs deals with the tobacco *N* gene which confers resistance to the tobacco mosaic virus (TMV) and codes for a TNL protein. Two miRNAs were shown to guide the cleavage of the *N* gene, namely nta-miR6019 and nta-miR6020 of 22 and 21 nt-long, respectively. In addition, a production of secondary siRNAs "in phase" with the miR6019 cleavage site of the *N* gene transcript was evidenced, and their biogenesis is dependent on DCL4 and RDR6 [95]. The co-expression of *N* with both of these nat-miRNAs led to reduced resistance against TMV confirming the importance of these nat-miRNA in the regulation of the *N* gene and N-dependent immune responses. The authors expanded these discoveries to tomato and potato, two species of the same Solanaceae family, finding that members of these miRNAs families are conserved across species as well as their potential for cleavage of NBS-LRR transcription products and the generation of secondary siRNAs [95].

During the infection of pine with the fusiform rust fungus *Cronartium quercuum*, a ta-siRNA (pta-22 ta-siRNA) targeting two disease resistance proteins was identified. In addition this study also reported and validated experimentally pta-miR946 and pta-miR948 and six of their targets which are predicted to encode for disease resistance-related transcripts, a transcript with similarity to *RPS2* and serine/treonine kinases [86]. Two other miRNAs (pta-miR950 and miR951) also target *R* genes [86]. In the response of poplar to the fungus *Dothiorella gregaria*, the targets of miRNAs pbe-miR482b, pbeSR3, pbe-SR23 and pbe-SR25 also include *R* genes. Other miRNAs previously identified in *Populus trichocarpa* include miR1447 which targets a related disease resistance-coding gene, while two other conserved miRNAs also targeting *R* genes (miR1447 and miRNA1448) are repressed [85]. Once again, these results highlight a complex network of several *R* genes whose regulation is coordinated by a huge collection of miRNAs belonging to different families and isoforms. Although miRNAs and their targets are not always validated experimentally, there is overall a clear consistency between the expression of disease resistance-related genes and their corresponding miRNAs.

R proteins play a pivotal role in triggering immune responses and should be able to recognize a broad spectrum of effector proteins (Jones and Dangl, 2006). The high repertory of plant immunity genes raises the question as to the control of their activity. Not surprisingly, several mechanisms were reported explaining the regulation of the expression and activity of this important type of genes. Because of the constitutive nature of many resistance genes expression, fitness costs translating into reduction of growth and productivity are significant. The regulation of *R* genes by miRNAs could have evolved as an alternative strategy for tight and cost effective regulation.

On the other hand, to achieve successful colonization, symbiotic microbes must be able to block plant immune responses triggered by non-self recognition. An expected strategy would be trough the control of plant immunity master regulators, such as *R* genes-encoded proteins and other immunity receptors. As a matter of fact, miR482 was reported to be induced during the establishment of symbiosis between soybean and *Bradyrhizobium japonicum*. Interestingly, bioinformatically-predicted targets of miR482 include various *R* genes of which two were validated experimentally. In addition, a considerable increase in the number of mature nodules was observed upon accumulation of miR482 conditionally expressed in roots under a *Rhizobium*-responsive promoter [96]. In *M. truncatula* challenged with *Shinorhizobium meliloti*, 14 targets predicted for 9 Mtr-miRNA candidates correspond to NBS-LRR coding genes [94]. Also, a high proportion of targets identified in a degradome library generated from *M. truncatula* plants infected with *Glomus intraradices* include *R* genes (27 genes) and transcription factors (33 genes). In particular it was established that miR1510a*, miR1507, miR2678 and miR5213 regulate the expression of a subset of *R* genes [84]. More recently a deep sequencing analysis of 21 sRNAs libraries generated from four legumes (*M. truncatula*, soybean, peanut and common bean) led to the identification of several phased siRNAs (potentially ta-siRNAs), most of them targeting NBS-LRR encoding genes. These findings were expanded to potato based on bioinformatic analysis [97]. Although none of the phased siRNAs were validated by alternative experiments, the deep sequencing and the high number of libraries support well these data.

Besides disease resistance-related genes, other genes involved in defense pathways signaling are regulated by miRNAs. In the above-mentioned study of *Pinus taeda* infected with the fusiform rust, potential targets of a few isolated miRNAs include transcripts encoding a MYB transcription factor (pta-miR159), laccase-like genes, (miR397), peroxidases (miR420) and glutation S-transferase (GST). All these genes play a role in plant responses to pathogens, notably in gene regulation and control of ROS production [86]. Consistently, it has been shown that some of these targets are also regulated by miRNAs to avoid their expression during symbiosis. miR5282 and miRc_275 were induced specifically in mycorrhizal roots. These miRNAs both target *MtGst1* which encodes for a GST [84].

In conclusion it appears obvious that a successful symbiosis requires suppression of host defenses. Altogether these reports stress that *R* genes are set under a multilayered and complex regulation network during interactions with microorganisms, meant to allow the establishment of beneficial interactions in favorable conditions and avoid in the mean time the invasion of pathogens.

11. Novel and specific miRNAs in beneficial interactions

sRNA studies for beneficial interactions have focused mainly on the study of legumes, and had benefited from the identification of sRNA loci in model legumes as *Lotus*, *Medicago*, *Glycine* and *Phaseolus* [98-100]. Some studies focused on the expression of specifically-irduced or repressed miRNAs during symbiosis at early [83] or late stages of the infection [72, 94]. For example, during the infection of soybean with *Bradyrhizobium japonicum*, miR168 and miR172 were induced during the first 3 hours but were gradually down regulated to reach basal levels at 12 hours. In contrast, the induction of miR159 and miR393 was sustained along the 12 hours, whereas miR160 and miR169 were down-regulated [83]. Interestingly, these studies allowed the identification of apparently specific miRNAs present or expressed only in plants able to form AM or in the symbiotic structures, respectively. Among the soybean mRNAs identified during interaction with *B. japonicum*, miR1507 seems to be legume-specific whereas miR1512, miR1515 and miR1521 were only reported in soybean [96].

miRNA expression was studied in soybean mutants nod49 (mutant for a Nod factor receptor *NFR1*) and nts382 (mutant for Nodule Autoregulation Receptor Kinase NARK) which are a non-nodulation and supernodulation mutants, respectively, as a result the expression of legume-specific miR1507, miR1511 and miR1512 was compromised in both mutants [96]. Another interesting and apparently specific symbiotic miRNA is miR5229a/g, which was identified in mycorrhizal roots of *M. truncatula* plants infected with *Glomus intraradices* [84]. MiR5229a/g which is the most strongly induced, was found by in situ hybridization to be exclusively expressed in arbuscules-containing cells in the root cortex, albeit with different signal intensities indicative of a specific function during different stages of the arbuscule development [84]. miR167 was localized in the differentiating peripheral vascular bundles and the novel miRNAs miR2586 and Mtr-s107 accumulate in the nodule meristem, leading the authors to conclude that miRNAs accumulate mainly in undifferentiated cells [94].

As stated before, rhizobacteria and AM share several elements of their symbiosis pathway [6] and miRNAs-mediated regulation is not an exception. miR169 which targets the CCAAT-binding transcription factor MtHAP2-1, was identified in the symbiotic interaction occuring between *Glycine max* and *B. japonicum* [83]. This transcription factor is highly induced during symbiosis and its degradation is mediated by miR169, causing a delayed nodule development and subsequent inability to fix N_2 [101]. This miRNA was also found to be up-regulated in *Medicago* interacting with AM, accumulating in the phloem and around fungal hyphae [84]. Another crosstalk must be established between symbiosis and nutrition pathways to determinate if the colonization of microbes occurs or not. An illustration of this comes from miR167 and miR5204 which are up-regulated in *Medicago* mycorrhizal roots under low phosphate conditions, as compared to nonmycorrhizal roots [84]. However these miRNAs were also regulated by phosphate, pointing to a direct connection between nutrition and symbiosis. Previous studies demonstrated an induction of miR399 under Pi depleted conditions in mycorrhizal *M. truncatula* and tobacco plants associated with a concomitant increase in Pi content [102]

12. Conclusions and perspectives

Research in sncRNAs is ultimately one of the most active and promising fields in plant biology, and it is expected to grow even more in importance in the near future, however many aspects of sncRNAs functions during plant-microbe interactions still remain unclear. How are these sncRNAs regulated? Are there common regulatory and feedback regulatory circuits between the different classes of sncRNAs? Are there core sncRNAs and targets for different class of pathogens and for different plant species? What is the evolutionary history of these different families of sncRNAs and how did they shape plant evolution? How did differences in sncRNA regulation across the plant kingdom arise? How are new sncRNA-specificities generated and how variable can these molecules be within species or populations? Studies reviewed here highlight the importance of sncRNAs in gene regulation in response of plants to pathogens as diverse as viruses, bacteria and fungi. Some sncRNAs have been shown to be induced or repressed in response to these diverse pathogens during incompatible and compatible interactions indicating a dual role of these RNAs as positive and negative regulators of plant immunity. This fact demonstrates the complex network of gene expression during plant-microbe interactions and should be considered in biotechnological programs focused to enhance the crop resistance to plant diseases. The notable repression of *R* genes during symbiotic interactions stresses the importance of these molecules during plant-microbe interactions and provides a bridge between pathogenic and beneficial interactions. The role that effectors have and its interaction with the plant silencing machinery reveals also the amazing and surprising mechanism that pathogens have evolved to surpass the plant immunity mechanisms. Deepening on all this knowledge surely will open new ways to improve resistance against biotic stress in several plants including crops of economical importance.

Author details

Camilo López and Álvaro L. Perez-Quintero

Universidad Nacional de Colombia, Bogotá, Departamento de Biología, Bogota D.C.,Colombia

Boris Szurek

UMR 186 IRD-UM2-Cirad, Résistance des Plantes aux Bioagresseurs (RPB), Institut de Recherche pour le Développement, Montpellier Cedex 5, France

13. References

[1] Jones-Rhoades MW, Bartel DP, Bartel B. MicroRNAS and their regulatory roles in plants. Annual review of plant biology. 2006; 57: 19-53.

[2] Chisholm ST, Coaker G, Day B, Staskawicz BJ. Host-microbe interactions: shaping the evolution of the plant immune response. Cell. 2006; 124(4): 803-14.

[3] Boller T, He SY. Innate immunity in plants: an arms race between pattern recognition receptors in plants and effectors in microbial pathogens. Science. 2009; 324(5928): 742-4.

[4] Mackey D, Holt BF, 3rd, Wiig A, Dangl JL. RIN4 interacts with Pseudomonas syringae type III effector molecules and is required for RPM1-mediated resistance in Arabidopsis. Cell. 2002; 108(6): 743-54.

[5] Tsuda K, Katagiri F. Comparing signaling mechanisms engaged in pattern-triggered and effector-triggered immunity. Current opinion in plant biology. 2010; 13(4): 459-65.

[6] Zamioudis C, Pieterse CM. Modulation of host immunity by beneficial microbes. Molecular plant-microbe interactions : MPMI. 2012; 25(2): 139-50.

[7] Popp C, Ott T. Regulation of signal transduction and bacterial infection during root nodule symbiosis. Current opinion in plant biology. 2011; 14(4): 458-67.

[8] Bonfante P, Requena N. Dating in the dark: how roots respond to fungal signals to establish arbuscular mycorrhizal symbiosis. Current opinion in plant biology. 2011; 14(4): 451-7.

[9] Napoli C, Lemieux C, Jorgensen R. Introduction of a Chimeric Chalcone Synthase Gene into Petunia Results in Reversible Co-Suppression of Homologous Genes in trans. The Plant cell. 1990; 2(4): 279-89.

[10] van der Krol AR, Mur LA, Beld M, Mol JN, Stuitje AR. Flavonoid genes in petunia: addition of a limited number of gene copies may lead to a suppression of gene expression. The Plant cell. 1990; 2(4): 291-9.

[11] Ramachandran V, Chen X. Degradation of microRNAs by a family of exoribonucleases in Arabidopsis. Science. 2008; 321(5895): 1490-2.

[12] Bartel DP. MicroRNAs: genomics, biogenesis, mechanism, and function. Cell. 2004; 116(2): 281-97.

[13] Zhu JK. Reconstituting plant miRNA biogenesis. Proceedings of the National Academy of Sciences of the United States of America. 2008; 105(29): 9851-2.

[14] Jamalkandi SA, Masoudi-Nejad A. Reconstruction of Arabidopsis thaliana fully integrated small RNA pathway. Functional & integrative genomics. 2009; 9(4): 419-32.

[15] Filipowicz W, Jaskiewicz L, Kolb FA, Pillai RS. Post-transcriptional gene silencing by siRNAs and miRNAs. Current opinion in structural biology. 2005; 15(3): 331-41.

[16] Llave C. Virus-derived small interfering RNAs at the core of plant-virus interactions. Trends in plant science. 2010; 15(12): 701-7.

[17] Allen E, Xie Z, Gustafson AM, Carrington JC. microRNA-directed phasing during trans-acting siRNA biogenesis in plants. Cell. 2005; 121(2): 207-21.

[18] Yoshikawa M, Peragine A, Park MY, Poethig RS. A pathway for the biogenesis of trans-acting siRNAs in Arabidopsis. Genes & development. 2005; 19(18): 2164-75.

[19] Guleria P, Mahajan M, Bhardwaj J, Yadav SK. Plant Small RNAs: Biogenesis, Mode of Action and Their Roles in Abiotic Stresses. Genomics, Proteomics & Bioinformatics. 2011; 9(6): 183-99.

[20] Borsani O, Zhu J, Verslues PE, Sunkar R, Zhu JK. Endogenous siRNAs derived from a pair of natural cis-antisense transcripts regulate salt tolerance in Arabidopsis. Cell. 2005; 123(7): 1279-91.

[21] Katiyar-Agarwal S, Gao S, Vivian-Smith A, Jin H. A novel class of bacteria-induced small RNAs in Arabidopsis. Genes & development. 2007; 21(23): 3123-34.

[22] Garcia-Ruiz H, Takeda A, Chapman EJ, Sullivan CM, Fahlgren N, Brempelis KJ, Carrington JC. Arabidopsis RNA-dependent RNA polymerases and dicer-like proteins in antiviral defense and small interfering RNA biogenesis during Turnip Mosaic Virus infection. The Plant cell. 2010; 22(2): 481-96.

[23] Voinnet O. Use, tolerance and avoidance of amplified RNA silencing by plants. Trends in plant science. 2008; 13(7): 317-28.

[24] Palauqui JC, Balzergue S. Activation of systemic acquired silencing by localised introduction of DNA. Current biology : CB. 1999; 9(2): 59-66.

[25] Molnar A, Melnyk CW, Bassett A, Hardcastle TJ, Dunn R, Baulcombe DC. Small silencing RNAs in plants are mobile and direct epigenetic modification in recipient cells. Science. 2010; 328(5980): 872-5.

[26] Voinnet O, Pinto YM, Baulcombe DC. Suppression of gene silencing: a general strategy used by diverse DNA and RNA viruses of plants. Proceedings of the National Academy of Sciences of the United States of America. 1999; 96(24): 14147-52.

[27] Dunoyer P, Lecellier CH, Parizotto EA, Himber C, Voinnet O. Probing the microRNA and small interfering RNA pathways with virus-encoded suppressors of RNA silencing. The Plant cell. 2004; 16(5): 1235-50.

[28] Lakatos L, Csorba T, Pantaleo V, Chapman EJ, Carrington JC, Liu YP, Lopez-Moya JJ, Burgyan J. Small RNA binding is a common strategy to suppress RNA silencing by several viral suppressors. The EMBO journal. 2006; 25(12): 2768-80.

[29] Thomas CL, Leh V, Lederer C, Maule AJ. Turnip crinkle virus coat protein mediates suppression of RNA silencing in Nicotiana benthamiana. Virology. 2003; 306(1): 33-41.

[30] Diaz-Pendon JA, Li F, Li WX, Ding SW. Suppression of antiviral silencing by cucumber mosaic virus 2b protein in Arabidopsis is associated with drastically reduced accumulation of three classes of viral small interfering RNAs. The Plant cell. 2007; 19(6): 2053-63.

[31] Chapman EJ, Prokhnevsky AI, Gopinath K, Dolja VV, Carrington JC. Viral RNA silencing suppressors inhibit the microRNA pathway at an intermediate step. Genes & development. 2004; 18(10): 1179-86.

[32] Omarov R, Sparks K, Smith L, Zindovic J, Scholthof HB. Biological relevance of a stable biochemical interaction between the tombusvirus-encoded P19 and short interfering RNAs. Journal of virology. 2006; 80(6): 3000-8.

[33] Lozsa R, Csorba T, Lakatos L, Burgyan J. Inhibition of 3' modification of small RNAs in virus-infected plants require spatial and temporal co-expression of small RNAs and viral silencing-suppressor proteins. Nucleic acids research. 2008; 36(12): 4099-107.

[34] Yu B, Chapman EJ, Yang Z, Carrington JC, Chen X. Transgenically expressed viral RNA silencing suppressors interfere with microRNA methylation in Arabidopsis. FEBS letters. 2006; 580(13): 3117-20.

[35] Bortolamiol D, Pazhouhandeh M, Marrocco K, Genschik P, Ziegler-Graff V. The Polerovirus F box protein P0 targets ARGONAUTE1 to suppress RNA silencing. Current biology : CB. 2007; 17(18): 1615-21.

[36] Fraile A, Garcia-Arenal F. The coevolution of plants and viruses: resistance and pathogenicity. Advances in virus research. 2010; 76: 1-32.

[37] Whitham S, Dinesh-Kumar SP, Choi D, Hehl R, Corr C, Baker B. The product of the tobacco mosaic virus resistance gene N: similarity to toll and the interleukin-1 receptor. Cell. 1994; 78(6): 1101-15.

[38] Vidal S, Cabrera H, Andersson RA, Fredriksson A, Valkonen JP. Potato gene Y-1 is an N gene homolog that confers cell death upon infection with potato virus Y. Molecular plant-microbe interactions : MPMI. 2002; 15(7): 717-27.

[39] Seo YS, Rojas MR, Lee JY, Lee SW, Jeon JS, Ronald P, Lucas WJ, Gilbertson RL. A viral resistance gene from common bean functions across plant families and is up-regulated in a non-virus-specific manner. Proceedings of the National Academy of Sciences of the United States of America. 2006; 103(32): 11856-61.

[40] Bendahmane A, Kanyuka K, Baulcombe DC. The Rx gene from potato controls separate virus resistance and cell death responses. The Plant cell. 1999; 11(5): 781-92.

[41] Cooley MB, Pathirana S, Wu HJ, Kachroo P, Klessig DF. Members of the Arabidopsis HRT/RPP8 family of resistance genes confer resistance to both viral and oomycete pathogens. The Plant cell. 2000; 12(5): 663-76.

[42] Takahashi H, Miller J, Nozaki Y, Takeda M, Shah J, Hase S, Ikegami M, Ehara Y, Dinesh-Kumar SP. RCY1, an Arabidopsis thaliana RPP8/HRT family resistance gene, conferring resistance to cucumber mosaic virus requires salicylic acid, ethylene and a novel signal transduction mechanism. The Plant journal : for cell and molecular biology. 2002; 32(5): 655-67.

[43] Palukaitis P, Carr JP, Schoelz JE. Plant-virus interactions. Methods Mol Biol. 2008; 451: 3-19.

[44] Marathe R, Guan Z, Anandalakshmi R, Zhao H, Dinesh-Kumar SP. Study of Arabidopsis thaliana resistome in response to cucumber mosaic virus infection using whole genome microarray. Plant molecular biology. 2004; 55(4): 501-20.

[45] Espinoza C, Vega A, Medina C, Schlauch K, Cramer G, Arce-Johnson P. Gene expression associated with compatible viral diseases in grapevine cultivars. Functional & integrative genomics. 2007; 7(2): 95-110.

[46] Uzarowska A, Dionisio G, Sarholz B, Piepho HP, Xu M, Ingvardsen CR, Wenzel G, Lubberstedt T. Validation of candidate genes putatively associated with resistance to SCMV and MDMV in maize (Zea mays L.) by expression profiling. BMC plant biology 2009; 9: 15.

[47] Gonzalez-Ibeas D, Canizares J, Aranda MA. Microarray analysis shows that recessive resistance to Watermelon mosaic virus in melon is associated with the induction of defense response genes. Molecular plant-microbe interactions : MPMI. 2012; 25(1): 107-18.

[48] Bazzini AA, Hopp HE, Beachy RN, Asurmendi S. Infection and coaccumulation of tobacco mosaic virus proteins alter microRNA levels, correlating with symptom and plant development. Proceedings of the National Academy of Sciences of the United States of America. 2007; 104(29): 12157-62.

[49] Csorba T, Bovi A, Dalmay T, Burgyan J. The p122 subunit of Tobacco Mosaic Virus replicase is a potent silencing suppressor and compromises both small interfering RNA- and microRNA-mediated pathways. Journal of virology. 2007; 81(21): 11768-80.

[50] Feng J, Wang K, Liu X, Chen S, Chen J. The quantification of tomato microRNAs response to viral infection by stem-loop real-time RT-PCR. Gene. 2009; 437(1-2): 14-21.

[51] Naqvi AR, Haq QM, Mukherjee SK. MicroRNA profiling of tomato leaf curl New Delhi virus (tolcndv) infected tomato leaves indicates that deregulation of mir159/319 and mir172 might be linked with leaf curl disease. Virology journal. 2010; 7: 281.

[52] Pacheco R, Garcia-Marcos A, Barajas D, Martianez J, Tenllado F. PVX-potyvirus synergistic infections differentially alter microRNA accumulation in Nicotiana benthamiana. Virus research. 2012; 165(2): 231-5.

[53] He XF, Fang YY, Feng L, Guo HS. Characterization of conserved and novel microRNAs and their targets, including a TuMV-induced TIR-NBS-LRR class R gene-derived novel miRNA in Brassica. FEBS letters. 2008; 582(16): 2445-52.

[54] Bazzini AA, Almasia NI, Manacorda CA, Mongelli VC, Conti G, Maroniche GA, Rodriguez MC, Distefano AJ, Hopp HE, del Vas M, Asurmendi S. Virus infection elevates transcriptional activity of miR164a promoter in plants. BMC plant biology. 2009; 9: 152.

[55] Havelda Z, Varallyay E, Valoczi A, Burgyan J. Plant virus infection-induced persistent host gene downregulation in systemically infected leaves. The Plant journal : for cell and molecular biology. 2008; 55(2): 278-88.

[56] Varallyay E, Valoczi A, Agyi A, Burgyan J, Havelda Z. Plant virus-mediated induction of miR168 is associated with repression of ARGONAUTE1 accumulation. The EMBO journal. 2010; 29(20): 3507-19.

[57] Llave C. MicroRNAs: more than a role in plant development? Molecular plant pathology. 2004; 5(4): 361-6.

[58] Lecellier CH, Dunoyer P, Arar K, Lehmann-Che J, Eyquem S, Himber C, Saib A, Voinnet O. A cellular microRNA mediates antiviral defense in human cells. Science. 2005; 308(5721): 557-60.

[59] Niu QW, Lin SS, Reyes JL, Chen KC, Wu HW, Yeh SD, Chua NH. Expression of artificial microRNAs in transgenic Arabidopsis thaliana confers virus resistance. Nature biotechnology. 2006; 24(11): 1420-8.

[60] Qu J, Ye J, Fang R. Artificial microRNA-mediated virus resistance in plants. Journal of virology. 2007; 81(12): 6690-9.

[61] Duan CG, Wang CH, Fang RX, Guo HS. Artificial MicroRNAs highly accessible to targets confer efficient virus resistance in plants. Journal of virology. 2008; 82(22): 11084-95.

[62] Lin SS, Wu HW, Elena SF, Chen KC, Niu QW, Yeh SD, Chen CC, Chua NH. Molecular evolution of a viral non-coding sequence under the selective pressure of amiRNA-mediated silencing. PLoS pathogens. 2009; 5(2): e1000312.

[63] Fahim M, Millar AA, Wood CC, Larkin PJ. Resistance to Wheat streak mosaic virus generated by expression of an artificial polycistronic microRNA in wheat. Plant biotechnology journal. 2012; 10(2): 150-63.

[64] Jelly NS, Schellenbaum P, Walter B, Maillot P. Transient expression of artificial microRNAs targeting Grapevine fanleaf virus and evidence for RNA silencing in grapevine somatic embryos. Transgenic research. 2012.

[65] Ai T, Zhang L, Gao Z, Zhu CX, Guo X. Highly efficient virus resistance mediated by artificial microRNAs that target the suppressor of PVX and PVY in plants. Plant Biol (Stuttg). 2011; 13(2): 304-16.

[66] Zhang X, Li H, Zhang J, Zhang C, Gong P, Ziaf K, Xiao F, Ye Z. Expression of artificial microRNAs in tomato confers efficient and stable virus resistance in a cell-autonomous manner. Transgenic research. 2011; 20(3): 569-81.

[67] Perez-Quintero AL, Neme R, Zapata A, Lopez C. Plant microRNAs and their role in defense against viruses: a bioinformatics approach. BMC plant biology. 2010; 10: 138.

[68] Naqvi AR, Choudhury NR, Mukherjee SK, Haq QM. In silico analysis reveals that several tomato microRNA/microRNA* sequences exhibit propensity to bind to tomato leaf curl virus (ToLCV) associated genomes and most of their encoded open reading frames (ORFs). Plant physiology and biochemistry : PPB / Societe francaise de physiologie vegetale. 2011; 49(1): 13-7.

[69] Boss IW, Renne R. Viral miRNAs: tools for immune evasion. Current opinion in microbiology. 2010; 13(4): 540-5.

[70] Moissiard G, Voinnet O. RNA silencing of host transcripts by cauliflower mosaic virus requires coordinated action of the four Arabidopsis Dicer-like proteins. Proceedings of the National Academy of Sciences of the United States of America. 2006; 103(51): 19593-8.

[71] Blevins T, Rajeswaran R, Aregger M, Borah BK, Schepetilnikov M, Baerlocher L, Farinelli L, Meins F, Jr., Hohn T, Pooggin MM. Massive production of small RNAs from

a non-coding region of Cauliflower mosaic virus in plant defense and viral counter-defense. Nucleic acids research. 2011; 39(12): 5003-14.

[72] Wang Y, Li P, Cao X, Wang X, Zhang A, Li X. Identification and expression analysis of miRNAs from nitrogen-fixing soybean nodules. Biochemical and biophysical research communications. 2009; 378(4): 799-803.

[73] Grant MR, Jones JD. Hormone (dis)harmony moulds plant health and disease. Science. 2009; 324(5928): 750-2.

[74] Navarro L, Dunoyer P, Jay F, Arnold B, Dharmasiri N, Estelle M, Voinnet O, Jones JD. A plant miRNA contributes to antibacterial resistance by repressing auxin signaling. Science. 2006; 312(5772): 436-9.

[75] Fahlgren N, Howell MD, Kasschau KD, Chapman EJ, Sullivan CM, Cumbie JS, Givan SA, Law TF, Grant SR, Dangl JL, Carrington JC. High-throughput sequencing of Arabidopsis microRNAs: evidence for frequent birth and death of MIRNA genes. PloS one. 2007; 2(2): e219.

[76] Zhang W, Gao S, Zhou X, Chellappan P, Chen Z, Zhang X, Fromuth N, Coutino G, Coffey M, Jin H. Bacteria-responsive microRNAs regulate plant innate immunity by modulating plant hormone networks. Plant molecular biology. 2011; 75(1-2): 93-105.

[77] Fahlgren N, Jogdeo S, Kasschau KD, Sullivan CM, Chapman EJ, Laubinger S, Smith LM, Dasenko M, Givan SA, Weigel D, Carrington JC. MicroRNA gene evolution in Arabidopsis lyrata and Arabidopsis thaliana. The Plant cell. 2010; 22(4): 1074-89.

[78] Li Y, Zhang Q, Zhang J, Wu L, Qi Y, Zhou JM. Identification of microRNAs involved in pathogen-associated molecular pattern-triggered plant innate immunity. Plant physiology. 2010; 152(4): 2222-31.

[79] Jay F, Renou JP, Voinnet O, Navarro L. Biotic stress-associated microRNAs: identification, detection, regulation, and functional analysis. Methods Mol Biol. 2010; 592: 183-202.

[80] Jin H. Endogenous small RNAs and antibacterial immunity in plants. FEBS letters. 2008; 582(18): 2679-84.

[81] Perez-Quintero AL, Quintero A, Urrego O, Vanegas P, Lopez C. Bioinformatic identification of cassava miRNAs differentially expressed in response to infection by Xanthomonas axonopodis pv. manihotis. BMC plant biology. 2012; 12: 29.

[82] Pruss GJ, Nester EW, Vance V. Infiltration with Agrobacterium tumefaciens induces host defense and development-dependent responses in the infiltrated zone. Molecular plant-microbe interactions : MPMI. 2008; 21(12): 1528-38.

[83] Subramanian S, Fu Y, Sunkar R, Barbazuk WB, Zhu JK, Yu O. Novel and nodulation-regulated microRNAs in soybean roots. BMC genomics. 2008; 9: 160.

[84] Devers EA, Branscheid A, May P, Krajinski F. Stars and symbiosis: microRNA- and microRNA*-mediated transcript cleavage involved in arbuscular mycorrhizal symbiosis. Plant physiology. 2011; 156(4): 1990-2010.

[85] Chen L, Ren Y, Zhang Y, Xu J, Zhang Z, Wang Y. Genome-wide profiling of novel and conserved Populus microRNAs involved in pathogen stress response by deep sequencing. Planta. 2012; 235(5): 873-83.

[86] Lu S, Sun YH, Amerson H, Chiang VL. MicroRNAs in loblolly pine (Pinus taeda L.) and their association with fusiform rust gall development. The Plant journal : for cell and molecular biology. 2007; 51(6): 1077-98.

[87] Xin M, Wang Y, Yao Y, Xie C, Peng H, Ni Z, Sun Q. Diverse set of microRNAs are responsive to powdery mildew infection and heat stress in wheat (Triticum aestivum L.). BMC plant biology. 2010; 10: 123.

[88] Ellendorff U, Fradin EF, de Jonge R, Thomma BP. RNA silencing is required for Arabidopsis defence against Verticillium wilt disease. Journal of experimental botany. 2009; 60(2): 591-602.

[89] Navarro L, Jay F, Nomura K, He SY, Voinnet O. Suppression of the microRNA pathway by bacterial effector proteins. Science. 2008; 321(5891): 964-7.

[90] Katiyar-Agarwal S, Morgan R, Dahlbeck D, Borsani O, Villegas A, Jr., Zhu JK, Staskawicz BJ, Jin H. A pathogen-inducible endogenous siRNA in plant immunity. Proceedings of the National Academy of Sciences of the United States of America. 2006; 103(47): 18002-7.

[91] Padmanabhan C, Zhang X, Jin H. Host small RNAs are big contributors to plant innate immunity. Current opinion in plant biology. 2009; 12(4): 465-72.

[92] Takken FL, Albrecht M, Tameling WI. Resistance proteins: molecular switches of plant defence. Current opinion in plant biology. 2006; 9(4): 383-90.

[93] Yi H, Richards EJ. A cluster of disease resistance genes in Arabidopsis is coordinately regulated by transcriptional activation and RNA silencing. The Plant cell. 2007; 19(9): 2929-39.

[94] Lelandais-Briere C, Naya L, Sallet E, Calenge F, Frugier F, Hartmann C, Gouzy J, Crespi M. Genome-wide Medicago truncatula small RNA analysis revealed novel microRNAs and isoforms differentially regulated in roots and nodules. The Plant cell. 2009; 21(9): 2780-96.

[95] Li F, Pignatta D, Bendix C, Brunkard JO, Cohn MM, Tung J, Sun H, Kumar P, Baker B. MicroRNA regulation of plant innate immune receptors. Proceedings of the National Academy of Sciences of the United States of America. 2012; 109(5): 1790-5.

[96] Li H, Deng Y, Wu T, Subramanian S, Yu O. Misexpression of miR482, miR1512, and miR1515 increases soybean nodulation. Plant physiology. 2010; 153(4): 1759-70.

[97] Zhai J, Jeong DH, De Paoli E, Park S, Rosen BD, Li Y, Gonzalez AJ, Yan Z, Kitto SL, Grusak MA. MicroRNAs as master regulators of the plant NB-LRR defense gene family via the production of phased, trans-acting siRNAs. Genes & development. 2011; 25(23): 2540-53.

[98] Sunkar R, Jagadeeswaran G. In silico identification of conserved microRNAs in large number of diverse plant species. BMC plant biology. 2008; 8: 37.

[99] Szittya G, Moxon S, Santos DM, Jing R, Fevereiro MP, Moulton V, Dalmay T. High-throughput sequencing of Medicago truncatula short RNAs identifies eight new miRNA families. BMC genomics. 2008; 9: 593.

[100] Arenas-Huertero C, Perez B, Rabanal F, Blanco-Melo D, De la Rosa C, Estrada-Navarrete G, Sanchez F, Covarrubias AA, Reyes JL. Conserved and novel miRNAs in

the legume Phaseolus vulgaris in response to stress. Plant molecular biology. 2009; 70(4): 385-401.

[101] Combier JP, Frugier F, de Billy F, Boualem A, El-Yahyaoui F, Moreau S, Vernie T, Ott T, Gamas P, Crespi M, Niebel A. MtHAP2-1 is a key transcriptional regulator of symbiotic nodule development regulated by microRNA169 in Medicago truncatula. Genes & development. 2006; 20(22): 3084-8.

[102] Branscheid A, Sieh D, Pant BD, May P, Devers EA, Elkrog A, Schauser L, Scheible WR, Krajinski F. Expression pattern suggests a role of MiR399 in the regulation of the cellular response to local Pi increase during arbuscular mycorrhizal symbiosis. Molecular plant-microbe interactions : MPMI. 2010; 23(7): 915-26.

Chapter 8

Micropropagation of *Anthurium spp.*

Çimen Atak and Özge Çelik

Additional information is available at the end of the chapter

1. Introduction

Micropropagation as an alternative method to conventional propagation, the culture of somatic cells, tissues and organs of plants under controlled conditions is a suitable way to produce a large number of progeny plants which are genetically identical to the stock plant in a short time. The important property of the plant cells is totipotency which is a capacity to produce the whole plant from different plant parts. Micropropagation has some features to be chosen in commercial production such as multiplicative capacity in a relatively short time, healthy and disease-free production capacity and ability to generate population during a year [1-5].

The genetic pattern of the plant is key element to select the propagation method. Using micropropagation techniques in plant biotechnology applications are costlier than conventional propagation methods. Propagation by using *in vitro* techniques instead of conventional methods offer some advantages like utilizing small pieces of plants called as explants to maintain the whole plant and increase their number. The main point is to evolve new strategies to lower the time and cost consumed per plant [2,3]. In tissue culture applications selection of initiating material is important in the beginning of the culture. Therefore it is easy to provide virus-free clones in a short time. Production of plants during all year long independent of seasonal changes, long storage periods make micropropagation preferable to propagate plants in short time. There are also some disadvantages of micropropagation. Adaptation of cultured plants to the environmental conditions need transitional period to allow the plants to produce organic matter by photosynthesis [2,4].

The main methods of *in vitro* propagation can be classified in two groups:

1. Propagation from axillary or terminal buds
2. Propagation by the formation of adventitious shoots or adventitious somatic embryos

The meristem and shoot tip cultures are used to establish virus-free plant culture. Many important horticulture crops were propagated by meristem culture for rapid growth and virus elimination. Adventitious shoots or adventitious somatic embryos are established directly or indirectly. Cultures are directly started with the excised explants from the mother plant tissues for organogenesis or embryogenesis. If shoots or embryos regenerate on previously formed callus or in cell culture, they are called as indirect organogenesis or embryogenesis [3,6,7].

When propagation occurs via an indirect callus phase, the genetic identity of the progenies decreases. This is an important problem in commercial propagation to affect the uniformity of progenies. Callus formation also increases the somaclonal variation. Increasing of somaclonal variation incidence is a crucial result of long term period of callus growth. Origin of the callus also causes somaclonal variation.

Propagation from axillary or terminal buds is the most ensurable method to have the highest genetic stability during *in vitro* propagation of plants.

George et al. [2008] described five stages of micropopagation which are mother plant selection and preparation [Stage 0], *in vitro* culture establishment [Stage 1], shoot multiplication [Stage 2], rooting of microshoots [Stage 3] and acclimatization [Stage 4]. These stages are necessary for a successful micropropagation.

Establishing aseptic culture conditions can be classified as Stage 0 which contains pre-surface sterilization applications of explants to reduce contamination of stock plants. The success of Stage 2 depends on different factors such as plant species, cultivar or genotype, plant growth regulators, the ingredients of the medium and physical culture conditions. Stage 3 is responsible of rooting of microshoots. It depends on the factors given in Stage 2. Transplantation of rooted shoots to the environment is the main step of Stage 4. This is also the important part of micropropagation. Acclimatization needs to be well controlled to avoid loss of propagated plants [4,5].

2. Propagation of *Anthurium*

The commercial production of ornamental pot plants has a great potential in international markets. In the global market, *Anthurium* cultivars with valued flowers are the second beside the Orchids among tropical cut flowers. *Anthurium* species and hybrids in Araceae family have an importance in monocotyledonous ornamental plants and they are commercially produced as cut flowers and potted plants in tropical and subtropical countries [2, 8-12].

The propagation rate of *Anthurium* by seeds is very low and it is not recommended. The cultivation has also been limited because of the inherent heterozygosity. The time between pollination and seed maturity and the development time take three years in a breeding program. To grow plants from seed may not provide a practical method of making new planting areas, in such circumstances vegetative propagation [stem cutting] seems the only way of multiplying a unique individual. Propagation method selection for a plant depends on its genetic potential and its intended use. Stem cutting methods are also not practical to propagate in large scale. Today *Anthurium* can be multiplied in large number by micropropagation. Application of biotechnology on *in vitro* propagation of *Anthurium* is important to increase the productivity of *Anthurium* [3, 6, 13].

2.1. Tissue culture of *Anthurium*

In vitro propagation methods have several advantages over conventional propagation like flexible adjustment of factors affecting regeneration such as explant type, nutrient and plant growth regulator levels and conditions of the environment, production of clones in desired rate, continued production during seasonal changes using tissue culture methods also increase the multiplication rate of plants [14].

Explant type

The success of tissue culture is related to the correct choice of explants. Shoot or shoot tips and node cultures are the most commonly used culture types in micropropagation of plants. Explants from shoot tips and nodal stem segments are suitable for enhanced axillary branching. *Anthurium* micropropagation from axillary buds, shoot tip, lamina explants, node, petiole, and microcuttings have been successfully utilized [15-18]. Among these plant parts, leaves are the most used explant source in *in vitro* culture of *Anthurium*.

The genotype of *Anthurium* plays an important role in *in vitro* propagation. The studies showed that different genotypes had different responses to the same culture conditions. For this reason, it is necessary to establish a suitable procedure for each varieties of *Anthurium* that can be adapted to commercial production [3,4,19,20].

Selection of explant type to induce callogenesis and orgonogenesis is important for plants. In direct and indirect orgonogenesis studies, using young leaf explants are important for the success of culture. Martin et al.[2003] observed higher number of shoots in the brown young lamina explants than young green lamina. Viégas et al.[2007] also indicated the importance of using new brown leaves for callus induction. Bejoy et al.[2008] reported that the explants excised from pale green leaves showed better callus development than pale brown leaves. Atak and Çelik [2009] also used young brown and green leaves of *Anthurium andreanum* to evaluate the effectiveness of callus formations. They achieved to decrease the callus formation time by using brown leaf explants and induced the callus formation percentage 50% more than performed by green leaves.

Establishing aseptic culture

The second important step in micropropagation is to obtain aseptic culture of plant material. Aseptic culture systems are effective to eradicate the bacterial, fungal and insect contaminants. The sterilization protocols used for different *Anthurium* explant sources were given in Table 1. NaOCl is the main disinfection material used in establishing aseptic culture conditions of *Anthurium*. NaOCl has been used for the concentrations differ from 1%-5% [Table 1]. The incubation times of the explants in sodium hypochloride showed differences due to its concentrations. There is also need to used extra disinfectant solutions to eradicate the fungal and bacterial contaminants. Benomyl [commercial name is Benlate], Cetrimite, gentamicin and streptomycin sulphate are effectively used for this aim [11,13,15,18,20,22].

A. species	Explant Source	Sterilization method	Reference
A.andreanum	Leaf	0.6% Benlate [30 min]+70% ethanol [30 sec]+1.5% NaOCl containing two drops of Tween 20 [20 min]	[15]
	Leaf	0.1% HgCl$_2$	[19]
	Leaf	70% ethanol [1 min]+gentamicin [30 min]+20% [v/v] commercial bleach [5% NaOCl,12 min]	[20]
A.andreanum L.	Apical shoot buds	Teepol+ antifungal solution Cetrimite [5min.] +NaOCl [5 min]+ 0.1% [w/v] HgCl$_2$ [5min.]	[11]
	Spadices	Washing under running tap water [30-60 min]+1% pesticide solution of 50% benomyl and 20% streptomycin sulphate [30 min] +5 times distilled water [5min each rinse.]+ 1% NaOCl [10 min] + 2% NaOCl [5min] +80% alcohol [30s.] +5-6 times distilled water [5min each rinse.]	[22]
	Leaf and spadix Segments	Washing under running tap water [30 min]+0.5% [v/v]Trix [Commercial detergent]+70% ethanol [1 min.] + 1.5% NaOCl containing 0.01% Tween 20 [8 min]	[23]
A.andreanum cv Rubrun	Seeds from plant spadix	1%NaOCl [20 min]	[7]
	Separate fruits from spadix Isolate seeds	3% [v/v]NaOCl [15 min]+3 times distilled water [5min.] 1% [v/v]NaOCl [20 min]+2 times distilled water [10min.]	[18]
A.andreanum Hort	Lamina segments	5% [v/v] Extran [5 min. with detergent]+0.1% [w/v] mercuric chloride [10-12 min]	[1]
	Leaf	15% [v/v] commersial bleach [20 min]+0.1%HgCl$_2$ [7 min]	[21]

Table 1. Sterilization methods used in *Anthurium* tissue culture

Culture medium

Culture medium influences the propagation efficiency in plant tissue culture applications. Organic compounds, vitamins and plant growth regulators are used to stimulate healthy growth. The rate of tissue growth and morphogenetic responses highly affected by the features of nutrients included.

There are several basal media such as Chu [N6] [24], Gamborg's B5 [25], Murashige and Skoog [MS] [26], Murashige and Tucker [MT] [27] and Nitsch and Nitsch [NN] [28]. These media are successfully used for establishing tissue cultures of different explants of various plants [22].

In plant tissue culture studies, different combinations of every medium based on different concentrations of macro and micronutrients have been used to develop efficient protocols. The rapid and efficient tissue culture protocols are important for micropropagation of *Anthurium* as much as in other plants.

The success of plant tissue culture depends on the composition of the medium used. Different combinations of macronutrients as nitrogen, potassium, calcium, phosphorus, magnesium and sulphur and micronutrients [trace elements] as iron, nickel, chlorine, manganase, zinc, boron, copper and molybdenum change the nature of the medium.

Each plant species has its own medium composition or it should be improved for better results. The modifications can be made up in macro and micronutrients, sugar content, plant growth regulators, vitamins and other nitrogen supplements.

MS media with some modifications have been frequently applied in tissue culture of *Anthurium*. The differences caused by using different concentrations of plant growth regulators in combination with MS organics used to obtain desired tissues [Table 2].

Nitrogen is an essential macronutrient in plant life. It is an important component of proteins and nucleic acids. Nitrate [NO_3^-] is the main source of nitrogen. NO_3^- is reduced to ammonium [NH_4^+] after uptake. Plants have ability to use the reduced form of nitrogen for their metabolism. Nitrate uptake happens effectively in an acidic pH. But after nitrate uptake, the medium are becoming less acid. When ammonium uptake, it makes the medium more acidic. The pH of the plant culture media is important because in a buffered media, existence of both ions affects efficient nitrogen uptake. The form and the amount of nitrogen in media have significant effects on cell growth and differentiation. pH controlling in the media is not the only reason of using both ions, excessive ammonium ions are toxic to the plants. Media containing high levels of NH_4^+ also inhibits chlorophyll synthesis [4].

It has been known that the root growth is induced by NO_3^- and reduced by NH_4^+. But morphogenesis is being controlled by total amount of nitrogen in the medium and it needs both of NO_3^- and NH_4^+. Because of using optimum NH_4^+: NO_3^- has a key role in morphogenesis, therefore the balance between NO_3^- and NH_4^+ differs for different plants and different kinds of cultures. This situation implies that this ratio should be specifically adjusted for each plant species and for different purposes. Changing the NO_3^- to NH_4^+ ratio by small alterations affects differentiation and growth.

Anthurium species	Explant source	Medium components	Aim	Reference
A.andreanum	Leaf	MS+2.2-4.4µM BA+0.9µM 2,4-D	Adventitious shoots	[33]
	Root	Modified MS+2.2µM BA	Multiple shoots	[34]
	Leaf	Modified Nitsch [200mg/l NH$_4$NO$_3$] +1mg/l BA+0.1mg/l 2,4-D	Callus initiation	[15]
		Nitsch [720mg/l NH$_4$NO$_3$] +0.5mg/l BA	Shoot development	
		Nitsch [720mg/l NH$_4$NO$_3$] +1.0mg/l IBA+0.04% AC	Roots	
	Leaf	½MS+0.6mg/l 2,4-D+1mg/l BAP	Callus induction	[20]
		½MS+250mg/l NH$_4$NO$_3$+0.1mg/l 2,4-D+1mg/l BA	Shoot regeneration	
		½MS+1mg/l IBA+0.04% AC	Roots	
	Leaf, spadix	¼MS+1mg/l BAP	Multiple shoots	[23]
		¼MS+1mg/l IBA	Roots	
	Seed	MS+2mg/l BA+0.5mg/l NAA	Callus proliferation	[18]
	Petiol	½MS+0.1mg/l 2,4-D+0.5 mg/l BA	Callus	[36]
		½MS+0.1mg/l 2,4-D+1.0 mg/l BA	Shoot	
		½MS+0.5mg/l 2,4-D	Root	
Anthurium ssp.	Leaf	½MS+1mg/l BA+0.08mg/l 2,4-D	Callus induction	[19]
		½MS+1mg/l BA	Callus multiplication	
		½MS[206 mg/l NH$_4$NO$_3$] +1mg/l BA	Shoot regeneration	
		¼MS+1g/l AC	Roots	
A.andreanum André cv.	Leaf, petiole	Modified Pietrik medium+0.36µM 2,4-D+4.4µM BA	Callus	[32]
Anthurium andreaum cv Rubun	Microcutting from germinated seed	MS+4.4µM BA+0.05µM NAA	Multiple shoots	[7]
A.andreanum Lind.	Apical shoot bud	MS+0.1mg/l NAA+0.25mg/l BAP	Multiple apical shoots	[11]
		MS+0.5mg/l BAP+60mg/l adenine sulphate	Multiple shoots	

Anthurium species	Explant source	Medium components	Aim	Reference
		MS+0.5mg/l IAA+2g/l AC	Roots	
	Half anther culture	NWT+0.25mg/l 2,4-D +0.02mg/l NAA+1.5mg/l TDZ + 0.75 mg/l BAP	Callus Shoot regeneration	[22]
		NWT+ 0. 2mg/l NAA+1.0 mg/lKIN	Roots	
A.andreanum Lindl.cv.	Nodal segments	MS+4.44mM BAP+2.89mM GA₃	Shoot induction	[16]
A.andreanum Hort	Lamina	½MS+1.11µM+BA+1.14µM IAA+0.46µM Kin	Shoot induction	[1]
		½MS+0.44µM BA	Multiple shoots	
		½MS+0.54µM+NAA+0.93µM Kin	Roots	
	Leaf	¼MS+0.88µM BA+0.9µM 2,4-D+0.46µM Kin	Callus	[35]
		¼MS+0.88µM BA+0.54µM NAA+0.46µM Kin	Multiple shoots	
		½MS+0.54µM NAA	Roots	
		½MS+0.5mg/l 2,4-D+1mg/l BAP	Adventitious shoots	[21]
	Leaf, petiole	½MS+0.90µM 2,4-D+8.88µM BA	Callus induction	[17]
		½MS+0.90µM 2,4-D+4.44µM BA	Callus proliferation	
		MS+5.71mM NAA	Roots	
A.scherzerianum	Leaf	½MS+0.08mg/l 2,4-D+1mg/l BAP+1mg/l 2-iP	Callus	[10]
		MS+0.5mg/l BAP	Shoots	
A.scherzerianum Schott	Leaf	Modified MS+2.5 mM NH₄NO₃+18µM 2,4-D+6% sucrose	Embryo induction	[6]

BAP, 6-benzylaminopurine; BA, N⁶-benzyladenine; 2,4-D, 2,4-dichlorophenoxyacetic acid, IAA, indole-3-acetic acid; IBA, indole-3-butyric acid; 2-iP, N⁶-[2-isopentenyl]adenine; Kin, kinetin; NAA, α-naphthalene acetic acid; TDZ, thidiazuron; AC, activated charcoal; AS, adenin sulphate; MS, Murashige Skoog [1962] medium; WM: Winarto-Teixeria medium, NWM: New Winarto-Teixeria medium GA₃, giberellic acid.

Table 2. *In vitro culture medium components for Anthurium cultivars (Modified from [22]).*

MS media used frequently for tissue culture of *Anthurium* and the ratio of NO_3^- to NH_4^+ is 66:34 at this medium. For this reason generally modified MS medium used at *Anthurium* organogenesis. The modifications of ammonium nitrate concentration have been studied at *Anthurium* media by researchers. Hamidah et al. [1997] used half-strength MS macroelements with 2.5 mM ammonium nitrate for *in vitro* stock cultures. While Puchooa [2005] used 200 mg/L reduced ammonium nitrate concentration for callus culture, they increased the amount to 720 mg/L for regeneration. Dufour and Guérin [2005] used different compositions of NO_3 and NH_4 to evaluate the developmental results. According to their results, the ratio of 0.37 showed better plant growth and development. Atak and Çelik [2009] preferred to use half-strength MS salts with NH_4NO_3 lowered to 250 mg/L for shoot regeneration. Winarto et al. [2011] were improved a protocol for callus induction and plant regeneration and NWT-3 media contains 750 mg/l NH_4NO_3.

In culture conditions, using synthetic chemicals with similar physiological activities as plant hormones have capabilities to induce plant growth as desired. Auxin and cytokinins are the most important hormones regulating growth and morphogenesis in plant tissue culture. Their combinative usage promote growth of calli, cell suspensions, root and shoot development and have capability to regulate the morphogenesis [4,29]. There are synthetic auxin and cytokinins beside naturals. Different combinations and concentrations of plant growth regulators such as 2,4-dichlorophenoxyacetic acid [2,4-D], naphthalene acetic acid [NAA], benzylaminopurine [BAP] and kinetin [Kin] were used to indicate callus formation from different kinds of explants of *Anthurium* cultivars. In preliminary studies, induction and regeneration of callus followed by shoot and root regeneration are the main steps of tissue culture of whole plants. As an important commercial plant, to develop a rapid and more effective tissue culture protocol to shorten the time is the main objective of *Anthurium* tissue culture [7,10, 22,23].

As given in Table 2, combination of 2,4-D and BA in culture media to induce callus initiation from leaf explants in different varieties of *Anthurium* is frequently used. Also, adding of BAP and 2-iP to the callus medium has been evaluated by different researchers. The concentrations of 2,4-D used in the callus medium is ranging from 0.08 mg/l to 1 mg/l 2,4-D. The BA concentrations are changing between 0.1 mg/l and 1 mg/l.

Micropropagated plants require a developed root system to resist the external environmental conditions. Rooting of the shoots take place *in vitro*. Therefore, determination of the appropriate auxin type and levels in the media required to promote rooting [4].

Activated charcoal [AC] is added to medium for promoting root growth [11, 13, 15, 19, 20]. AC is composed of carbon and it is often used in plant tissue culture to absorb gases and dissolved solids. It is not a growth regulator but it has an ability to modify medium composition [4].

There are several advantageous uses of charcoal on the type of culture. These are adsorption of secreted compounds from cultured tissues, decrements in the pherolic oxidations, pH changes of the medium to optimize for morphogenesis, prevention of unwanted callus growth, simulation of soil conditions because of the ability to promote root formation, capability to use in production of secondary plant products in culture conditions [4, 30].

The most important effect of using AC to the medium is the rigorous decrease in the concentrations of plant growth regulators and other organic supplements. AC shows greater adsorptive capacity to phenolics commonly produced by wounded tissues, plant hormones like IAA, NAA, IBA, BA, kinetin, zeatin and other hormones [30,31]. The adsorptive property of AC changes with purity, pH and density [3]. The *Anthurium* seedlings propagated by Atak and Çelik [2009] were rooted in medium containing AC and given in Figure 1.

Figure 1. *In vitro* propagation of *Anthurium* cultivars [Arizona]. The shoots with root were growth inplant tissue culture medium with AC [20].

3. The importance of subculturing in micropropagation

In plant propagation applications, subculturing has an importance to prolong the life of plants and expand the number of cultured seedlings. At *in vitro* propagation of *Anthurium andreanum* cultivars, the number of shoots per explants was increased subsequent subcultures. Atak and Çelik [2009] observed that shoot multiplication for two *Anthurium andreanum* cultivars Arizona and Sumi was increased in the next multiplication stage. At every subculture, shoot numbers regenerated form nodal explants gradually increased [Table 3]. Bejoy et al.[2008] reported that multiplication was enhanced in the next multiplication stage. They succeeded to increase the rate of shoot production in the second multiplication stage.

Subculture	Arizona		Subculture	Sumi	
	Number of Explants	Number of Shoots per Explant [Mean±SE]		Number of Explants	Number of Shoots per Explant [Mean±SE]
Va1	50	15. 64±1.69[a]	Vs1	50	12.24±1.18[a]
Va2	50	22.70±1.46[b]	Vs2	50	15.98±1.36[b]
Va3	50	26.76±1.30[c]	Vs3	50	21.82±1.87[c]
Va4	50	33.70±1.09[d]	Vs4	50	26.96±1.46[d]

Data presented as means with different letters within a column indicating significant differences at P<0.05 according to Duncan's Multiple range test. Each mean represented 5 replications.

Abbreviations: **Va1** =initial shoot [Shoot regeneration from callus cultures of Arizona variety] **Va2, Va3, Va4**= subcultures of Arizona variety, **Vs1** = initial shoot [Shoot regeneration from callus cultures of Sumi variety], **Vs2, Vs3, Vs4**= subcultures of Sumi variety.

Table 3. Shoot multiplication of *Anthurium andreanum* cultivars [20].

4. Acclimatization

In micropropagation studies, the last and the critical step is acclimatization of the rooted seedlings to the environment. In this stage, plant losses have been due to different reasons [37,38]. Directly rooted shoots in soil show higher survival rate in the field than rooted under *in vitro* conditions. Therefore, there are several methods to high the survival rate of *in vitro* rooted shoots.

Cultured plants must adapt to low humidity, high light intensities and large temperature fluctuations with *ex vitro* acclimatization techniques. However these methods are expensive, time consuming and labor-intensive, *in vitro* acclimatization techniques have been improved. Using growth chambers which have relative humidity, controlled ventilation and possibility to change the components of the media make it possible to reduce the steps need for the process [37,39].

The success of acclimatization of *in vitro* cultured plants depends on the nutrients reserved in the leaves during development [40]. The important point in acclimatization is to keep the rooted plants in incubator in order to keep the humidity high.

Different acclimatization protocols for *in vitro* regenerated *A.andreanum* plantlets have been reported. Soilrite-perlite with the rate of 10:1, vermicompost and sand mixture [1:3], vermiculite and perlite [1:1], soil and organic humus [1:1] are the most used acclimatization mediums with the high survival ratios ranging from 60% to 98% [1,10,13].

During *in vitro* development stage, the cultural conditions such as humidity, air turbulence, CO_2 concentration, sugar content in medium effect acclimatization ability of plants to *ex v-vo* conditions. Therefore, for each *Anthurium* varieties, efficient acclimatization protocols have to be improved to prolong the success of micropropagation.

5. Discussion and conclusion

In micropropagation studies, the success of the protocols depends on the variety of *Anthurium*, explant type, the components of the media used for shoot and root regenerations. Different combinations of plant growth regulators and additives used in relation to increase the regeneration potential of the explants should be evaluated for each cultivars to determine the efficient tissue culture protocol. In this chapter, we compared the explant types and tissue culture components for *Anthurium* species which is an important ornamental pot plant.

Stages of the leaves show different response to propagation by indirect organogenesis. Explants prepared from brown leaves have higher callus formation rates in a shorter time than green leaf explants. Therefore, selection and using the right leaf explants at the appropriate leaf stage is the first step of establishing a successful tissue culture. Using different combinations of plant growth regulators and nitrogen additives should be evaluated to control the organogenesis for *Anthurium* varieties. NO_3^- : NH_4^+ balance in the growth medium has to be adjusted for each *Anthurium* varieties to obtain desired differentiation and growth. Developing an ideal acclimatization condition is important to increase the survival rate of micropropagated and rooted seedlings to adapt to *ex vivo* conditions.

In conclusion, the primary point to be remembered is the effects of genotypical differences on culture efficiencies. Different genotypes of varieties show different organogenesis responses in explant cultures. Therefore for each *Anthurium* varieties suitable micropropagation methods should be determined.

Author details

Çimen Atak and Özge Çelik

Istanbul Kultur University, Faculty of Science and Letters,
Department of Molecular Biology and Genetics, Ataköy, Istanbul, Turkey

6. References

[1] Martin KP, Joseph D, Madassery J, Philip VJ. Direct shoot regeneration from lamina explants of two commercial cut flower cultivars of *Anthurium andreanum* Hort. *In Vitro* Cell. Dev. Biol-Plant 2003; 39 500-504.

[2] Rout GR, Mohapatra A, Jain SM. Tissue culture of ornamental pot plant: A critical review on present scenario and future prospects. Biotechnology Advances 2006; 24 531-560.

[3] Harb EM, Talaat NB, Weheeda BM, El-Shamy M, Omira GA. Micropropagation of *Anthurium andreanum* from shoot tip explants. Journal of Applied Sciences Research 2010; 6(8) 927-931.

[4] George EF, Hall MA, Klerk JD. Plant propagation by tissue culture, Volume 1. The 9-Background, Springer, 2008.

[5] Dobranszki J, Silva JAT. Micropropagation of apple – A review. Biotechnology Advances 2010; 28 462-488.

[6] Hamidah M, Karim AGA, Debergh P. Somatic embryogenesis and plant regeneration in *Anthurium scherzerianum*. Plant Cell, Tissue and Organ Culture 1997; 48 189-193.

[7] Vargas TE, Mejias A, Oropeza M, Garcia E. Plant regeneration of *Anthurium andreanum* cv Rubrun. Electronic Journal of Biotechnology 2004;7(3) 285-289.

[8] Dufour L, Guerin V. Growth, developmental features and flower production of *Anthurium andreanum* Lind. in tropical condition. Scientia Horticulturae 2003; 98 25-35.

[9] Dufour L, Guerin V. Nutrient solution effects on the development and yield of *Anthurium adreanum* Lind. in tropical soilless conditions. Scientia Horticulturae 2005; 105 269-282.

[10] Viegas J, Rosa da Rocha MT, Ferreira-Moura I, Lairia da Rosa D, Almeida de Souza J, Correa MGS, Telxelra da Silva JA. *Anthurium andraeanum* [Linden ex Andre] culture: *in vitro* and ex vitro. Floricult. Ornamental Biotechnology 2007;1 61-65.

[11] Gantait S, Mandal N, Bhattacharyya S, Das PK. *In vitro* Mass Multiplication with pure genetic identity in *Anthurium andreanum* Lind. Plant Tissue Cult. Biotechnology 2008;18(2) 113-122.

[12] Gantait S, Sinniah UR. Morphology, flow cytometry and molecular assessment of ex-vitro grown micropropagated anthurium in comparison with seed germinated plants. African Journal of Biotechnology 2011;10(64) 13991-13998.

[13] Gantait S, Mandal N. Tissue culture of Anthurium andreanum: Asignificant review and future prospective. İnternational Journal of Botany 2010;6(3) 207-219.

[14] Silva JAT, Nagae S, Tanaka M. Effect of physical factors on micropropagation of *Anthurium andreanum*. Plant Tissue Culture 2005;15(1) 1-6.

[15] Puchooa D. *In vitro* mutation breeding of Anthurium by gamma radation. International Journal of Agricultural Biology 2005;7 11-20.

[16] Lima FC, Ulisses C, Camara TR, Cavalcante UMT, Albuquerque CC, Willadino L. *Anthurium andraeanum* Lindl. cv. Eidibel *in vitro* rooting and acclimation with arbuscular mycorrhizal fungi. Rev. Bras. Cienc. Arar. Recife 2006;1 13-16.

[17] Yu Y, Liu L, Liu J, Wang J. Plant regeneration by callus-mediated protocorm-like body induction of *Anthurium andreanum*. Hod. Agric. Sci. China 2009; 8 572-577.

[18] Maira 0, Alexander M, Vargas TE. Micropropagation and organogenesis of Anthurium andraeanum Lind cv. Rubun. Jain SM, Ochatt SJ.[Eds.] Protocols for *in vitro* propagation of ornimental plants, Methods in Molecular Biology 2009;589 3-14.

[19] Nhut DT, Duy N, Vy NNH, Khue CD, Khiem DV, Vinh DN. Impact of *Anthurium* spp. genotype on callus induction derived from leaf explants and shoot and root regeneration capacity from callus. Journal of Applied Horticulture 2006;8(2) 135-137.

[20] Atak C, Celik O. Micropropagation of *Anthurium andreanum* from leaf explants. Pakistan Journal of Botany 2009;41 1155-1161.

[21] Bejoy M, Sumitha VR, Anish NP. Foliar Regeneration in *Anthurium andreanum* Hort. cv. Agnihothri. Biotechnology 2008;7(1)134-138.

[22] Winarto B, Rachmawati F, Silva JAT. New basal media for half –anther culture of *Anthurium andreanum* Linden ex Andre cv. Tropical. Plant Growth Regulation 2011;65 513-529.

[23] Jahan MT, Islam MR, Khan R, Mamun ANK, Ahmed G, Hakim, L. *In vitro* clonal propagation of *Anthurium* [*Anthurium andreanum* L.] using callus culture. Plant Tissue Cult. Biotech. 2009;19 61-69.

[24] Chu CC, Wang CC, Sun CS, Hsu C, Yin KC, Chu CY, Bi FY. Establishment of an efficient medium for anther culture of rice through comparative experiments on the nitrogen sources. Science Sinica 1975;18 659-668.

[25] Gamborg O, Miller R, Ojimo K. Nutrient requirement suspensions cultures of soybean root cells. Experimental Cell Research 1968;50(1) 151-158.

[26] Murashige T, Skoog F. A revised medium fo rapid growth and bio-assays with tobacco tissue cultures. Physiologica Plantarum 1962;15 473-497.

[27] Murashige T, Tucker DPH. Growth factor requirements of Citrus tissue culture. Proc. 1st In. Citrus Symp. 1969;3 1155-1161.

[28] Nitsch JP and Nitsch C. Haploids plants from Pgrains. Science 1969;163 85-87.

[29] Bajguz A, Piotrowska A. Conjugates of auxin and cytokinin. Phytochemistry 2009;70(8) 957-969.

[30] Thomas TD. The role of activated charcoal in plant tissue culture. Biotechnology Advences 2008;26 618-631.

[31] Asaduzzaman M, Asao T. Autotoxicity in beans and their allelochemicals. Scientia Horticulturae 2012;134(1) 26-31.

[32] Kuehnle AR, Sugii N. Callus induction and plantlet regeneration in tissue culture of Hawaiian *Anthuriums*. Hort Science 1991;26 919-921.

[33] Teng WL. Regeneration of *Anthurium* adventitious shoots using liquid or raft culture. Plant Cell,Tissue and Organ Culture 1997;49 153-158.

[34] Chen FC, Kuehnle A, Sugii N. *Anthurium* roots for micropropagation and *Agrobacterium tumefaciens*-mediated gene transfer. Plant Cell, Tissue and Organ Culture 1997;49 71-74.

[35] Joseph M, Martin KP, Mundassery J, Philip VJ. *In vitro* propagation of three commercial cut flower cultivars of *Anthurium andreanum*. Hortic. Indian Journal of Experimental Biology 2003;41 154-159.

[36] Zhao Q, Jing J, Wang G, Wang JH, Feng YY, Xing HW, Guan CF. Optimization in Agrobacterium-medium transformation of *Anthurium andreanum* using GFP as a reporter. Electronic Journal of Biotechnology 2010;13(5)1-11.

[37] Diaz LP, Namur JJ, Bollati SA, Arce OEA. Acclimatization of Phalaeropsis and Cattleya obtained by micropropagation. Rev. Colomb. Biotechnol. 2010;12(2) 27-40.

[38] Hazarika BN. Acclimatization of tissue cultured plants. Current Science 2003;85 1704-1712.

[39] Lavanya M Venkateshwarlu B, Devi BP. Acclimatization of neem microshoots adaptable to semi-sterile conditions. Indian Journal of Biotechnology 2009;8 218-222.

[40] Premkumar A, Mercado JA, Quesada MA. Effects of *in vitro* tissue culture conditions and acclimatization on the contents of Rubisco, leaf soluble proteins, photosynthetic pigments and C/N ratio. Journal of Plant Physiology 2001;158 835-840.

Transient Virus Expression Systems for Recombinant Protein Expression in Dicot- and Monocotyledonous Plants

Gregory P. Pogue and Steven Holzberg

Additional information is available at the end of the chapter

1. Introduction

Plants have long been a source for traditional medicinal products. Indeed, greater than four billion people utilize plants to meet their primary health care [1-2]. There are >120 distinct drugs derived from plant sources representing >70% of the approved drugs in the past 20 years [3, 4]. The manner to exploit the scale and cost advantages of agriculture while diversifying the product offerings made available by plants has been under intense investigation since the early 1980s. Traditional transgenic approaches were initially pursued, but the challenges associated with the transformation and regeneration of viable recombinant crops delayed the appearance of initial products of medicinal promise until 1989 with the production of antibodies [5] and 1990 with the production of human serum albumin [6]. During this time, the concept of using plant virus genomes as expression vectors emerged. In early investigations, researchers recognized the natural capability of virus systems to change the translational priorities within infected cells such that virally encoded proteins were produced preferentially. This ability suggested that expression vectors could be constructed from viral nucleic acids to produce recombinant proteins throughout infected plants [7]. However, for this hypothesis to be tested, the genomes of viruses, starting with positive (+) strand RNA viruses, had to be cloned and characterized [8-10]. Soon after the first full-length "infectious" clones of a (+) strand RNA plant virus were constructed, and preceding traditional transgenic systems, the virus genome was converted into an expression vector [11]. Although limited with regards *to in planta* expression, this first vector revealed the promise of virus genomes to be efficient expression systems for plants. The advantages revealed in these early studies, continue to be present: cDNA "infectious clones" offer facile subcloning vehicles allowing rapid prototyping of genetic expression constructs, and recombinant protein expression levels that exceed that offered by transgenic systems.

The rapid replication cycle of the virus systems provided amplification of messenger RNA and the resulting proteins providing for a "burst" of recombinant expression that can provide impressive yields (reviewed in [12-15]). While these early vectors were useful in plant cell systems to produce recombinant protein products with potential market value [16], these early systems could not support large scale manufacturing nor did they exploit the advantages of agriculture to provide cost-effective products. This review will provide an overview of plant virus-based expression vectors, and provide select examples how virus expression systems have evolved to offer valuable tools for the production of medically important products [17] and support the study of plant structure and metabolic function (reviewed in [18]) in dicot- and monocotyledonous plants. The growing biomedical and agricultural markets have encouraged great creativity in the construction and testing of plant virus expression systems.

2. Biomedical market for recombinant proteins

As an example of the market drivers for plant-based expression vectors, the biopharmaceutical industry market will be briefly reviewed. Recombinant proteins, including monoclonal antibodies (mAbs), enzymes, hormones, cytokines and growth factors, and vaccine antigens, are the source for new medical therapies and the pharmaceutical market. The global pharmaceutical market continues to prove to be robust >$850B in 2010, in spite of generic pressure and biosimilars appearance [19]. Recombinant protein drugs, known as biologics, expanded their market to $149B, including $48B in sales of the top selling monoclonal antibody (mAb) products treating cancer and other disorders [19, 20]. The global cancer therapeutic market is projected to continue growth at 12.6% compound annual growth rate through 2014 [21]. Monoclonal antibody immunotherapy has revolutionized the treatment of many diseases – most notably cancer where the nondestructive nature of mAb treatment synergizes with many existing therapies to result in improved efficacy. These molecules make up the most promising part of product portfolios for biopharmaceutical companies and this market is predicted to grow by 11-14% compounded annual growth rate (CAGR) in the next five years compared with the rather tepid growth of ~3% CAGR of small molecule drugs [22, 23]. Indeed, in the midst of an economic downturn (2008-2010), the growth in mAbs continues to occur with sales of therapeutic mAbs being $48B in 2010 compared with $40B in 2009 and $37B in 2008 [24]. Adding the $10B of sales for mAbs used for diagnosis and research reagents raises the total mAb market to $58B in 2010. The clinical development of immunotherapy has been revived after several breakthroughs that have led to the approval of drugs and treatment for cancer. Indeed, four of the top 10 mAbs in terms of sales, are used for cancer treatment [24]. New drug targets and associated drug interventions are under investigation that will provide therapeutic options for traditionally underserved populations.

Recent successes and the growing market demand for more innovative biologic products to treat chronic patients has continued to fuel interest and investment to identify tools and strategies to accelerate discovery and product validation in immunotherapy fields. Further, clinical success is ultimately determined by established clinical endpoints indicative of survival. However, these results often require monitoring for months to years after the therapy has been given to the patient. These timelines are not conducive to the iterative and experimental process that is cancer therapy. Therefore, surrogate markers are sought: more rapidly appearing measurements that correlate with longer-term clinical endpoints. These functions require access to relevant clinical samples from diseased and healthy patients as well as adapting laboratory assays to more clinical formats. Surrogate markers require specific assays used to demonstrate efficacy of immunologic therapies and, as noted above, will fuel extraordinary market growth in the coming decade. These assays are highly empirical and require well trained staff, highly controlled conditions and consistency in procedure to ensure trends in data can be validated as statistically significant. Diagnostic assays incorporating recombinant proteins or exploiting mAb for detecting and assessing medical conditions was $776M in 2010 and is expected to grow at a CAGR of 47% through 2015 [25]. This growth in the use of biologics for immunotherapy and diagnostic products continues, in part, because of recent FDA approvals and physician implementation of several new immunology tools and immunotherapeutic products to diagnose and treat patients.

3. Plant production of bio-pharmaceutical proteins

Plants have steadily gained acceptance as alternative production systems for biologics. The recent United States Food and Drug Administration approval of Protalix Biotherapeutic's Elelyso (taliglucerase alfa) [20] represents the successful realization of the goal for plant-produced human biologics, initiated soon after the formation of Agrigenetics in 1981. This product, produced in engineered carrot cells, benefits from the simpler culture conditions required by plant cells compared with Chinese hamster ovary (CHO), insect, or yeast systems. However, it still requires the capital intensive production methods, requiring multiplicative costs for increased scale [26]. Nevertheless, the approval of Elelyso, demonstrates that plants can be used to source biologics that meet the stringent demands for high quality in human products, at competitive scale and costs. Indeed, Protalix indicates that the lower production costs associated with carrot cell systems will allow pricing of Elelyso to be 75% of Cerezyme, the leading product in this market sector [20]. This approval strengthens the regulatory case for plant-based production systems that was established by the 2006 approval of a DowAgrosciences, LLC vaccine by the US Department of Agriculture for the prevention of Newcastle disease in chickens [27]. As with the Protalix product, this recombinant vaccine was derived from a transgenic plant cell culture system.

Agriculture offers several advantages as a biologic production system. Plants allow capital-efficient design of upstream manufacturing capacity at various scales providing cost savings that cannot be easily matched by fermentation technologies. The market opportunities provided by follow-on biologics and the rising capital costs associated with production using traditional systems make plants particularly attractive. Considerable capital and time is required to construct the upstream facilities for cell culture production. The upstream facility must be linked with downstream capabilities supporting product purification and characterization. Although outdated, the published costs associated with these facilities are \$300–\$500 million and require from four to five years to complete construction, validation, and to gain regulatory approval [28]. Agriculture-based production requires less specialized upstream facilities, typically controlled growth chambers, linked with similar downstream production capabilities. The use of plants therefore reduces capital expenditures and also provides for more flexible use of space and capital.

The handling of plant biomass and its initial extraction requires unique biomanufacturing solutions [14]. Virus vector expression systems offer significant advantages at considerably reduced costs to current cell-based manufacturing systems, such as employed by Protalix and Dow AgroSciences, while avoiding concerns associated with stable plant transformation [29, 30]. Virus based expression systems have been extensively tested and shown safe and environmentally-friendly in both indoor and outdoor tests since 1991 [12] and multiple products completing early stage human clinical investigations [31, 32]. Additional advantages also exist compared with traditional cell-based fermentation approaches include: 1) speed and low cost of genetic manipulation; 2) rapid manufacturing cycles; 3) no mammalian pathogen contamination; 4) minimal endotoxin concentrations and 5) economical production [12-15, 33, 34].

4. Types of plant virus-based expression vectors

Many different types of plant viruses have been converted into vectors for the production of recombinant proteins or peptides (for complete review, see [12-15, 33, 34]). As different viruses have distinct biological limitations and gene expression potential, each vector system has its own unique opportunities. This review focuses on virus vectors that have been particularly useful to produce recombinant proteins for biomedical, therapeutic and

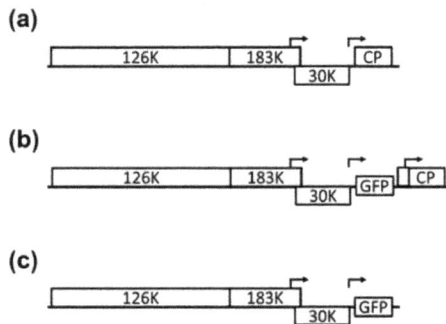

Figure 1. The genomic structure of wild type Tobacco Mosaic *tobamovirus* RNA (a) and derivative independent (b) and minimal (c) expression vectors. Boxes represent the open reading frames on each RNA and are labeled with the the viral protein they encode. The 126K and 183K proteins are required for virus replication. The 30K gene is the virus movement protein required for cell-to-cell transport. The CP is the virus coat protein that is required for encapsidation and systemic movement. The green fluorescent protein (GFP) is used as a placeholder for a gene for protein overexpression *in planta*. Arrows indicate the position of subgenomic promoters used to express the downstream proteins.

research, use. Many groups have sought to categorize virus based expression systems. For this review, we will organize our thinking by using two categories: "independent-virus" or "minimal-virus". Independent-virus vectors are replication competent vectors that can be principally inoculated to plants as virus particles or viral RNA, multiply in initially infected cells and exploit virus encoded cell to cell and systemic movement activities to infect the majority of the phloem sink tissue of a host. In contrast, minimal virus systems are replication competent systems that have be modified in order to possess greater expression capabilities. The modifications are typically replacement of a virus-encoded open reading frame, not essential for genome replication, with the gene of interest such that the minimal-virus systems lack the ability to systemically infect a host. Examples of each system will be provided in the following sections with Figure 1 illustrating the genetic structures of independent and minimal virus systems derived from the tobacco mosaic *tobamovirus*. (TMV) genome.

4.1. Independent-virus vectors

Initial RNA virus vectors were functionally minimal-virus vectors that utilized a "gene replacement" strategy where a foreign gene of interest replaced the capsid protein (CP) gene of a virus [11]. These early vectors expressed foreign genes, but, as with other minimal-virus systems, lacked certain virus functions thereby limiting activities. For example, brome mosaic virus (BMV) CP replacement vectors could not even move from cell to cell in an infected leaf [11], and although TMV-based CP replacement vectors could move from cell to cell, they could not move systemically in inoculated plants [35-37]

With greater understanding of virus function, plant RNA virus vectors were constructed to express a foreign gene product in addition to all required viral proteins [36, 38]. These vectors were the first independent-virus system that expressed recombinant products while moving systemically in a host plant. To construct independent-virus systems for (+) strand RNA viruses, vectors exploit subgenomic mRNA production to express foreign genes by using an additional subgenomic promoter inserted into the virus [38-40]. For viruses that used polyprotein processing, the foreign gene was inserted in translational frame with the existing virus open reading frame (ORF) and peptide sequences that facilitate the proteolytic processing of the fusion protein were present to insure release of the recombinant protein. Some independent-virus vectors are designed to express potential products as fusions to viral proteins, such as the potatovirus X (PVX) CP (reviewed in [41]). Often the fusion methodology employs the foot and mouth virus 2A translational cleavage sequence (see references in [42]). The apparent pausing of the ribosome, and the discontinuity of the peptide bond that results, allows proteins upstream and downstream of the 2A sequence to be differentially targeted, such as a single chain antibody accumulating in the plant apoplast while the CP was sequestered in its normal cytosolic localization [43]. Using these strategies, independent-virus systems have been derived from the genomes of *potexviruses* (including potato virus X; PVX), *tobamoviruses* (including TMV), *comoviruses* (including cowpea mosaic virus), *potyviruses, tobraviruses, closteroviruses* and others [12, 13, 15].

Most independent-virus vectors are functional in *Nicotiana* or other herbacious species. This follows from their ease of inoculation and the receptivity of species to virus expression systems. Recently, the adaptation of *closterovirus* vectors to a non-herbacious plant system was accomplished by adaption of Grapevine leafroll-associated virus-2 into an expression vector [44]. This vector showed characteristic phloem-associated expression in inoculated grapevines involving the roots, stems, petioles, leaves and berries. A grapevine-A vector was also shown to induce inhibitory RNA (RNAi) of model genes in micropropogated grapevines [45]. These vectors for use in expression and RNAi-based functional genomics studies open new non-transgenic strategies for researchers in woody plants.

As a more detailed example of an independent-virus system, *Tobamoviruses*-based vectors have been commonly used for recombinant protein expression. These viruses have a (+) sense single stranded RNA genome of ~6400 nucleotides helically encapsidated by many ~2,100 copies of a 17.5 kDa CP in rigid rod shaped particles [12]. The viral replication-associated proteins are directly transcribed from the genomic RNA directly, whereas expression of internal genes is through the production of subgenomic RNAs. Sequences in the *tobamovirus* genome function as subgenomic promoters regulating the production of subgenomic RNAs. The virus movement protein MP and CP are translated from two separate, but co-terminal, subgenomic RNAs, with the CP being among the most abundant protein and RNA produced in the infected cell [46]. In a *tobamovirus* infected plant there are several milligrams of CP produced per gram of infected tissue.

Tobamoviruses-based vectors are readily constructed as independent-virus systems, including cell-to-cell and systemic movement activities mediated by MP and CP, respectively (Figure 1). These vectors benefit from the strength of the viral subgenomic promoter's activity to reprogram the translational activities of infected plant cells such that virus-encoded proteins are synthesized at high levels, often similar to the TMV CP [40]. A foreign gene encoding the protein for overexpression is added in place of the virus CP so it will be expressed from the endogenous virus CP promoter [38, 40]. A second CP promoter from a different *tobamoviruses* strain, of sequence divergent to the first CP promotor, is placed downstream of the heterologous coding region and a virus CP gene is then added 3′ terminal to the heterologous subgenomic promoter. *Tobamoviruses*-based vectors infect various species, but most commonly tobacco-related species (genus *Nicotiana*), including *tabacum* and *benthamiana*. For the vector to express foreign proteins, the infectious vector RNA enters plant cells via wounds induced by an abrasive. The virus RNA is released from the CP subunits, translated to produce replication-associated proteins and is replicated in the initial cell. The progeny RNA is moved to adjacent cells in association with the MP to produce infection foci. A proportion of the RNA, complexed with CP, enters the plants vascular system for transport to phloem sink tissues in the aerial leaves. This movement produces the systematic infection and the foreign gene is expressed in all cells that express other virus protein products. Within the cell, the foreign protein is deposited in the site dictated by its protein sequence, either naturally or purposely engineered [12, 46].

Tobamoviruses-based vectors have been used by literally hundreds of researchers to produce a range of human enzymes, antimicrobials, cytokines, subunit vaccine and immunoglobulin proteins. Several reviews have provided surveys of these products [12, 14]. In general, expression results were obtained in *Nicotiana* hosts and proteins were extracted from leaf tissues using total homogenization and clarification methods or leaf infiltration and isolation of interstitial fluids [46, 47]. Several products have been purified using differential separation and standard chromatographic separations and tested in various model systems of human disease. TMV-based vectors, especially the well-developed GENEWARE® system, have been used to express a large range of recombinant proteins under research and Current Good Manufacturing Practice (cGMP) compliance [14]. One class of products has been successfully tested in human clinical trials, as described below.

Active vaccination of Non-Hodgkin's Lymphoma (NHL) patients with cancer antigens, in this case the idiotypic antibody expressed by the tumorogenic B-cells, has been shown to induce clinical remissions in human clinical trials [48]. However, more efficient and effective vaccines are sought. Full antibodies contain both idiotype-specific elements as well as constant sequences, shared by many antibodies, which may reduce the immunogenicity of the vaccine. In order to provide higher antigen content to vaccines, single chain antibodies (scFvs) were constructed from tumor-derived idiotypic antibodies to provide simpler and more sequence-focused vaccines for clinical testing. This vaccine strategy was shown to be effective in murine models of NHL [49] and a GENEWARE®-based production methodology was developed that could produce >80% of scFvs from human tumor samples [50]. Sixteen patients were enrolled in a Phase I clinical trial under the regulatory oversight of the US Food and Drug Administration [31, 51]. Vaccines were successfully produced for all patients and applied in two dosing groups with and without granulocyte-macrophage colony-stimulating factor adjuvant. The primary endpoint of the study was safety which, as the first parenteral administration of a plant-made vaccine, was an important outcome to monitor. The study results confirmed the safety of plant-derived vaccines, including plant-specific glycoforms present on 15 of the 16 vaccines. The secondary endpoint was determination of the immunogenicity of the vaccines in human subjects. Overall, 70% of the patients developed cellular or humoral immune responses to the scFv vaccines, with the adjuvant improving the frequency of responses, as predicted. The majority of the responses was shown to be vaccine specific and did not cross react with control idiotype proteins. These results demonstrated the flexibility of the TMV-based expression systems as well as the safety and effectiveness of the plant derived products [31, 51].

4.2. Minimal-virus vectors

In contrast to independent-virus vectors, minimal-virus systems are capable of functions supporting RNA replication, yet are lacking in one or more functions necessary for systemic infection. Although this vector was the first type constructed, researchers moved away from this approach in favor of the independent-virus vectors. However, as limitations emerged from independent-virus systems, including the size of genes that can be expressed, host range limitations and problems with systemic movement, researchers revised minimal-virus systems with new energy. The resulting vectors were found to be incapable of systemic movement in inoculated plants, thus they must be delivered to each and every plant leaf to allow cell-to-cell movement activities allow infection of all inoculated leaves. Standard abrasion methods are too tedious to deliver inoculum to each leaf, so new methods were developed. The most common method is Agro-infiltration of host plants to launch the infection process [33, 52-54]. This process introduces a DNA plasmid, containing the virus vector under the control of an appropriate transcriptional unit within normal Ti plasmid integration sites, into *Agrobacterium tumefaciens* cells to create an inoculum. The *Agrobacterium* strain containing the DNA expression constructs are grown in overnight cultures and diluted for inoculation. Plants are inverted, submerged in inoculum solution and a vacuum is applied removing the air trapped in the leaves. As the vacuum is released, the inoculum replaces the air in the leaf spaces providing the *Agrobacterium* strain access to cells throughout the submerged leaves for invasion. The bacteria then exploits the transfer DNA mechanisms to introduce the DNA copy of the virus expression vector into the nucleus of infected plants. RNA transcription produces infectious RNA or mRNA transcripts that are processed by nuclear enzymes into an intact and capped transcript and exported to the cytoplasm. These transcripts are then translated and replication protein expression initiates similar to independent-virus systems.

The activity of viral movement proteins move the vectors from the initially infected cells to adjacent cells creating a more rapid and synchronous infection of inoculated leaves than independent-virus vectors. This eliminates the delays associated with systemic plant movement and can yield greater amounts of recombinant proteins in a shorter period of time than independent-virus systems [33, 54]. This approach increases the genetic load carried by the minimal-virus systems allowing efficient expression of larger recombinant proteins [55]. Minimal-virus systems have been developed from the genomes of *potexviruses, tobamoviruses, bromoviruses, comoviruses* and *geminiviruses* [12, 13].

Nicotiana benthamiana is an ideal host for minimal-virus expression since it highly susceptible to *Agrobacterium* infection to mediate initial entry and introduction of the viral expression vectors. The expression of a defective form of RNA dependent RNA polymerase in *N. benthamiana* [56] makes it nearly universally susceptible to plant viruses and the great experience with this host has led it to be the common host for independent-virus as well for the expression of many recombinant proteins. The flexibility of Agroinfiltration inoculation procedures allow more than one expression vector into a host plant in a given treatment. The co expression of silencing suppressor proteins has been shown to be a key factor for optimized yields [57-59]. Such methods have been used to produce a range of biopharmaceutical proteins [60-62] and offer strategies to modify the plant enzymatic machinery, producing more stable and "human" like recombinant proteins, including glycan structures [34, 60] which will be discussed later in this review.

Examples of minimum-virus vectors include the systems developed using TMV genomes include those developed by (Figure 1)[63,64]. Two TMV variants were developed – one, actually an independent-virus system, employed the full TMV virus capable of systemic movement [63] and a second – lacked the virus coat protein as minimum-virus system [64]. The minimum-virus system (TRBO vector) produced significantly higher levels of the green fluorescent protein (up to 5.5 mg/g FW). This vector did not require the co-expression of a silencing suppressor and worked with very high inoculum dilution in infiltration medium. The RNA2 of cowpea mosaic virus (CPMV) was also adapted as a minimum-virus system which overcame the historic limitation of the insert size in CPMV vectors and allows expression hetero-oligomeric proteins from a single vector [65, 66]. However, the system also requires co-expression of silencing suppressors for optimal expression.

DNA viruses have been adapted to minimal virus systems. Both single and bipartite plant geminivirus systems have modified to produce recombinant proteins – usually at the expense of expression of the capsid or key movement or transmission proteins. Maize streak virus is an example of a single component virus which has been converted into an expression vector [67]. Bean yellow dwarf virus (BeYDV) has been developed into a single and dual-component replicon system that permits simultaneous, efficient replication of two DNA replicons and thus high-level accumulation of one or two recombinant proteins in the same plant cell [68, 69]. This system has been used to produce express immunoglobulin proteins and human papilloma virus HPV-16 and the p24 protein of HIV-1 [70]. The system requires co-expression of the silencing suppressors. Some geminivirus systems have been adapted to express recombinant proteins in non-*Nicotiana* species, including lettuce [71].

The most advanced minimum-virus system is magnICON®. This technology has been used to express a large number of recombinant proteins, including cytokines, interferon, bacterial and viral antigens, growth hormone, single chain antibodies (reviewed in [54, 72]). The ability of Agroinfiltration to introduce more than one expression vector into a host plant in the same inoculation allows the use of two magnICON® vectors to produce heteromeric recombinant proteins, such as mAbs. For production of mAbs, two non-competitive virus vectors are used: one based on turnip vein clearing tobamovirus (TVCV) and the second, potatovirus X (PVX [55, 73]). In mAb production, two magnICON® virus expression vectors each contain a separate mAb chain, heavy or light, and are co-delivered by Agroinfiltration. Each vector replicates independently and expresses mAb chains in the same cells that self-assemble functional mAbs at yields up to 1 g/kg fresh weight [33, 54, 55]. These vectors have been used for efficient large scale production of multi-gram batches of mAbs under Current Good Manufacturing Practices that have been tested in several challenge model systems, including non-human primates [14, 74].

5. Synergy of transgenic plants and virus-based expression vectors

Standard integrative plant expression vectors allow transformation of plant lines using *Agrobacterium tumefaciens*-mediated transfer-DNA delivery methods. These methods allows for great flexibility and synergy when mixed with transient, virus-based expression vectors. The mixture of these two approaches allows for the efficient expression of the recombinant biopharmaceutical protein of interest and provision of the required co-factors improving pharmaceutical protein yield and processing. As noted above, the co-expression of silencing suppressor proteins has been shown to be a key factor for optimized yields with some minimal-virus expression systems. Further, entire protein processing systems can be introduced into plants to produce more stable and "human" like recombinant proteins, including glycan structures [60, 34]. Concerns over the potential immunogenicity of plant-specific glycan linkages on recombinant proteins (the presence of $\beta1,2$-xylosylation and core α1,3-fucosylation) have been mitigated by the use of RNAi technologies to down-regulate endogenous beta1,2-xylosyltransferase and alpha1,3-fucosyltransferase genes [75-77] in *N. benthamiana* plant lines. Proteins produced from these lines show almost homogeneous N-glycan species without detectable beta 1,2-xylose and alpha1,3-fucose residues providing a host that produces humanized glycan structures. Indeed, production of antibody products using the magICON® system using the RNAi plants have demonstrated the synergy of the systems – rapid production of high quantity production of antibodies that show humanized glycan structures [14]. Indeed, the lack of fucose glycans on mAb products produced in the *N. benthamiana* producing humanized glycans have shown enhanced potency in the treatment of Ebola infection in a non-human primate model, anti-tumoral investigations of a plant-derived trastuzumab in murine xenograph models and HIV-1 neutralization studies [74, 78, 79]. The mechanism responsible for this enhanced *in vivo* activity appears to be the improved Fcγ and FcγRIII receptors of nonfucosylated, plant produced mAbs. This advantage based on glycan engineering offers great promise for plant derived biosimilar antibody products.

The present of sialic acid terminal sugars on glycan structures of human plasma proteins is correlated with their long half-life and the pharmacokinetic properties of effective recombinant therapeutics which must function in human plasma [80]. The capability to sialylate plant proteins has been demonstrated in transient and transgenic *Arabidopsis* systems by [81-83]. This effort required transformation events providing enzymatic synthesis of the sialic acid metabolic precursor, which is normally not synthesized in plants, in addition to transferase activities. Efforts are now ongoing to introduce this multi-genic modification into *N. benthamiana*. This glyco-modified host in combination with transient virus-based expression of recombinant proteins benefiting from sialylation offers potential single-step production of human plasma enzymes with similar pharmacokinetic properties as the native proteins. By eliminating any need for post-production enzyme modification, successful development and deployment of these plant lines would contribute to both faster speed of delivery and lower cost of goods. From these examples, the synergy between tailored expression hosts, genetically engineered for appropriate protein post-translational modifications, and virus-based expression strategies to provide recombinant products that meet the biological function and production rigors of modern biotechnology.

6. Transient virus-based expression: Transitioning from dicot- to monocotyledonous plants

The focus of the previous material was virus expression systems providing recombinant protein expression primarily in plant dicotyledonous plant species. This emphasis comes from the historical emphasis on dicot expression due to the availability of more facile systems. Yet monocotyledonous species, especially the cereals, are the most important of crops for feeding humans and livestock in many parts of the world. Further, maize, sugarcane, bamboo and other monocot species, are used extensively in the production of biofuels and other industrial products. Monocots possess unique morphological features and seed biology suggesting gene functions not present in dicots. Monocot species have also developed unique mechanisms for tolerance of adverse environmental conditions like drought and high salinity. In the past, most protein functions were proposed based on homology to better characterized dicot systems, such as *Arabidopsis thaliana*, but functional assays are required to truly ascertain gene function. Overexpression and RNAi gene silencing strategies are key to these direct studies [84]. The construction of transgenic plant lines is not an adequate solution due to the time consuming nature of the approach and the frequency of lethal phenotypes. Therefore, a more rapid, transient strategy is needed. RNA virus expression vectors offer such powerful tools for understanding the biology of these species. In the remaining sections of this review, we will examine the contributions that monocot viral expression systems have made and provide a few detailed examples.

6.1. Use of transient expression vectors as functional gene discovery tools in moncot species

One of the challenges in developing any plant based expression system is to achieve high level expression without triggering the post translational gene silencing (PTGS) and related RNAi mechanisms that plants and other organisms have evolved (reviewed in [18]). These mechanisms have been observed to operate in transgenic plants, and even exploited to generate pathogen-derived resistance to viruses in cases where the silencing of viral transgenes prevents related viruses from infecting the transgenic host [85, 86]. The identification of plant viral proteins that are able to suppress these silencing mechanisms suggested that these mechanisms have evolved in part to prevent or slow viral infection.

Expression of proteins using viral vectors can also trigger PTGS, and is referred to as viral induced gene silencing (VIGS). Using VIGS, an endogenous plant gene can be silenced by inserting only a small portion of the target gene (100-500 nt in length) which creates loss of function phenotypes to study gene function. A wide range of viruses have been developed as VIGS vectors, originally, and most extensively, for dicot hosts [18, 87]. Shortly after the first dicot examples, a barley stripe mosaic virus (BSMV) vector, was used to silence the endogenous *phytoene desaturase (pds)* gene systemically in barley [88].

BSMV, the type member of the hordeivirus family, infects many agriculturally important monocot species such as barley, rice, corn, oat and wheat [89-92]. It is known to be highly seed transmitted in barley, which could be a potential advantage for assessing gene function in seeds and early development. As showen in Figure 2, BSMV has a tripartite (+) sense RNA genome consisting of RNAs α, β and γ [93]. The virus uses well characterized subgenomic promoters for gene expression from each of its genomic RNAs [94].

As a more detailed example of a VIGS vector, BSMV was constructed by inserting fragments of the silencing target so that they would be expressed only as untranslated RNA on a subgenomic promoter following the γb ORF [88]. Although the strongest silencing in barley plants was observed using the barley *pds* fragments, fragments from *pds* taken from maize and rice caused a degree of silencing that was proportional to their sequence similarity with the barley homolog. Further modification of the virus to delete the coat protein enhanced the suppression of the endogenous *pds*. Since this first demonstration of VIGS in barley, BSMV silencing vectors have been successfully used to demonstrate gene functions related to pathogen resistance, aphid defense, development and mophogenesis in a variety of monocots including oat, rice, and the model grass *Brachypodium distachyon* [87]. BSMV mediated VIGS was recently shown to operate in root, leaf and meiotic tissues of wheat, along with efforts to optimize its efficacy in this important agricultural host [95]. In addition, BSMV mediated VIGS can be inherited and has been observed for up to 6 generations in wheat and barley, which not only enhances the range of phenotypes that can be explored for reasons related to the timing of developmental events, but also because progeny frequently have fewer viral symptoms [87, 96]. Most recently, the β RNA has been modified to incorporate untranslated foreign gene fragments and was used with the modified γ RNA vector described above to silence two endogenous genes simultaneously, or to achieve enhanced silencing of a single gene [97]. In addition to BSMV, a strain of BMV has also been engineered as a VIGS vector and has been used successfully in barley, rice and maize [98, 99].

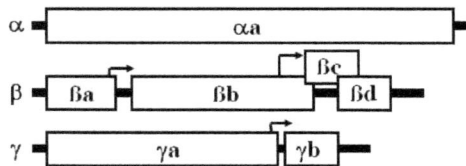

Figure 2. The general structure of BSMV genomic RNAs α, β and γ. Boxes represent the open reading frames on each RNA and are labeled with the the viral protein they encode. Arrows indicate the position of subgenomic promoters used to express the downstream proteins.

Coupling VIGS with high-throughput cloning and sequencing technologies has additionally allowed these viral vectors to be used in functional genomics. In this approach, cDNA libraries are constructed within virus expression vectors and gene function to be assessed by screening infected plant hosts for phenotypic or metabolic changes measured by various input and output focused screening assays. TMV and PVX VIGS vectors were the first to employed in this manner in the dicot *N. benthamiana*, with BSMV used shortly after in the monocot barley [84, 100]. Since these first reports, vector systems for functional genomics have continued to be refined and optimized in an expanding number of hosts [18, 87, including systems amenable for high throughput screens in monocots based on BSMV [101, 102].

6.2. Use of transient virus expression systems for recombinant protein expression in monocot species

As with dicot expression vectors, monocot vectors include both minimal and independent virus vectors [103]. As noted above, minimal type BMV vectors have been constructed by replacing the coat protein ORF with a foreign gene. For example, French *et al.* [11] used this strategy to express the CAT protein in barley protoplasts. In another example, involving BSMV, substitution of the open reading frame (ORF) for the βb triple gene block 1 (TG31) movement protein was used to express the luciferase reporter gene [104]. An example of an independent vector is one based on the potyvirus, wheat streak mosaic virus (WSMV). In this system the foreign ORF is inserted in frame within the virus polyprotein and flanked by cleavage sequences that allow for its release after translation [105] These systems have the limitations of low foreign gene insertion efficiency and the smaller number of hosts that can be infected and tested for recombinant protein expression.

6.3. Using barley stripe mosaic virus vectors for recombinant protein expression

BSMV has been developed to express recombinant proteins as fusions to several individual virus proteins expressed from subgenomic RNAs. To date these systemic vectors have not incorporated any sequences to liberate the foreign protein from the fusion protein. For example Lawrence and Jackson [106] expressed GFP as a fusion to the N terminus of the βb (TGB1) protein at levels sufficient to explore the function of the viral protein in cell-to-cell movement. Higher levels of expression however were achieved with fusions to the C-terminus of the BSMV γb protein. This includes systemic expression of a γb::GFP fusion protein used to study viral movement in barley [107, 108]. More recently, C-terminal γb fusions have been use to test the fungal ToxA protein activity in wheat, barley and *N. benthamiana* [109]. In the latter case, the recombinant virus genome, containing the ToxA gene, was sufficiently stable that Agroinfiltrated *N. benthamiana* was used as the inoculum source for systemic expression in barley and wheat. ToxA phenotypes were replicated in all species. The systemic expression by BSMV of non-fusion recombinant proteins in barley and wheat has been achieved by substituting most of the γb ORF with the gene for either the wild type and mutant ToxA protein [110]. Again, the resulting phenotypes indicate the ToxA proteins were expressed appropriately.

These efforts to develop independent viral expression vectors in monocots are extended by previously unpublished work described here. Our aim in this work was to improve foreign gene stability, increase the level of foreign gene expression, and generate free, non-fusion, foreign proteins that could function and localize independently of viral proteins.

Expression of GFP::γb Fusion Protein from RNA γ

To test whether insert stability was related to the fusion orientation, we constructed a vector with GFP fused (indicated by the "::" in construct name) to N-terminus of the γb protein (γ.GFP::γb, Figure 3A). Co-inoculation of protoplasts and barley leaves with α and β RNAs (BSMV-GFP::γb) resulted GFP fluorescence and fusion protein accumulation to levels that were indistinguishable from co-inoculations with α, β and γ.γb::GFP (BSMV-γb::GFP) as shown in Table 1 (compare #3 and #4). However, as the infection in plants progressed beyond 7 dpi, GFP fluorescence and fusion protein accumulation were observed longer in γ GFP::γb infected plants than in those infected with γ.γb::GFP. Specifically, GFP::γb expression is regularly observed in leaves 1-4 above the inoculated leaf by BSMV-GFP::γb, compared to 1-2 leaves for BSMV-γb::GFP (data not shown). Thus, the GFP gene is generally maintained and expressed by BSMV-GFP::γb to 18 dpi. We believe that the increased stability of fusions to the N-terminus of γb results from more restrictive requirements for the deletion of the GFP gene in order to gain a competitive advantage. For example, internal deletions in the GFP ORF have a one in three chance of maintaining the continuity of the ORF with γb, which is critical because γb is an important virulence factor for BSMV [111]. Thus, those deletions which result in a γb frame shift are likely to be less competitive than those viruses which maintained the full GFP ORF. In contrast, for C-terminal fusions of GFP to γb, all deletions within the GFP ORF would maintain γb expression and would presumably be more competitive than the parental virus due the reduced genetic load.

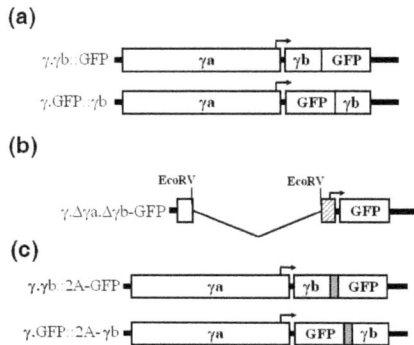

Figure 3. Genomic organization of BSMV γ RNAs engineered to express GFP. Vector components γ.γb::GFP and γ.GFP:: γb as designed to produce a fusion of GFP to the γb protein C and N terminus, respectively (a). The vector component γ.Δγa. Δγb-GFP designed to produce only free GFP (b). Vector components γ.γb::2A-GFP and γ.GFP::2A-γb as designed to produce a free GFP protein following cleavage by the FMDV 2A cleavage sequence from the γb protein C or N terminus, respectively (c). Open boxes indicate ORFs, hatched boxes indicate untranslated ORF sequences, grey boxes indicate FMDV 2A cleavage sequence, arrows indicate subgenomic promoters. EcoRV restriction sites indicate the position of the Δγa sequences that have been deleted from the wild type γ RNA.

Inoculum				GFP expression		Systemic symptoms of BSMV infection
Inoculation#	Vector RNA components			Tobacco Protoplasts	Barley Plants	
						Barley Plants
1	α	β	none	γ	-	-	yes
2	α	β	TMV-GFP		*****	-	no
3	α	β	γ.γb::GFP		****	systemic	yes
4	α	β	γ.GFP:: γb	γ	****	systemic	yes
5	α	β	γ.Δγa. Δγb.GFP		-	-	no
6	α	β	γ.Δγa. Δγb.GFP	γ	*****	-	yes

Table 1. Comparative expression of GFP in tobacco protoplasts and barley plants using BSMV vector constructs

Expression of native GFP from RNA γ

To develop a BSMV vector able to express a free foreign protein, we tested a variety of novel vectors with the most promising of which described here. In each case, the BSMV γ RNA was modified to express GFP in combination with wild type BSMV α and β RNA. A three component vector with RNA α, β, and γ.Δγa.Δγb-GFP (a γ RNA containing a deletion of γa, the replicase protein ORF, and GFP in place of the γb ORF, Figure 3C) did not express GFP in protoplasts (#5, Table 1). However, the addition of wild type γ RNA created a four component vector, which was equivalent in GFP expression levels to TMV 30B (TMV-GFP [40]) in protoplasts (compare #6 with #2, Table 1). In contrast to its behavior in protoplasts, the four component vector did not express GFP systemically in barley, even though systemic viral symptoms were observed (#6, Table 1), apparently due to the loss of the GFP bearing component during systemic movement. This system may be amenable to expression in whole plants if delivered using an Agroinoculation strategy, or in transgenic plants expressing the γa and γb proteins to complement RNA α, β, and γ.Δγa.Δγb-GFP.

In a different approach, BSMV vectors expressing N- and C-terminus fusions of GFP to BSMV γb were modified to release GFP *in vivo* after autocleavage of the fusion protein. Cleavage of the fusion protein was achieved by inserting the foot and mouth disease virus (FMDV) 2A co-translational cleavage sequence [112, 113] between the fused ORFs (Figure 3C). Western analysis of γ.GFP::2A::γb and γ.γb::2A::GFP infected tobacco protoplasts and barley plants revealed that the FMDV 2A cleavage sequence was at least >90% effective in both cases (data not shown). Viral spread and GFP fluorescence in barley plants infected with γ.GFP::2A::γb was indistinguishable from those containing γ.GFP::γb. In contrast, the systemic spread of the BSMV-γb::2A::GFP virus was slightly delayed compared to BSMV-γb::GFP (data not shown). Both BSMV-γb::2A::GFP and BSMV-GFP::2A:: γb were able to infect and express GFP in *N. benthamiana* plants in a manner similar to that reported for γ.γb::GFP [108].

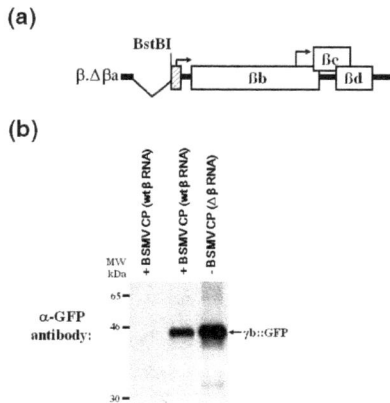

(a)

(b)

Figure 4. BSMV Coat protein deletion vector and GFP expression. Genomic organization of BSMV β RNA engineered with a deletion of the BSMV coat protein, βa (a). Open boxes indicate ORFs, hatched boxes indicate indicate untranslated ORF sequences, arrows indicate subgenomic promoters. BstB1 restriction site indicates the position of the βa sequences that have been deleted from the wild type β RNA. Western blot analysis of γb::GFP protein in barley leaves (b).

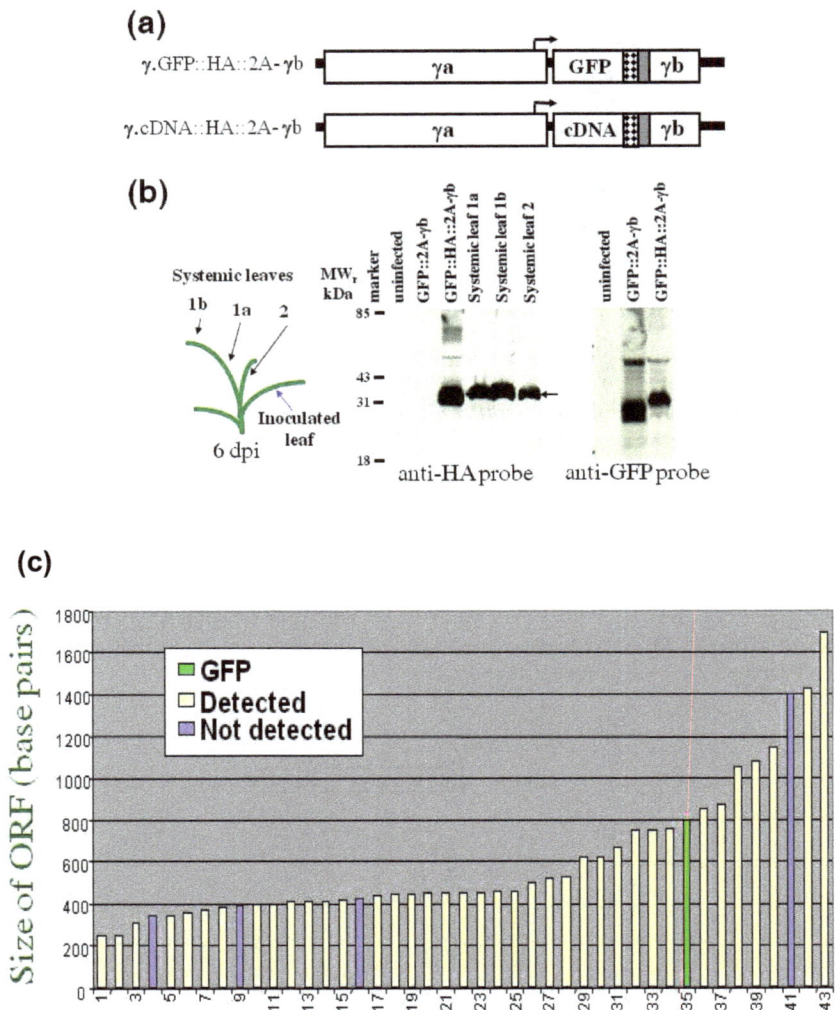

Figure 5. Genomic organization of BSMV γ RNAs and expression of free GFP or heterologous proteins from a cDNA library. Vector components γ.GFP::HA::2A-γb and γ.cDNA::HA::2A-γb as designed to produce a fusion of GFP or a heterologous protein with an HA epitope tag and the FMDV 2A sequence. The 2A sequence is designed to release the γb protein is during translation (a). Open boxes indicate ORFs, checkered boxes indicate the HA epitope sequence, grey boxes indicate FMDV 2A cleavage sequence, arrows indicate subgenomic promoters. Western blot analysis of GFP::HA fusion protein in barley leaves (b). Graph of rice cDNA sequences arranged according to size (c). cDNAs whose predicted proteins were detected by Western blot analysis are unshaded, while blue shaded bars indicate the lack of detectable HA tagged protein. Green bar represents the internal GFP::HA::2A control.

Coat Protein Deletion Enhances Expression from γb Subgenomic Promoter

The coat protein (βa) of BSMV is not required for systemic infection. To determine the impact of βa deletion on βb subgenomic expression, a β genomic RNA containing a deletion of the βa ORF was constructed (β.Δβa, Figure 4A). In tobacco protoplasts, using β.Δβa increased GFP accumulation from the γ RNA 2-5 fold for viruses bearing the γb fusions (γ.γb::GFP, γ.GFP:: γb, γ.γb::2A::GFP, and γ.GFP::2A:: γb) or GFP substitutions of γb (γ.Δγa.Δγb.GFP) (Figure 4B and data not shown). Using β.Δβa also increased GFP fluorescence and accumulation for γb::GFP, GFP:: γb, γb::2A::GFP, and GFP::2A:: γb, in systemic barley tissue (data not shown).

Expression of Epitope Tagged Heterologous Proteins from RNA γ Derived Vectors

To detect heterologous proteins, γ.GFP::2A-γb was modified to contain the hemagglutinin (HA) epitope tag fused between GFP and the FMDV 2A cleavage sequence (γ.GFP::HA::2A-γb, Figure 4A). In barley plants infected with BSMV RNAs α, β.Δβa, and γ.and γ.GFP::HA::2A-γb, Western blot analysis detected the HA tagged GFP (Figure 5B). The addition of the HA tag had no effect on the efficiency of the 2A cleavage sequence. These results suggested that the HA epitope tags would allow for the efficient detection of BSMV expressed heterologous proteins.

To test the flexibility of BSMV to express heterologous proteins in barley, plant genes, from a variety of subcellular locations were expressed as N-terminal fusions to HA::2A-γb (γ.cDNA::HA::2A-γb, Figure 5A). A total of 42 full-length rice genes were amplified from sequenced, cDNA library clones and inserted in frame with the HA::2A-γb ORF. All 42 γ.cDNA::HA::2A-γb RNAs were infectious when co-inoculated onto barley plant with BSMV RNAs α and γ.Δβa. For each vector, systemically infected barley tissue was tested in replicate for recombinant protein expression using two anti-HA antibodies. From the 42 genes, 38 genes showed confirmed expression of protein product in each plant tested (Figure 5C). In each case, the size of the protein detected in the Western blot was consistent with the size predicted for the post-cleavage heterologous protein, with the addition of the HA epitope tag and the 2A sequence. The 2A cleavage differed between the different heterologous genes, but cleavage rates of 60-95% were usually observed.

In our survey of 42 full length rice ORFs between 200 and 1800 nucleotides in size. The resulting gene products including proteins of 38, 46, 54 and 64 kDa proteins. Due to the average insert size of the library, proteins of 20-30 kDa were most commonly detected in our study. The flexibility of the BSMV expression vector was further demonstrated is ability to successfully express proteins of cytosolic localization (GFP and BMV CP) and those that are matured through the plant secretory pathway (including a lysozyme, interferon, human growth hormone and protease inhibitor). The maturation of these proteins demonstrates the ability of the 2A cleavage system to deliver proteins to distinct subcellular fates and still retain activity. In addition, these data also demonstrate the ability of the vector to express genes from both plant and animal sources successfully.

In summary, to develop a monocot vector capable of expressing free, heterologous proteins, we have tested a variety of strategies based on the BSMV γ RNA. All of the strategies relied on expression from the γb subgenomic RNA facilitating expression of >90% (38/42) recombinant proteins tested. The recombinant proteins were of a variety of sizes, ranging from 11 to 64 kDa. These data demonstrate that monocot recombinant expression vectors can be developed that show many of the flexile and attractive features of traditional dicot expression systems. These vectors can be deployed for testing the function of plant genes in both monocot and dicot species as well as express proteins of recombinant proteins of biomedical importance.

Dicot Vectors	Monocot Vectors
Primarily monopartite genomes with some multipartite examples	Primarily multipartite genomes with a monopartite example
Independent and minimal-virus vectors demonstrated for whole plant expression	Independent-virus vectors only demonstrated for whole plant expression; minimal-virus vectors restricted to cell culture expression
Vector delivery as infectious RNA transcripts or Agroinfiltration of DNA-based expression vectors	Vector delivery as infectious RNA transcripts
Successfully used for gene silencing and gene overexpression	Successfully used for gene silencing and gene overexpression
Systemic expression of foreign proteins primarily through non-genetic fusion strategies	Systemic expression of foreign proteins primarily through genetic fusions to virus proteins and inclusion of cleavage sequences
Expression of single gene cistrons	Potential expression of multiple cistrons as β and γ gene fusions
Systemic expression of wide range of gene sizes and classes of proteins	Systemic expression of wide range of gene sizes and classes of proteins
Successful integration into cGMP recombinant protein production environment	Not integrated to date into cGMP production environment

Table 2. Comparison of properties of virus vectors for expression in dicot and monocotyledonous plant species.

7. Conclusion

The last few decades have seen tremendous progress in developing tools and expertise to produce recombinant proteins in plants. Although conceptually straightforward, the technical hurdles included not only improving our understanding of plant biology, development of expression systems, but also the perfection of purification and analytical methods to meet the specifications of research, industrial and medical applications. The successes described in this chapter involved a convergence of economic incentives, market forces and regulatory acceptance, the latter being particularly important for biomedical products. Throughout much of this time, transient viral expression systems have played a significant role. The adaptability of virus systems and their ease of use continue to help push the boundaries of recombinant protein expression in plants. Beginning with a few examples highlighted in this review, a diverse array of viral vector systems have emerged capable of delivering target genes to a wide array of host species, and compliant with a wide range of regulatory and technical constraints. As shown in this chapter, expression successes initially observed in dicotyledonous plants have now been extended to monocotyledonous plants through the use of the BSMV genome. A comparison of the properties of virus vectors designed for recombinant protein expression in dicot and monocot plant species is provided in Table 2. Research continues to improve their effectiveness and ease of use. Indeed, the advantages of low development costs, flexible design, and relatively quick turnaround from conception to proof of principle to scale up, will continue to drive innovation and application of viral expression vectors. The synergy of customizing specific plant hosts for post-translational modifications offers a strategy to produce plant-sourced products which match the needs of the proposed end uses. Further, the unexpected discovery of viral vectors as tools to silence specific genes in plants has also been of tremendous value to the plant research community, and has effectively complemented their use in overexpression in a way that has only begun to be appreciated and applied in the last decade. The use of these two complementary approaches to address functional genomics in a high throughput fashion, and on a broad range of hosts, will likely emerge in the coming years.

Author details

Gregory P. Pogue
Kentucky BioProcessing, LLC, Owensboro, KY, USA
IC² Institute, The University of Texas at Austin, Austin, Texas, USA

Steven Holzberg
Folsom Lake College, Folsom, CA, USA

Acknowledgement

The authors would like to acknowledge the support of Kentucky BioProcessing, LLC in the preparation of this manuscript and their continued leadership in the development and application of plant-virus expression systems in recombinant protein manufacturing processes. The excellent technical efforts of Mr. Paul Brosio and Dr. Long Nguyen are also acknowledged for their contributions to the novel BSMV overexpression vectors described within this chapter.

8. References

[1] Farnsworth NR, Akerele O, Bingel AS, Soejarto DD, Guo Z. Medicinal Plants in Therapy. Bulletin - World Health Organization 1985;63: 965-81.

[2] Dias DA, Urban S, Roessner U. A Historical Overview of Natural Products in Drug Discovery. Metabolites 2012;2: 303-36.

[3] Taylor L. The Healing Power of Rainforest Herbs. City Park, NY: Square One Publishers, Inc; 1996.

[4] Newman DJ, Cragg GM. Natural Products as Sources of New Drugs Over the Last 25 Years. Journal of Natural Products 2007;70: 461-77.

[5]   Hiatt A, Cafferkey R, Bowdish K. Production of Antibodies in Transgenic Plants. Nature 1989;342: 76-78.

[6]   Sijmons PJ, Dekker BM, Schrammeijer B, Verwoerd TC, van den Elzen PJ, Hoekema A. Production of Correctly Processed Human Serum Albumin in Transgenic Plants. Biotechnology 1990;8(3): 217-21.

[7]   Siegel A. RNA Viruses as Cloning Vehicles. Phytopathology 1983;73: 775.

[8]   Ahlquist P, French R, Janda M, Loesch-Fries LS. Multicomponent RNA Plant Virus Infection Derived from Cloned Viral cDNA. Proceedings of the National Academy of Sciences USA 1984;81(22): 7066-70.

[9]   Ahlquist P, Dasgupta R, Kaesberg P. Nucleotide Sequence of the Brome Mosaic Virus Genome and Its Implications for Viral Replication. Journal of Molecular Biology 1984;172(4): 369-83.

[10] Haseloff J, Goelet P, Zimmern D, Ahlquist P, Dasgupta R, Kaesberg P. Striking Similarities in Amino Acid Sequence Among Nonstructural Proteins Encoded by RNA Viruses that Have Dissimilar Genomic Organization. Proceedings of the National Academy of Sciences USA 1984;81(14): 4358-62.

[11] French R, Janda M, Ahlquist P. Bacterial Gene Inserted in an Engineered RNA Virus: Efficient Expression in Monocotyledonous Plant Cells. Science 1986;231: 1294-97.

[12] Pogue GP, Lindbo JA, Garger SJ, Fitzmaurice WP. Making an Ally from an Enemy: Plant Virology and the New Agriculture. Annual Review of Phytopathology 2002;40: 45-74.

[13] Lico C, Chen Q, Santi L. Viral Vectors for Production of Recombinant Proteins in Plants. Journal of Cellular Physiology 2008;216: 366-77.

[14] Pogue GP, Vojdani F, Palmer KE, Hiatt E, Hume S, Phelps J, Long L, Bohorova N, Kim D, Pauly M, Velasco J, Whaley K, Zeitlin L, Garger SJ, White E, Bai Y, Haydon H, Bratcher B. Production of Pharmaceutical-Grade Recombinant Aprotinin and a Monoclonal Antibody Product Using Plant-Based Transient Expression Systems. Plant Biotechnology Journal 2010;8(5): 638-54.

[15] Lico C, Santi L, Twyman RM, Pezzotti M, Avesani L. The Use of Plants for the Production of Therapeutic Human Peptides. Plant Cell Reports 2012;31(3): 439-51.

[16] De Zoeten GA, Penswick JR, Horisberger MA, Ahl P, Schultze M, Hohn T. The Expression, Localization, and Effect of a Human Interferon in Plants. Virology 1989;172 213-22.

[17] Paul M, van Dolleweerd C, Drake PMW, Reljic R, Thangaraj H, Barbi T, Stylianou E, Pepponi I, Both L, Hehle V, Madeira L, Inchakalody V, Ho S, Guerra T, Ma JKC. Molecular Pharming: Future Targets and Aspirations. Human Vaccines 2011;3: 375-82.

[18] Lacomme C. Milestones in the Development and Applications of Plant Virus Vector as Gene Silencing Platforms. Current Topics in Microbiology and Immunology 2011;28.

[19] Maggon K. Top Ten/Twenty Best Selling Drugs 2010: World Best Selling Human Medicinal Brands 2010.
      http://knol.google.com/k/krishan-maggon/top-ten-twenty-best-selling-drugs-2010/5fy5e owy8suq3/141 (accessed 10 November 2011).

[20] Maxmen A. Drug-Making Plant Blooms. Nature 2012;485: 160.

[21] TechNavio. Global Cancer Drug Market 2010-2014.
http://top10drugs.wordpress.com/2012/03/26/top-20-drugs-2010/ (accessed 27 May 2012).

[22] Nelson AL, Reichert JM. Development Trends for Therapeutic Antibody Fragments.
Nature Biotechnology 2009;27: 331-7.

[23] Datamonitor. Monoclonal Antibodies Report: 2008. London: Business Information
Center; 2008.

[24] Maggon K. Top Ten Monoclonal Antibodies 2010: Global Market Analysis &
Blockbuster mAbs.
http://monoclonalantibodies.wordpress.com/2012/03/07/top-10-mabs- 2010 (accessed 17
March 2012).

[25] BCC Research. Next-Generation Cancer Diagnostics: Technologies and Global Markets.
http://www.bccresearch.com/report/BIO081A.html (accessed 20 May 2012).

[26] Odum JN. Biotech Manufacturing: Is the Crisis Real? Pharmaceutical Engineering
2001;21: 22-33.

[27] Metzler NT. The First Plant-Derived Vaccine Approved for Chickens.
http://www.pharmexec.com/pharmexec/The-First-Plant-Derived-Vaccine-Approved-for
-Chick/ArticleStandard/Article/detail/307471 (accessed 21 May 2012).

[28] Thiel KA. Biomanufacturing, from Bust to Boom....to Bubble? Nature Biotechnology
2004;22: 1365-72.

[29] Belson NA. US Regulation of Agricultural Biotechnology: An Overview. AgBioForum
2000;3: 268-80.

[30] European Food Safety Authority. Scientific Opinion on Guidance for the Risk
Assessment of Genetically Modified Plants Used for Non-Food or Non-Feed Purposes.
European Food Safety Authority Journal 2009;1164: 1-42.

[31] McCormick AA, Reddy S, Reinl SJ, Cameron TI, Czerwinkski DK, Vojdani F, Hanley
KM, Garger SJ, White EL, Novak J, Barrett J, Holtz RB, Tusé D, Levy R. Plant Produced
Idiotype Vaccines for the Treatment of Non-Hodgkin's Lymphoma: Safety and
Immunogenicity in a Phase I Clinical Study. Proceedings of the National Academy of
Sciences USA 2008;105: 10131-36.

[32] Yusibov V, Streatfield SJ, Kushnir N. Clinical Development of Plant-Produced
Recombinant Pharmaceuticals: Vaccines, Antibodies and Beyond. Human Vaccines and
Immunotherapeutics 2011;7(3): 313-21.

[33] Gleba Y, Marillonnet S, Klimyuk V. Plant Virus Vectors (Gene Expression Systems).
Encyclopedia of Virology 2008;4: 229-37.

[34] Vézina LP, Faye L, Lerouge P, D'Aoust MA, Marquet Blouin E, Burel C, Lavoie PO,
Bardor M, Gomord V. Transient Co-Expression for Fast and High Yield Production of
Antibodies with Human Like N-Glycans in Plants. Plant Biotechnology Journal 2009;7:
442-55.

[35] Dawson WO, Bubrick P, Grahtham GL. Modifications of the Tobacco Mosaic Virus Coat
Protein Gene Affecting Replication, Movement and Symptomatology. Phytopathology
1988;78: 783-89.

[36] Dawson WO, Lewandowski DJ, Hilf ME, Bubrick P, Raffo AJ, Shaw JJ, Grantham GL,
Desjardins PR. A Tobacco Mosaic Virus-Hybrid Expresses and Loses an Added Gene.
Virology 1989;172: 285-92.

[37] Takamatsu N, Ishikawa M, Meshi T, Okada Y. Expression of Bacterial Chloramphenicol Acetyltransferease Gene in Tobacco Plants Mediated by TMV-RNA. The EMBO Journal 1987;6: 307-11.

[38] Donson J, Kearney CM, Hilf ME, Dawson WO. Systemic Expression of a Bacterial Gene by a Tobacco Mosaic Virus-Based Vector. Proceedings of the National Academy of Sciences USA 1991;88: 7204-8.

[39] Chapman S, Kavanagh T, Baulcombe D. Potato Virus X as a Vector for Gene Expression in Plants. The Plant Journal 1992;2: 549-57.

[40] Shivprasad S, Pogue GP, Lewandowski DJ, Hidalgo J, Donson J, Grill LK, Dawson WO. Heterologous Sequences Greatly Affect Foreign Gene Expression in Tobacco Mosaic Virus-Based Vectors. Virology 1999;255: 312-23.

[41] Lacomme C, Pogue GP, Wilson TMA, Santa Cruz S. Plant Viruses as Gene Expression Vectors. In: Ring CJA, Blair ED (eds.) Genetically Engineered Viruses. Oxford, UK: BIOS Science; 2001. p59-99.

[42] Santa Cruz S, Chapman S, Roberts AG, Roberts IM, Prior DA, Oparka KJ. Assembly and Movement of a Plant Virus Carrying a Green Fluorescent Protein Overcoat. Proceedings of the National Academy of Sciences USA 1996;93: 6286-90.

[43] Smolenska L, Roberts IM, Learmonth D, Porter AJ, Harris WJ, Wilson TM, Santa Cruz S. Production of a Functional Single Chain Antibody Attached to the Surface of a Plant Virus. FEBS Letters 1998;441: 379-82.

[44] Kurth EG, Peremyslov VV, Prokhnevsky AI, Kasschau KD, Miller M, Carrington JC, Dolja VV. Virus-Derived Gene Expression and RNA Interference Vector for Grapevine. Journal of Virology 2012;86(11): 6002-9.

[45] Muruganantham M, Moskovitz Y, Haviv S, Horesh T, Fenigstein A, Preez J, Stephan D, Burger JT, Mawassi M. Grapevine Virus A-Mediated Gene Silencing in Nicotiana Benthamiana and Vitis Vinifera. Journal of Virological Methods 2009;155(2): 167-74.

[46] Turpen TH. Tobacco Mosaic Virus and the Virescence of Biotechnology. Philosophical Transactions of the Royal Society B: Biological Sciences 1999;354: 665-73.

[47] Pogue GP, Lindbo JA, Dawson WO, Turpen TH. Tobamovirus Transient Expression Vectors: Tools for Plant Biology and High Level Expression of Foreign Proteins in Plants. In: Gelvin SB, Schilperoot RA (eds.) Plant Molecular Biology Manual. The Netherlands: Kluwer Academic Publishers; 1998. p1-27.

[48] Brody J, Levy R. Lymphoma Immunotherapy: Vaccines, Adoptive Cell Transfer and Immunotransplant. Immunotherapy 2009;1(5): 809-24.

[49] McCormick AA, Kumagai MH, Hanley K, Turpen TH, Hakim I, Grill LK, Tusé D, Levy S, Levy R. Rapid Production of Specific Vaccines for Lymphoma by Expression of the Tumor-Derived Single-Chain Fv Epitopes in Tobacco Plants. Proceedings of the National Academy of Sciences USA 1999;96(2): 703-8.

[50] McCormick AA, Reinl SJ, Cameron TI, Vojdani F, Fronefield M, Levy R, Tusé D. Individualized Human scFv Vaccines Produced in Plants: Humoral Anti-Idiotype Responses in Vaccinated Mice Confirm Relevance to the Tumor Ig. Journal of Immunological Methods 2003;278(1-2): 95-104.

[51] McCormick AA. Tobacco Derived Cancer Vaccines for Non-Hodgkin's Lymphoma: Perspectives and Progress. Human Vaccines and Immunotherapeutics 2011;7(3): 305-12.

[52] Marillonnet S, Giritch A, Gils M, Kandzia R, Klimyuk V, Gleba Y. In Planta Engineering of Viral RNA Replicons: Efficient Assembly by Recombination of DNA Modules Delivered by Agrobacterium. Proceedings of the National Academy of Sciences USA 2004;101: 6852-7.

[53] Gleba Y, Klimyuk V, Marillonnet S. Magnifection: A New Platform for Expressing Recombinant Vaccines in Plants. Vaccine 2005;23: 2042-8.

[54] Gleba Y, Klimyuk V, Marillonnet S. Viral Vectors for the Expression of Proteins in Plants. Current Opinion in Biotechnology 2007;18: 134-41.

[55] Giritch A, Marillonnet S, Engler C, van Eldik G, Botterman J, Klimyuk V, Gleba Y. Rapid High Yield Expression of Full Size IgG Antibodies in Plants Coinfected with Noncompeting Viral Vectors. Proceedings of the National Academy of Sciences USA 2006;103: 14701-6.

[56] Yang SJ, Carter SA, Cole AB, Cheng NH, Nelson RS. A Natural Variant of a Host RNA Dependent RNA Polymerase is Associated with Increased Susceptibility to Viruses by Nicotiana Benthamiana. Proceedings of the National Academy of Sciences USA 2004;101: 6297-302.

[57] Azhakanandam K, Weissinger SM, Nicholson JS, Qu R, Weissinger AK. Amplicon-Plus Targeting Technology (APTT) for Rapid Production of a Highly Unstable Vaccine Protein in Tobacco Plants. Plant Molecular Biology 2007;63: 393-404.

[58] Hellens RP, Allan AC, Friel EN, Bolitho K, Grafton K, Templeton MD, Karunairetnam S, Gleave AP, Laing WA. Transient Expression Vectors for Functional Genomics, Quantification of Promoter Activity and RNA Silencing in Plants. Plant Methods 2005;1: 13.

[59] Mallory AC, Parks G, Endres MW, Baulcombe D, Bowman LH, Pruss GJ, Vance VB. The Amplicon Plus System for High Level Expression of Transgenes in Plants. Nature Biotechnology 2002;20: 622-5.

[60] Benchabane M, Saint Jore Dupas C, Faye L, Gomord V, Michaud D. Nucleocytoplasmic Transit of Human Alpha1 Antichymotrypsin in Tobacco Leaf Epidermal Cells. Plant Biotechnology Journal 2009;7: 161-71.

[61] Joh LD, VanderGheynst JS. Agro Infiltration of Plant Tissues for Production of High Value Recombinant Proteins: An Alternative to Production in Transgenic Crops. Journal of the Science of Food and Agriculture 2006;86: 2002-4.

[62] Sourrouille C, Marshall B, Liénard D, Faye L. From Neanderthal to Nanobiotech: From Plant Potions to Pharming with Plant Factories. Methods in Molecular Biology 2009;483: 1-23.

[63] Lindbo JA. TRBO: A High-Efficiency Tobacco Mosaic Virus RNA-Based Overexpression Vector. Plant Physiology 2007;145(4): 1232-40.

[64] Lindbo JA. High-Efficiency Protein Expression in Plants from Agroinfection-Compatible Tobacco Mosaic Virus Expression Vectors. BMC Biotechnology 2007;7: 52.

[65] Sainsbury F, Lavoie PO, D'Aoust MA, Vézina LP, Lomonossoff GP. Expression of Multiple Proteins Using Full-Length and Deleted Versions of Cowpea Mosaic Virus RNA-2. Plant Biotechnology Journal 2008;6(1): 82-92.

[66] Sainsbury F, Liu L, Lomonossoff GP. Cowpea Mosaic Virus-Based Systems for the Expression of Antigens and Antibodies in Plants. Methods in Molecular Biology 2009;483: 25-39.

[67] Palmer KE, Rybicki EP. Investigation of the Potential of Maize Streak Virus to Act as an Infectious Gene Vector in Maize Plants. Archives of Virology 2001;146: 1089-104.

[68] Mor TS, Moon YS, Palmer KE, Mason HS. Geminivirus Vectors for High-Level Expression of Foreign Proteins in Plant Cells. Biotechnology and Bioengineering 2003;81(4): 430-7.

[69] Chen Q, He J, Phoolcharoen W, Mason HS. Geminiviral Vectors Based on Bean Yellow Dwarf Virus for Production of Vaccine Antigens and Monoclonal Antibodies in Plants. Human Vaccines and Immunotherapeutics 2011;7: 331-8.

[70] Regnard GL, Halley-Stott RP, Tanzer FL, Hitzeroth II, Rybicki EP. High Level Protein Expression in Plants through the Use of a Novel Autonomously Replicating Geminivirus Shuttle Vector. Plant Biotechnology Journal 2010;8(1): 38-46.

[71] Lai H, He J, Engle M, Diamond MS, Chen Q. Robust Production of Virus-Like Particles and Monoclonal Antibodies with Geminiviral Replicon Vectors in Lettuce. Plant Biotechnology Journal 2012;10(1): 95-104.

[72] Gleba Y, Giritch A. Plant Viral Vectors for Protein Expression. In: Carole Caranta MAA, Tepfer M, Lopez-Moya JJ (eds.) Recent Advances in Plant Virology. Norfolk: Caister Academic Press; 2011. p387-412.

[73] Hiatt A, Pauly M. Monoclonal Antibodies from Plants: A New Speed Record. Proceedings of the National Academy of Sciences USA 2006;103(40): 14645-6.

[74] Zeitlin L, Pettitt J, Scully C, Bohorova N, Kim D, Pauly M, Hiatt A, Ngo L, Steinkellner H, Whaley KJ, Olinger GG. Enhanced Potency of a Fucose-Free Monoclonal Antibody Being Developed as an Ebola Virus Immunoprotectant. Proceedings of the National Academy of Sciences USA 2011;108(51): 20690-4.

[75] Strasser R, Stadlmann J, Schähs M, Stiegler G, Quendler H, Mach L, Glössl J, Weterings K, Pabst M, Steinkellner H. Generation of Glyco-Engineered Nicotiana Benthamiana for the Production of Monoclonal Antibodies with a Homogeneous Human-Like N-Glycan Structure. Plant Biotechnology Journal 2008;6(4): 392-402.

[76] Castilho A, Gattinger P, Grass J, Jez J, Pabst M, Altmann F, Gorfer M, Strasser R, Steinkellner H. N-Glycosylation Engineering of Plants for the Biosynthesis of Glycoproteins with Bisected and Branched Complex N-Glycans. Glycobiology 2011;21: 813-23.

[77] Castilho A, Gattinger P, Pabst M, Grass J, Kunert R, Strasser R, Altmann F, Steinkellner H. Humanization of the Plant N-Glycosylation Pathway for the Production of Therapeutically Relevant Proteins. Glycoconjugate Journal, 2011;28: 240.

[78] Komarova TV, Kosorukov VS, Frolova OY, Petrunia IV, Skrypnik KA, Gleba YY, Dorokhov YL. Plant-Made Trastuzumab (Herceptin) Inhibits Her2/Neu+ Cell Proliferation and Retards Tumor Growth. PLoS One 2011;6(3): e17541.

[79] Forthal DN, Gach JS, Landucci G, Jez J, Strasser R, Kunert R, Steinkellner H. Fc-Glycosylation Influences Fcγ Receptor Binding and Cell-Mediated Anti-HIV Activity of Monoclonal Antibody 2G12. Journal of Immunology 2010;185: 6876-82.

[80] Byrne B, Donohoe GG, O'Kennedy R. Sialic Acids: Carbohydrate Moieties that Influence the Biological and Physical Properties of Biopharmaceutical Proteins and Living Cells. Drug Discovery Today 2007;12: 319-26.

[81] Castilho A, Pabst M, Leonard R, Veit C, Altmann F, Mach L, Glossl J, Strasser R, Steinkellner H. Construction of a Functional CMP-Sialic Acid Biosynthesis Pathway in Arabidopsis. Plant Physiology 2008;147: 331-9.

[82] Castilho A, Strasser R, Stadlmann J, Grass J, Jez J, Gattinger P, Kunert R, Quendler H, Pabst M, Leonard R, Altmann F, Steinkellner H. In Planta Protein Sialylation Through Overexpression of the Respective Mammalian Pathway. Journal of Biological Chemistry 2010;285: 15923-30.

[83] Jez J, Castilho A, Stadlmann J, Grass J, Antes B, Vorauer-Uhl K, Strasser R, Altmann F, Steinkellner H. The Cherry on the Cake: Expression of Sialylated Human Erythropoietin in Plants. Glycoconjugate Journal 2011;28: 290.

[84] Fitzmaurice WP, Holzberg S, Lindbo JA, Padgett HS, Palmer KE, Wolfe GM, Pogue GP. Epigenetic Modification of Plants with Systemic RNA Viruses. OMICS 2002;6(2): 137-51.

[85] Lindbo JA, Dougherty WG. Pathogen-Derived Resistance to a Potyvirus: Immune and Resistant Phenotypes in Transgenic Tobacco Expressing Altered Forms of a Potyvirus Coat Protein Nucleotide Sequence. Molecular Plant Microbe Interactions 1992;5: 144-53.

[86] Lindbo JA, Silva-Rosales L, Proebsting WM, Dougherty WG. Induction of a Highly Specific Antiviral State in Transgenic Plants: Implications for Regulation of Gene Expression and Virus Resistance. Plant Cell 1993;5: 1749-59.

[87] Senthil-Kumar M, Mysore KS. New Dimensions for VIGS in Plant Functional Genomics. Trends in Plant Science 2011;16(12): 1360-85.

[88] Holzberg S, Brosio P, Gross C, Pogue GP. Barley Stripe Mosaic Virus-Induced Gene Silencing in a Monocot Plant. The Plant Journal 2002;30(3): 315-27.

[89] McKinney HH, Greeley LW. Biological Characteristics of Barley Stripe Mosaic Virus Strains and Their Evolution. Technical Bulletin U.S. Department of Agriculture 1965; 1324.

[90] Zhang L, French R, Langenberg WG, Mitra A. Accumulation of Barley Stripe Mosaic Virus is Significantly Reduced in Transgenic Wheat Plants Expressing a Bacterial Ribonuclease. Transgenic Research 2001;10(1): 13-9.

[91] Weiland JJ, Edwards MC. Evidence that the Alpha a Gene of Barley Stripe Mosaic Virus Encodes Determinants of Pathogenicity to Oat (Avena Sativa). Virology 1994;201(1): 116-26.

[92] Mottinger JP, Johns MA, Freeling M. Mutations of the Adh1 Gene in Maize Following Infection with Barley Stripe Mosaic Virus. Molecular and General Genetics 1984;195(1-2): 367-9.

[93] Palomar MK, Brakke MK, Jackson AO. Base Sequence Homology in the RNAs of Barley Stripe Mosaic Virus. Virology 1977;77(2): 471-80.

[94] Johnson JA, Bragg JN, Lawrence DM, Jackson AO. Sequence Elements Controlling Expression of Barley Stripe Mosaic Virus Subgenomic RNAs in Vivo. Virology 2003;313: 66-80.

[95] Bennypaul HS, Mutti JS, Rustgi S, Kumar N, Okubara PA, Gill KS. Virus-Induced Gene Silencing (VIGS) of Genes Expressed in Root, Leaf, and Meiotic Tissues of Wheat. Functional Integrative Genomics 2012;12: 143-56.

[96] Bruun-Rasmussen M, Madsen CT, Jessing S, Albrechtsen M. Stability of Barley Stripe Mosaic Virus–Induced Gene Silencing in Barley. Molecular Plant-Microbe Interactions 2007;20(11): 1323-31.

[97] Kawalek A, Dmochowska-Boguta M, Nadolska-Orczyk A, Orczyk W. A New BSMV-Based Vector with Modified β Molecule Allows Simultaneous and Stable Silencing of Two Genes. Cellular & Molecular Biology Letters 2012;17(1): 107-23.

[98] Ding XS, Schneider WL, Chaluvadi SR, Mian MAR, Nelson RS. Characterization of a Brome Mosaic Virus Strain and Its Use as a Vector for Gene Silencing in Monocotyledonous Hosts. Molecular Plant Microbe Interactions 2006;19(11): 1229-39.

[99] van der Linde K, Kastner C, Kumlehn J, Kahmann R, Doehlemann G. Systemic Virus-Induced Gene Silencing Allows Functional Characterization of Maize Genes During Biotrophic Interaction with Ustilago Maydis. New Phytologist 2011;189: 471-83.

[100] Lu R, Malcuit I, Moffett P, Ruiz MT, Peart J, Wu A, Rathjen JP, Bendahmane A, Day L, Baulcombe DC. High Throughput Virus-Induced Gene Silencing Implicates Heat Shock Protein 90 in Plant Disease Resistance. The EMBO Journal 2003;22(21): 5690-9.

[101] Meng Y, Moscou MJ, Wise RP. Blufensin1 Negatively Impacts Basal Defense in Response to Barley Powdery Mildew. Plant Physiology 2009;149: 271-85.

[102] Yuan C, Li C, Yan L, Jackson AO, Liu Z, Han C, Yu J, Li D. A High Throughput Barley Stripe Mosaic Virus Vector for Virus Induced Gene Silencing in Monocots and Dicots. PLoS ONE 2011; 6(10): 1-16.

[103] Hensel G, Himmelbach A, Chen W, Douchkov DK, Kumlehn J. Transgene Expression Systems in the Triticeae Cereals. Journal of Plant Physiology 2011;168: 30-44.

[104] Joshi RL, Joshi V, Ow DW. BSMV Genome Mediated Expression of a Foreign Gene in Dicot and Monocot Plant Cells. The EMBO Journal 1990;9(9): 2663-9.

[105] Choi I, Stenger DC, Morris TJ, French R. A Plant Virus Vector for Systemic Expression of Foreign Genes in Cereals. The Plant Journal 2000;23(4): 547-55.

[106] Lawrence DM, Jackson AO. Interactions of the TGB1 Protein during Cell-to-Cell Movement of Barley Stripe Mosaic Virus. Journal of Virology 2001;75(18): 8712-23.

[107] Haupt S, Duncan GM, Holzberg S, Oparka KJ. Sieve-Element Unloading in Sink Leaves of Barley is Symplastic. Plant Physiology 2001;125: 209-18.

[108] Lawrence DM, Jackson AO. Requirements for Cell-To-Cell Movement of Barley Stripe Mosaic Virus in Monocot and Dicot Hosts. Molecular Plant Pathology 2001;(2): 65-75.

[109] Manning VA, Chu AL, Scofield SR, Ciuffetti LM. Intracellular Expression of a Host-Selective Toxin, ToxA, in Diverse Plants Phenocopies Silencing of a ToxA-Interacting Protein, ToxABP1. New Phytologist 2010;187: 1034-47.

[110] Tai Y, Bragg J. Dual Applications of a Virus Vector for Studies of Wheat-Fungal Interactions. Biotechnology 2007;6(2): 288-91.

[111] Donald RGK, Jackson AO. The Barley Stripe Mosaic Virus CB Gene Encodes a Multifunctional Cysteine-Rich Protein that Affects Pathogenesis. Plant Cell 1994;6: 1593-606.

[112] Ryan MD, King AM, Thomas GP. Cleavage of Foot-and-Mouth Disease Virus Polyprotein is Mediated by Residues Located within a 19 Amino Acid Sequence. Journal of General Virology 1991;72(Pt 11): 2727-32.

[113] Donnelly ML, Hughes LE, Luke G, Mendoza H, ten Dam E, Gani D, Ryan MD. The 'Cleavage' Activities of Foot-and-Mouth Disease Virus 2A Site-Directed Mutants and Naturally Occurring '2A-Like' Sequences. Journal of General Virology 2001;82(Pt 5): 1027-41.

Chapter 10

Mutational Analysis of Effectors Encoded by Monopartite Begomoviruses and Their Satellites

Muhammad Shafiq Shahid, Pradeep Sharma and Masato Ikegami

Additional information is available at the end of the chapter

1. Introduction

The geminiviruses are plant-infecting viruses with genomes consisting of circular, single-stranded DNA (ssDNA) geminate particles [86]. Members of the family *Geminiviridae* have been grouped into four genera (*Begomovirus, Curtovirus, Mastrevirus and Topocuvirus*) based on genome organization, host range and insect vector [29, 87]. The majority of geminiviruses belong to the genus Begomovirus, are transmitted by whiteflies (*Bemisia tabaci*: Gennadius), and infect dicotyledonous plant species [85]. The monopartite begomovirus genome is ~2.8 kb nucleotides in length and encode genes both in complementary and virion sense from a non-coding intergenic region that contains promoter sequences and the origin (*ori*) of virion-strand DNA replication. The *ori* consists of a predicted hairpin structure that contains the absolutely conserved (for geminiviruses) nonanucleotide (TAATATTAC) loop sequence and repeated motifs upstream known as iterons.

2. Functions of effectors encoded by monopartite begomoviruses

2.1. Complementary-sense

The complementary-sense strand encodes the Rep proteins, also known as C1, AC1 and AL1, is a multifunctional protein and the only viral protein absolutely required for virus replication. Rep is encoded on the complementary sense strand (Fig. 1 DNA A). This protein is involved in several biological processes: initiation and termination of rolling circle replication (RCR) by nicking and religating the replication origin of viral DNA [45] and repression of its own gene transcription [19]. The Rep proteins of geminiviruses are closely related and show substantial sequence conservation. Four functional domains have been delineated for begomovirus Rep : the N-terminal domain (amino acids 1 to 120), which is involved in initiation by geminiviruses [63], AC1 protein initiates rolling circle replication by a site-specific cleavage within the loop of the conserved nonamer sequence, TAATATTAC [30]. The AC1 protein binding site is located between the TATA box and the transcription start site for the *Rep* gene and acts as the origin recognition sequence and as a negatively regulatory element for *AC1* gene transcription [19], the oligomerization domain (121 to 180 aa), leading to interactions with itself and with host factors [28]. The AC1 protein alone can initiate RCR without requiring other accessory viral factors [34]. AC1 protein also has DNA helicase activity which depends upon the oligomeric state of the protein [14].

The transcriptional activator protein (TrAP); is also known as AC2, C2 an AL2. AC2 is a ~15-KD a transcriptional activator protein unique to begomoviruses because it is absent in mastreviruses and a related protein in curtoviruses, AC2 protein, seems to play a different role. In mastreviruses, AC1 protein provide the functions of AC2 [51]. TrAP is necessary for transactivation of late genes [90]. Recently, several researchers have shown that the AC2 gene of *Cabbage leaf curl virus* (CaLCuV) activates the CP promoter in mesophyll and acts to derepress the promoter in vascular tissue, similar to that observed for TGMV [44]. Further, since AC2 1-100 is as effective a suppressor as the full-length AC2 protein, activation and silencing suppression appear to be independent activities. For example Gopal et al. [26] showed that AC2 of *Bhindi yellow vein mosaic virus* (BYVMV) is involved in transactivation and only mildly in suppression of gene silencing of monopartite begomoviruses viruses and not in transmission.

The replication enhancer protein (REn); also named as AC3/AL3. AC3 is a ~16 KD a protein in curtoviruses and in begomoviruses. The AC3 protein greatly enhances viral DNA accumulation of curtoviruses and begomoviruses [22, 92] by interacting with Rep [81]. Experimental observations suggested that AC3 protein might increase the affinity of Rep for the origin. Complementation studies revealed that AC3 could act on heterologous viruses [93].

The C4 protein, for which the function remains unclear but for some viruses is a pathogenicity determinant and a suppressor of PTGS [73]. AC4 is highly variable among begomoviruses, which is expressed from an open reading frame (ORF) embedded in the Rep ORF.

2.2. Virion-sense

The virion-sense strand encodes the genes required for insect transmission and movement in plants, coat protein (CP) and V2 protein. Monopartite begomovirus capsids are composed of a single CP, encoded by the *V1* gene or (also known as AV1), depending on the geminivirus [107]. For monopartite geminiviruses, CP is essential for systemic spread through the plant [12]. Besides the encapsidation function, CP is also required for transmission of the virus between the plants. The CP of the monopartite begomoviruses facilitates the transfer of infecting viral DNA into the host cell nucleus and is essential for systemic virus movement [5, 46, 50, 104]. The CP also determines the vector specificity [10, 32, 33] and protects the viral ssDNA from degradation during transmission by the insect vector [3], or mechanical inoculation [24].

In contrast to New World (NW) begomoviruses, Old World (OW) begomoviruses have AV2/V2 and this is involved in the movement of monopartite viruses. A recent report shown that the V2 (a homolog of AV2) of a monopartite begomovirus is involved in overcoming host defenses mediated by post-transcriptional gene silencing as well as in movement [114, 115]. V2 targets a step in the RNA silencing pathway which is subsequent to the Dicer-mediated cleavage of dsRNA [109, 70].

3. Role of effectors encoded by satellites

3.1. Betasatellite

Recently, the majority of the begomoviruses originating from the OW have been shown to be monopartite and to associate with a class of ssDNA satellites known as betasatellites (earlier known as DNAβ) [9]. Betasatellites are approximately half the size of their helper begomoviruses (approx.1.4 kb) and are required by the helper virus to induce typical disease symptoms in their original hosts. The success of begomovirus-betasatellite disease complexes appears to be due to the promiscuous nature of betasatellites that allows them to be *trans*-replicated by several distinct begomoviruses [53, 59]. These begomovirus-betasatellite disease complexes are widespread throughout OW and outnumber bipartite begomoviruses whereas in the NW only bipartite begomoviruses are native. There have been recent reports which showed that betasatellite can complement the function of DNA B, suggesting that the betasatellite may provide movement functions to its helper begomovirus [74] Betasatellite can be associated with distinct begomoviruses and it can interact and make new complex with diverse monopartite begomoviruses [110-112].

Tomato leaf curl virus (ToLCV), originating from Australia, was shown to be associated with a single-stranded DNA satellite molecule [18]. The ToLCV satellite (ToLCV-sat) is approximately 682 nt in length and sequence unrelated to ToLCV and it depends on ToLCV for replication and encapsidation. It has no discernable effects on viral replication or symptoms caused by ToLCV. Betasatellites have three structural features: a approx.115 bp highly conserved region, *βC1* gene and a region rich in adenine, [76, 108] (Fig. 1 betasatellite). This gene has the capacity to encode a 13-14-kDa protein comprising 118 amino acids, although some betasatellites have additional N-terminal amino acids [79, 108]. Recently it has been shown that betasatellite to be pathogenicity determinant and suppressor of RNA silencing [16, 66]. It also induced abnormal cell division in *N. benthamiana* [17]. Betasatellites do not contain the iterons of their helper begomoviruses, although betasatellite clearly must possess sequences that are recognized by the begomovirus-encoded Rep in order to allow transreplication of the betasatellite [76].

All the reported betasatellites [54] or defective betasatellites (half size of wild type betasatellite) [7] contain the A-rich region, the A-rich region may play biological role in betasatellites [95]. A-rich region is not required for *trans*-replication of betasatellite and not related with encapsidation also. However, the A-rich region deleted mutant caused milder symptom [95]. The begomovirus accumulates to normal levels in Ageratum in the presence of betasatellite suggesting that the satellite functions either by facilitating the replication or movement of the begomovirus or by suppressing a host defense mechanism such as gene silencing. Recently it has been shown that a betasatellite can override the AC4 pathogenicity phenotype of TLCV and it can complement the function of DNA B [73]. Despite its importance to the disease phenotype, there is still no information available concerning even the most fundamental properties of the satellite.

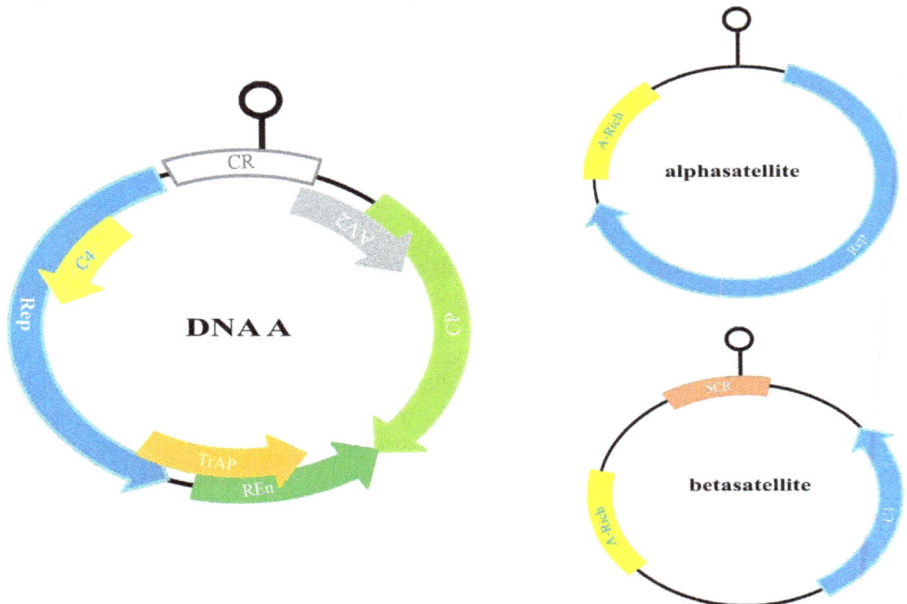

Figure 1. Genome organization of monopartite begomoviruses-satellite complex. DNA-A (encoding replication-associated protein [Rep], coat protein [CP], replication enhancer protein [REn], transcriptional activator protein [TrAP] and proteins possibly involved in virus movement [AV2], pathogenicity determinant and a suppressor of RNA silencing [AC4], viral genome replication [AC5]) Alphasatellites are self-replicating molecules encoding their own Rep. Betasatellites are dependent on their helper viruses for their replication and encode a single protein, βC1, which upregulate replication of helper virus and suppress host defense. Both satellites have an A-rich region and in addition to this betasatellites have a region of sequence conserved between all examples known as the satellite conserved region (SCR).

In Arabidopsis, these pathways are affected by the DICER- like proteins (DCL1, DCL2, and DCL3) that are nuclear localized and are required for miRNA and siRNA biogenesis. Thus, βC1 protein may affect the activity of the DICER-like proteins in plants during nuclear activities that function in silencing suppression. The other possibilities are that βC1 protein could down-regulate transcription of a host protein that acts in the PTGS pathway in the cytoplasm or that βC1 protein could activate transcription of a host PTGS inhibitor [15].

4. Alphasatellites

Many begomovirus betasatellite complexes are also associated with a third ssDNA component for which the collective term alphasatellite (earlier known as DNA 1; R.W. Briddon, manuscript in preparation). However, alphasatellites are dispensable for virus infection and appear to play no significant role in the etiology of the diseases with which they are associated [56]. Alphasatellite components are satellite-like, circular ssDNA molecules approx.1375 nucleotides in length (Fig.1 alphasatellite). They encode a single gene, a rolling circle replication initiator protein (Rep), and are capable of autonomous replication in plant cells. Closely related to the replication associated protein encoding components of nanoviruses (a second family of plant infecting ssDNA viruses), from which they are believed to have evolved, they require a helper begomovirus for movement within and between plants [56, 80].

Several alphasatellites are capable of replicating and systemically infecting their plant host in the presence of a helper begomovirus without a visible effect on symptom development or virulence [6, 40]. However AYVSGA a different type of 'DNA-2'-class alphasatellite that ameliorates symptom severity in an infected host and also capable of reducing virulence and the relative accumulation of its associated Tomato leaf curl betasatellite (ToLCB) [1]. Alphasatellites have been acquired by helper begomoviruses to restrain virulence to achieve increased viral fitness [76, 105].

Recently, two 'DNA-1-type' alphasatellites Gossypium mustelinium symptomless alphasatellite (GMusSLA) and Gossypium darwinii symptomless alphasatellite (GDarSLA) phylogenetically divergent from the DNA-2-type alphasatellite have each been shown to attenuate symptoms caused by their helper begomovirus [60]. However [35] hypothesize that symptom attenuation and a relative reduction in betasatellite accumulation might result from DNA-2-mediated modulation of betasatellite activity. Possibly alphasatellite modulates begomovirus-betasatellite pathogenicity by interfering with βC1, a key virulence factor [8]. Also alphasatellite rep can interact with C4 of CLCuRaV that might be providing an additional possible mechanism for symptom amelioration by alphasatellites. Furthermore alpha-Rep down regulate betasatellite replication (In the field), and thus down-regulation of the manifestation of the pathogenicity determinant βC1 [60], moreover alpha-Rep proteins GMusSLA and GDarSLA can act as a strong suppression of posttranscriptional gene silencing (PTGS) [60].

5. Post-transcriptional gene silencing (PTGS)

Post-transcriptional gene silencing (PTGS) which is initiated by double stranded RNA (dsRNA) is common in plant–virus interactions and is an evolutionarily conserved mechanism that protects host cells against invasive nucleic acids, such as viruses, transposons and transgenes [100]. As a counter to this host defense, most plant viruses encode proteins which act as suppressors of PTGS [71]. Viral suppressors of PTGS interfere with various steps of this pathway including initiation, maintenance or systemic silencing which are mainly downstream of dsRNA production [52, 57].

RNA silencing in plants operates as an antiviral defense response; to establish infection, viruses must suppress RNA silencing by the host [100]. Begomoviruses have been shown to induce PTGS in infected plants by producing virus specific siRNAs (21, 22 and 24 nt) [97]. To counteract this host defence, geminiviruses encode RNA silencing suppressors [4]. However, depending on each intrinsic virus and its interaction with plant host factors, the efficacy of virus-induced PTGS may vary [99]. At least three RNA-silencing suppressors have been reported in TYLCD-associated or related begomoviruses. Thus, the V2 protein of TYLCV functions as an RNA-silencing suppressor; it counteracts the innate immune response of the host plant by interacting with SlSGS3, the tomato homolog of the Arabidopsis SGS3 protein involved in the RNA-silencing pathway. The TrAP protein of the related monopartite begomovirus. *Tomato yellow leaf curl China virus* (TYLCCNV) is also involved in suppression of RNA silencing [98], probably by activating transcription of host genes that control silencing [97]. The C4 protein of the monopartite begomoviruses ToLCV, *Ageratum yellow vein virus* (AYVV), and *Bhendi yellow vein mosaic virus* (BiYVMV) also have the ability to suppress RNA silencing [26, 84].

6. Mutagenesis of effectors encoded by monopartite begomoviruses

Little is known about gene function in monopartite begomoviruses. However, gene function has been studied extensively in other types of geminiviruses which share organization and nucleotide sequence similarities with TYLCV. Mutational analysis of few monopartite begomoviruses like TYLCV define similarities and differences between this single component geminivirus and bipartite geminiviruses in functions essential for systemic spread and infectivity [103]. The CP appeared to be required for systemic movement of TYLCV in *N. benthamiana* and tomato, consistent with those of mutation analyses obtained with other monopartite geminiviruses such as MSV, BCTV, and TLCV [5, 12, 69, 61] have shown that *Tomato yellow leaf curl virus*-Sardinia (TYLCV-Sar) C2 can form stable complexes with ssDNA (and less preferably with dsDNA) and that the binding is sequence nonspecific. AC2 of TYLCV have also been involved in the activation of other viral genes and was considered as a transcriptional activator [91]. However, transcription factors usually show high sequence specificity.

TYLCV ORF V1 truncated either 133 nt upstream or 19 nt downstream of the initiation codon of ORF V2 would altered the viral DNA forms, it suggested that the V1 protein may participate in the switch from dsDNA to ssDNA synthesis. Indeed, interaction between V2 and the CP has already been proposed, in view of the concerted evolution of these two protein sequences following a geo-graphical gradient of similarity [39], and the synergistic reduction in ssDNA levels of a TYLCV V1-V2 double mutant compared to single mutants [69]. Although TYLCV V1 mutants did not greatly overproduce dsDNA, the similarity of phenotype between BCTV V2 and TYLCV V1 mutants may indicated that the two corresponding gene products serve a related function. It has shown [69] (Table 1) has shown that disruption of the V1 gene in the monopartite Australian isolate of TLCV did not affect its ability to spread in tomato, although the infection was asymptomatic and the DNA levels reduced.

Virus	Accession number	Mutated ORF	Mutation Position	Type of mutation	References
TLCV-[AU]	AF084006	V1	N-terminal/_Bam_HI	frameshift	Rigden et al., 1993
		V2	N-terminal/_Bgl_II	frameshift	
		V/V2	N-terminal/_Bgl_II/_Bam_HI	frameshift	
		V/V2	C-terminal/_Bgl_II/_Bam_HI	deletion	
		V/V2	C-terminal/_Bam_HI/_Bgl_II	inversion	
TLCV-[AU]	AF084006	C4	N-terminal (at 2457&2463)	deletion	Rigden et al., 1993
		C4	N-terminal (at 2457&2463)	revertion	
TYLCV-Sar[ES:Psp95:93]	Z25751	C4	C>T at 2432	point	Jupin et al., 1994
		C4	C>T at 2423		
TYLCV-Sar[ES:Psp95:93]	Z25751	C2	ΔCC2>31	deletion	Noris et al., 1996
		C2	ΔNC2>33	deletion	
		C2	ΔC2>33-104	deletion	
TYLCV-Sar[FR:98]	X61153	C2	1523+GATC	frameshift	Wartig et al.,1997
		V1	156+GATC	frameshift	
		V1	324C>T	stop	
		V2	748>CTAG	stop	
TYLCV-Sac[IT:pSic36:95]	Z28390	CP	H134	substitution	Noris et al., 1998
		RepC1	L198	substitution	
TYLCV-Sar[ES:Psp95:93]	Z25751	CP	Q134H	site-directed	Noris et al., 1998
TYLCV-DO[DO:99]	AF024715	CP	ΔNCP>180	deletion	Rojas et al., 2001
		CP	ΔCCP>150	deletion	
		V1	ΔNV1>63	deletion	
		V1	ΔCV1>84	deletion	
		C4	ΔNC4>54	deletion	
		C4	ΔCC4>60	deletion	
AYVV-[SG:pHN419:97]	X74516	C4	A>T at 2419 C4mut	substitution	Saunders et al., 2004

Virus	Accession number	Mutated ORF	Mutation Position	Type of mutation	References
PaLCuV-[PK:02]	AJ436992	V2	N-terminal (1-32)	deletion	Mubin et al., 2010
		V2	N-terminal (1-60)	deletion	
		V2	C-terminal (130-149)	deletion	
		V2	C-terminal (111-149)	deletion	
		V2	C-terminal (101-149)	deletion	
		V2	C-terminal (91-149)	deletion	
		V2	C-terminal (71-149)	deletion	
		V2	C-terminal (30-149)	deletion	
TYLCV-IL[IL:89]	X15656	CP	Lys-Thr CPmut3	substitution	Yaakov et al., 2011
		CP	Arg-Pro CPmut4	substitution	
		CP	Arg-Leu CPmut19	substitution	
TYLCV-Mld[ES:72:97]	AF071228	C4	C>G at 9 C4mut	substitution	Tomas et al., 2011
ToLCJV-A[ID]	AB100304	CP	CPΔ191-257	deletion	Sharma et al., 2009
		CP	CPΔ1-190	deletion	
		CP	CPΔ31-257	deletion	
		CP	CPΔ1-30/Δ50-257	deletion	
		CP	CPΔ1-30/Δ191-257	deletion	
		CP	CPΔ1-31	deletion	
		CP	CPΔ16-20	deletion	
		CP	CPΔ1-190/Δ245-250	deletion	
		CP	CPΔ1-30/Δ62-257	deletion	
ToLCJV-A[ID]	AB100304	V2	N-terminal (58aa)		Sharma et al., 2010
		V2	C-terminal (58aa)		
ToLCJV-A[ID]	AB100304	V2	pGEMV2ΔC		Sharma et al., 2011

Table 1. List of published studies reporting deletion mutants of monopartite begomoviruses

For example Noris et al. [62] suggested that the region of the CP between amino acids 129 and 134 is essential for both the correct assembly of virions and transmission by the insect vector. The genome of the SicRcv (infectious) had the same size as the original Sic DNA 9 (non-infectious) differed by only 2 nt. One change was at nt 2025 (A instead of T in the plus strand), determining a CAC-to-CUC codon change in the RepC1 mRNA and an H198L replacement in the RepC1 protein. The other mutation located at nt 708 (C instead of G), determining a CAG-to-CAC codon change in the CP mRNA and a Q134H replacement in the CP. This indicated that the Q134H mutation changed a viral DNA, only capable of replicating in single cells (Sic), into one that was systemically infectious, but not insect transmissible (SicRcv). Comparative analysis of Sic, SicRcv, and the hybrid genomes and showed that the mutation in the CP gene, not in the Rep gene, was responsible for restoring infectivity in SicRcv; however, it still did not result in a whitefly-transmissible TYLCV. In TYLCV-Sar, the two capsid protein alterations resulted in the same either non-infectious or non-transmissible phenotype. Mutants containing the combinations QQ, QH, and PH at positions 129 and 134 were infectious in plants, whereas those with PQ are not. The PQ mutants can replicate and accumulate CP and V2 protein in leaf discs, but appear unable to produce virus particles. Mutants having the PH combination at positions 129 and 134 infect plants and form apparently normal virions, but are not transmissible by whiteflies. Changing the amino acid at position 152 (D or E) does not influence the phenotype. Requirement of the CP for infection has been demonstrated previously [62] suggested that accurate particle assembly is also necessary. In fact, the PQ mutants, which are unable to assemble virions, accumulate CP in leaf discs, showing that its expression and stability were not altered. Another TYLCV protein, V2, for which a role in virus assembly has recently been, suggested [103].

For example Rojas et al. [70] has shown that C4, V1, and CP gene may function in TYLCV-DC movement. The CP localized to nuclei and nucleoli and was found to act as a nuclear shuttle, mediating the import and export of DNA [70]. It was consistent with results obtained for the TYLCV CP in heterologous experimental systems [43, 68]. Recently, Liu et al [49] also showed the same behavior for the CP of the monopartite mastrevirus, MSV. TYLCV CP was found to accumulate in the nucleolus and the absence of the N-and C-terminal CP mutants from the nucleolus implicates CP motifs in this localization. As the nucleolus is the site of rRNA synthesis and packaging of ribosomal proteins, it may also serve as the site of geminiviral replication/gene expression [70]. The TYLCV C4 targeted to the cell periphery and/or cell wall, consistent with a role in cell-to-cell movement of viral DNA [65, 75, 101].

Disruption of the AYVV C4 ORF (A>T at position 2419nt) alters the phenotype in agroinoculated N. benthamiana from upward leaf roll and vein swelling to downward leaf curl [79] (Table 2). Previously,[88] also found the identical functions of BCTV C4 ORF mutant in this host. The AYVV C4 proteins may perform partially redundant functions involving convergent pathways and the behavior of ACMV AC4 and TYLCCNV C4 is consistent with such a function [98].

For example Stanley and Latham [58] have shown that V2 protein of *Papaya leaf curl virus* (PaLCuV) is potentially involved in the elicitation of cell death response. The deletion mutants (having deletions of 32 and 60 amino acids, respectively, at the N-terminal end of V2) exhibited a systemic HR in *Nicotiana benthamiana* plants. While C-terminal end deletions of 60, 80 and 119 amino acids abolished the induction of HR, however 50 amino acids deletion induced local necrosis, but not systemic. The mutants with 20 and 40 amino acids deletion produce HR both at the inoculated and in newly emerged leaves, although the systemic symptoms for the 40-amino-acid deletion mutant were delayed and were milder. The amino acid sequences between positions 92 and 101 are essential for the elicitation of HR, whereas those between 102 and 115 affect the timing and severity of the response. V2 of PaLCuV at amino acid positions 116 and 118 contains a conserved CxC. Mutations of this motif have been shown to abolish both the pathogenicity and suppressor of RNA silencing activities of the protein [64, 109]. Phosphorylation of MPs may also play a role in controlling the switch from viral replication to translation [36, 37]. Few earlier studies showed that for PaLCuV V2, deletion of sequences encompassing this motif abrogates the ability to induce HR [58].

The first 30 N-terminal amino acids of the TYLCV-IL CP are needed for nuclear import of the protein into the plant cell, suggesting the CP's involvement in nuclear shuttling of the virus genome [43]. This was confirmed by the finding of a strong interaction between the CP and the plant nuclear import receptor karyopherin $\alpha 1$ (Kap $\alpha 1$) [94]. The TYLCV CP has been found to inter act with itself (CP–CP or homotypic interaction) which may be important for capsule assembly as it is made up solely of CP units serving as building blocks. Mutations in the TYLCV-IL V1 gene coding for the TYLCV-IL CP by replacing Lys with Thr, Arg with Pro, and Arg with Leu, according to the positions of amino acids mutated [31]. TYLCV CP mutated failed to interact with the w.t. CP, while the w.t. protein showed strong homo typic interaction. As the CP has been suggested to be a shuttle protein for the viral genome into the plant cell nucleus [43, 70], its interaction with the nuclear-transport mediator Kap $\alpha 1$ is an important step and has been shown to occur at high affinity [94]. A mutation in the NLS domain, in particular at Arg19, disrupts the CP's interaction with proteins that are known to interact with the w.t. CP [106.]. Earlier Sharma et al. [113-115] demonstrated by the constructed a series of single and double deletions into the coding sequence of *Tomato leaf curl Java virus* ToLCJAV-A[ID] CP and found that, amino acids (aa)16KVRRR20 in the N-terminal region of CP functioned as nuclear/nucleolar localization signals (NLSs). Further, the region from aa 52RKPR55 contained basic amino acid cluster was capable to redirect the CP to the nucleus. Deletion mutant analysis revealed that this property was attributed to a nuclear export signal (NES) sequence consisted of aa (245LKIRIY250) reside at C-terminal part of CP. Additionally ToLCJV V2 is a target of host defense responses. Deletion of 58 amino acids (aa) from the N-terminus did not affect the HR, suggesting that this region has no role in the HR, while deletion of 58 aa from the C-terminus of V2 abolished both the HR response and V2 silencing suppressor activity, suggesting that these sequences are required for the HR-like response and suppression of PTGS. He also demonstrated that ToLCJV V2 is a pathogenicity determinant that elicits an HR-like response. Further deletion analysis that fusion of Nterminal part of the V2 containing the nuclear export signals (NES), directed the accumulation of fluorescence towards the cell cytoplasm. Also V2 enhances the coat protein-mediated nuclear export of ToLCJV and is consistent with the model in which V2 mediates viral DNA export from the nucleus to the plasmodesmata.

Satellite	Accession number	region (nt.)	Type of mutation	References
AYVB-[SG:pBS-beta:99]	AJ252072	AT>TA at 547/ 548 βC1mut1 G>T at 486 βC1mut2		Saunders et al., 2004
TYLCCNB-[CN:Y10:01]	AJ421621	ATG>ATC (CIM-F)	site-derected	Cui et al., 2004
		ACT>TGA (ACIM-S)	site-derected	
		ATG (2) >ATC (2) (CIM-B)	site-derected	
		GAA>TAG (CIM-T)	site-derected	
TYLCCNB-[CN:Y10:01]	AJ421621	742-952	deletion	Xiaorong et al., 2004
TYLCCNB-[CN:Y10:01]	AJ421621	ΔC1β	truncation	Qian and Zhou 2005
CLCuVβ-[PK:00]	AJ298903	195-484	site-derected	Saeed et al., 2005
		504-596	stop	
		586	frame-shift	
		ATG>ATA	stop	
BYVMB-[IN:Muth:01]	AJ308425	51-140 ΔNβC1	deletion	Kumar et al., 2006
		1-80 ΔCβC1	deletion	
TYLCV-satDNA-[AU:96]	U74627	Δnt 35-146 (112nt)	deletion	Li et al., 2007
		Δnt 146-296 (151nt)	deletion	
		Δnt 35-296 (262nt)	deletion	
		Δnt 296-420 (1251nt)	deletion	
		Δnt 296-492 (197nt)	deletion	
		Δnt 492-540 (49nt)	deletion	
		Δnt 540-641 (105nt)	deletion	
AYVJB-[ID:04]	AB162142	4>ATGtga	stop	Kont et al., 2007
AYVB-[SG:pBS-beta:99]	AJ252072	794-795	deletion	Saunders et al., 2008
		118-119	deletion	
		804-806	deletion	
		801-1047	deletion	
		1048-1051	deletion	
		1146-1147	deletion	
		1146-1150	deletion	
		1269-1271	deletion	
		1229-1234	deletion	
TꓛCSVB-[CN:Y35:01]	AJ420318	DNAΔC1β		Qian et al., 2008

Satellite	Accession number	region (nt.)	Type of mutation	References
CLCuMA-[PK:2:99]	AJ132345	915-1117	deletion	Shahid et al., 2009
CLCuMuB-[PK:09]	FJ607041	Δ150-840	deletion	Nawaz-ul-Rehman et al., 2009
		Δ1130-116	deletion	
		Δ995-1095	deletion	
CLCuMB-[PK:00]	AJ298903	ΔC1β	deletion	Kharazmi et al., 2012
TYLCCNB-[CN:Y10:01]	AJ421621	N-terminal (NTG)	deletion	Cheng et al., 2011
		C-terminal (CTG)	deletion	
		βC1-ΔCTD	deletion	

Table 2. List of published studies reporting deletion mutants of DNA Satellites

7. Mutational analysis of effectors encoded by satellites

Betasatellite molecules have been associated with numerous monopartite begomoviruses in China, including *Tobacco curly shoot virus* (TbCSV) and TYLCCNV that infect tomato and tobacco field plants [108, 47]. TbCSB is not essential for infection but increases symptoms in some hosts [17, 47]. However in case of TYLCCNB which is essential for symptom induction, The βC1 gene of TYLCCNB is required for symptom induction but not for the replication of betasatellite. Also a mutated βC1 deleted is stably maintained in few hosts by TYLCCNV [17, 67]. TbCSB with the complete βC1 deleted (ΔβC1) returns to a size comparable to that of the intact betasatellite in few systemically infected *N. glutinosa*, *N. tabacum Samsun* and *P. hybrida* plants plants. The levels of accumulation of the size revertant betasatellite were similar to those of ΔβC1 in same hosts (*N. benthamiana* and *N. glutinosa*) plants showing size reversion of the betasatellite developed viral symptoms similar to those induced by TbCSV and DNAΔβC1 [67]. A βC1 gene frame-shift mutant of TYLCCNVB was unable to induce disease symptoms and consequently, did not play a role in silencing suppression [67]. The complete coding region of Y10βC1 (TYLCCNB), followed by N- and C-terminal deletion mutants showed multimerization mediated by amino acids between positions 60 and 100 [13]. Karyopherin-α, a transport receptor involved in nuclear import were reported to interact with the C-terminal sequences of BYVMB- βC1 [42]. A myristoylation-like motif (GMDVNE) positioned at the C-terminal of CLCuMB-βC1 (103 to 108aa) interacted with a ubiquitin-conjugating enzyme involved in targeting proteins for degradation by the 26S proteasome [20, 78] identified sequences on AYVB by deletion mutagenesis required for trans-replication by AYVV. βC1 of Cotton leaf curl Multan betasatellite (CLCuMuB) has been shown to have possible virus movement function [74]. Generally, sequences between the *βC1* gene and the A-rich region are not essential for trans-replication by begomoviruses. Nevertheless, deletion of these sequences abolish the ability of the betasatellite to upregulate virus levels in plants and the symptoms expression [59]. For geminiviruses hairpin structure that contains the nonanucleotide sequence is an essential part of the virion sense origin of replication that is recognized and nicked (within the nonanucleotide sequence) by Rep to initiate rolling-circle replication of the virion strand. Similarly, deletion of betasatellite sequences from 1130 to 116 that is conserved (between all betasatellites) stopped the betasatellite's ability to be trans-replicated and maintained by helper viruses both from OW (CLCuRaV) and New world and *Cabbage leaf curl virus* (CbLCuV). Trans-replication of CLCuMuB remained unaffected by deletion of the sequence between coordinates 995 and 1095 by CLCuRaV [59].

ToLCJAV alone can cause infection and displayed leaf curl symptoms. But, symptom expression of ToLCJAV in the presence of ToLCJAB is enhanced. In contrast, ToLCJAV and AYVB (mutated βC1) restored mild symptoms. It suggested that the βC1 protein was required for symptom induction and is a determinant of pathogenicity, βC1 protein expression in *N. benthamiana* plants and as a suppressor of PTGS [41].

For example Li et al. [47] have shown the deletion mutant of TYLCV sat-DNA (from 296-641nt) lacked the ability to replicate or replicated poorly by deleting of (region nt 35-296). Also sequence from nt 296-35 is to be essential for sat-DNA replication. The deletion of a 112 nt region downstream of the stem-loop from nt 35-146 and 151nt from 146-296 cannot effect on the replication of sat-DNA but reduced significantly. However, the deletion from nt 35-296 regions diminished sat-DNA replication   these deletions loss of genomic sequences required for replication or due to changes in genome size. Heterologous non-viral DNA fragments can restore the wild-type 682 nt sat-DNA size and of replication when the replacement occurred in the region between nt 35 and 296. However, the sequence replacements in the region nt 35 to 296 of the sat-DNA improved the accumulation of sat-DNA considerably relative to the deleted constructs in this region. The sequence elements distributed within the entire sat-DNA molecule contribute to replication activity, but that sequence elements within the region from nt 35 to 296 are dispensable for replication.

For example Saeed et al. [72] used mutagenesis study of CLCuMB and tobacco was used as the host plant rather that cotton, the natural host of CLCuB. Few studies showed that it was symptomless when inoculated with *Cotton leaf cur Multan virus* (CLCuMV) alone but showed drastic symptoms when coinoculated with CLCuMB [9]. *Nicotiana benthamiana* showed a severe symptom on inoculated with CLCuMV with or without CLCuMB. Evidence for the involvement of the βC1 ORF in modulation of symptom expression also provided by [108] demonstrated few DNA β species associated with tomato and tobacco infecting begomoviruses and found that in-frame mutation of the βC1 initiation codon resulted in loss of symptom severity in *N. benthamiana*.

In recent studies Saunders et al. [79] have proved that disruption of the βC1 ORF prevented infection of the AYVB complex in ageratum and altered their phenotype in *N. benthamiana* to that produced by AYVV alone. For example Kumar et al. [42] tested the infectivity of two βC1 mutant constructs, first carrying a stop codon at amino acid position 41 and second with two stop codons at positions 9 and 41, and both resulted in loss of pathogenicity in tobacco plants on coinoculated with TLCV as helper virus. These mutation studies indicated that the βC1 ORF is involved in pathogenicity and that the expression of its N terminal 40 amino acids is not sufficient for its function.

Disruption of the βC1 ORF of AYVB by introducing an internal in-frame nonsense codon (G>T) did not prevent transreplication and systemic movement of the βC1 mutant by AYVV in lab host (*N.benthamiana*). The mutated βC1 removed the influence of the satellite on symptom development in this host and prevented symptomatic infection of ageratum. That suggested the βC1 protein is an important pathogenicity factor that plays an essential role in the proliferation of the AYVV-betasatellite complex in its real host. For example Saunders et al. [79] also shown that βC1 ORF initiation codon (AT) to a nonsense codon (TA) did not completely eliminate betasatellite activity. A similar mutation in the βC1 ORF of a satellite associated with TYLCCNB was shown previously [108]. The βC1 ORF encodes a pathogenicity determinant that suppressed a host defense mechanism [76].

For example [78] have demonstrated that the region of AYVB between the introduced nt 114 and 1047 sites is not required for betasatellite replication. This region includes the βC1 open reading frame (ORF), which encodes a gene essential for pathogenicity [79] and an A-rich region that may serve to maintain the size integrity of the satellite [76]. For example [78] found that the entire ORF is dispensable and is consistent with the findings of [67] for the betasatellite associated with TYLCCNV. In addition, removal of the A-rich region from TYLCCNB was tolerated, although the deletion mutant was associated with milder infection than those produced by the wild-type satellite [95]. In contrast, deletion of this region in AYVB did not affect the phenotype, at least in *N. benthamiana*. Maximum deletions within non coding regions of the begomovirus genome were not tolerated and the deletion mutants revert to wild type size by both intra-and intermolecular recombination during systemic movement [23, 25,]. For example Saunders et al. [78] also demonstrated removal of 361 nt of betasatellite representing 27% of the satellite and the region between nt 1047 and 1146 is important for betasatellite replication. It contained an inverted repeat flanking a sequence that is identical to the ToLCV iteron ToLCV sat-DNA [18]. Protein binding assays followed by mutagenesis have demonstrated that this motif in both ToLCV and sat-DNA represents a high affinity Rep binding site, although it is not required for replication of either the begomovirus or its satellite [48]. Saunders and associates [80] found that the region between nt 1146-1229 and sequences across the nt 1268 of AYVB are also required for replication.

SCR is highly conserved nature between distinct satellites [typically above 65% sequence identity with blocks of absolutely conserved sequence [7] strongly suggests that it also plays an important role in the virus replication cycle. In addition, the adjacent stem-loop and conserved nonanucleotide sequence would be expected to participate in replication. Approximately the 386 nt upstream of the stem–loop structure in ToLCV sat-DNA, as well as the stem–loop structure itself, are essential for replication [47].

βC1 is a multi-functional protein encoded by betasatellites that are associated with the majority of monopartite begomoviruses [11]. For example Cheng et al. [13] proved by deletion mutants of Y10βC1 that multimerization was mediated by amino acids between positions 60 and 100. Previous studies say that the C-terminal sequences of BYVMB-βC1 were interact with karyopherin α, a transport receptor involved in nuclear import [42]. A myristoylation-like motif (GMDVNE) located at the C-terminal of CLCuMB-βC1 (103-108aa) interacted with a ubiquitin-conjugating enzyme involved in targeting proteins for degradation by the 26S proteasome [21]. It also seems to indicate interference with a functionality associated with the C terminus of Y10βC1. βC1 protein of AYVB, CLCuMB or BYVMB with GFP fused at the N-terminus also presented as granular spots in the cytoplasm and around the nucleus [42, 84].

TYLCCNB presumably has one or more *cis*-acting elements needed for replication and binds to TYLCCNV replication protein (Rep) for replication, and these elements are most probably located in the 115-nucleotide highly conserved region of betasatellite upstream of its stem-loop structure. Recently, Astroga [2] showed that a 5-bp core sequence (GGN1N2N3) is a typical constituent of Rep-binding iterons. Conserved GG motifs occur upstream of the115-nucleotide highly conserved region of betasatellite. One or more of these GG motifs, combined with the 115 nucleotide highly conserved region, possibley responsible for Rep binding to betasatellite. However, the Rep binding activity of the TLCV-sat from Australia seems much less specific: TLCV-sat contains an A-rich region but lacks a βC1 gene and is believed to be a defective betasatellite molecule [48]. The effect of mutation of the conserved βC1 gene of TYLCCNVB indicated that the βC1 protein plays a key role in symptom induction.

The position and size of the βC1 gene of the betasatellite molecules are conserved in all betasatellite molecules, and the mutation of the start codon of C1 gene in TYLCCNB showed that it's a pathogenicity determinant [108, 6]. Few studies has been also shown that the βC1 protein of betasatellite associated with TYLCCNV or AYVV is an essential pathogenicity determinant [17, 79], it may act as suppressors of post-transcriptional gene silencing that interfering the host defense system, thus, the presence of C1 protein facilitates efficient infection of the virus in hosts [102]. For example Tau and Zhou [96] showed that *βC1* gene was not required for the TYLCCNV and betasatellite replication and truncated betasatellite molecules with the deletion of the entire *βC1* gene were stable in infected plants. Defective DNAs, betasatellite and alphasatellite associated with begomoviruses are maintained at approximately half the size of the genomic components [83, 89, 55, 80, 77]. Some proofs have been displayed that geminiviruses CP can encapsidate circular ssDNA molecules of about half or quarter the size of the genomic DNA [18, 51]. Immunocapture PCR indicated that the truncated TYLCCNB of about 1 kb in length may be encapsidated with TYLCCNV coat protein in vivo [67].

βC1of BYVMB have a nuclear export or peripheral localization function and βC1 interacts with itself, also with CP and the tomato protein karyopherin α. Mutagenesis of βC1 protein showed that the domain of βC1 interacting with CP is at the N terminal half whereas the domain(s) of βC1interacting with itself and karyopherin α are at the C terminal half and the role of BYVMD βC1 as a suppressor of posttranscriptional gene silencing was explored [42]. Karyopherins are soluble transport receptors that interact with basic NLS sequences and help in nuclear import [27]. Full length betasatellite of CLCuMB can substitute for the movement function of the DNA B of a bipartite begomovirus *Tomato leaf cur New Dehli virus DNA-B* (ToLCNDV DNA-B). However, the betasatellite containing a disrupted βC1 ORF did not mobilize the DNA A for systemic infection, suggested that the βC1 protein was required for movement [74].

8. Potential of mutated satellites using a virus induced gene silencing vectors

8.1. Betasatellite

Betasatellites have about 200nt sequences (known as a-Rich region) conserved among all that indicating may be these sequences have some biological roles in satellites. The role of A-Rich sequence may be to increase the required size of the molecule that is essential for encapsidation or systemic movement by the coat protein or movement protein encoded by begomovirus. TYLCCNB-Y10 could be infectious and mutant betasatellite (deleted a-rich region) could be encapsidated in the coat protein encoded by DNA-A that suggested may be A-Rich region is not required for trans-replication of TYLCCNB but only for size maintaining [95]. For example [20] reported that only a small region of the nucleotide sequence of CLCuMB upstream of the start codon of $\beta C1$ (a 68-nt fragment), which contains a G-box, was important for $\beta C1$ promoter activity. In addition to $\beta C1$ ORF of CLCuMB delete a larger region (complete $\beta C1$) to make it a gene delivery vector for plants. It can potentially tolerate the insertion of larger foreign sequences without affecting promoter activity [38]. Putative promoter and TATA box are located upstream of the $\beta C1$ gene. Thus, the $\beta C1$ gene of betasatellite could be replaced by a foreign gene and be modified to convert it into an expression vector [17]. The modified betasatellite might be an candidate gene silencing vector to study functional genomics in plants [54]. Also leaf curl symptoms in Nicotiana species can be brought by transgenic expression of the $\beta C1$ gene of TYLCCNB that the severity of the symptoms parallels the level of $\beta C1$ transcript in the transgenic plants and their ability to induce symptoms is abolished by mutation of the $\beta C1$ gene. Possibly $\beta C1$ gene of betasatellite may be replaced with a foreign gene and used as an expression vector for gene function analysis in plants [20].

Evidence has been shown that TYLCCNB modified by deletion of its $\beta C1$ gene but retaining the $\beta C1$ promoter and terminator, can be turned into a gene silencing vector. Also, insertion in the vector with fragments of endogenous plant genes or a transgene, in either the sense or antisense direction, can result in effective silencing of the cognate gene in plants [95]. The size of the mutated satellite DNA molecule significantly influences replication efficiency. TLCV sat-DNA can be used as a potential gene expression/silencing vector [47].

8.2. Alphasatellite

Also alphasatellite is a small molecule and easy to manipulate and have a wide host range and can apparently be maintained by a large number of distinct Begomovirus species. It has some sequences (A-rich approx.200 nt.), similar to betasatellite which can, potentially, be removed and still it can replicate autonomous. The A-rich deleted sequences of CLCuMA can not affect its ability to replicate autonomously and move, in trans, by a helper begomovirus that provide a space suitable for insertion of foreign sequences to increase its capacity to accept and maintain foreign gene sequences [82] (Table 2). This ability to amplify itself is useful for construction of VIGS vectors it will increase the copy number (and thus also expression) of inserted sequences [82]. Rolling-circle replication initiator protein of GmusSA and GDarSLA act as a strong suppressor of PTGS.

9. Conclusions

The monopartite begomovirus associated with DNA-satellites (Betasatellite and Alphasatellite) complex is in the norm throughout the Old World, particularly in South Asian countries. The epidemiology and evolution of this complex has been extensively analyzed since its first description. Monopartite begomovirus encoded all the genes needed to cause a successful infection. Many of these genes are coding for multifunctional proteins, adding another level of complexity in their interaction with host proteins, and their de novo creation. This shows the ability of begomoviruses and their associated satellites to rapidly evolve in response to selection pressures such as host plant resistance.

Author details

Muhammad Shafiq Shahid and Masato Ikegami
NODAI Research Institute, Tokyo University of Agriculture, Tokyo, Japan

Pradeep Sharma *
Division of Crop Improvement, Directorate of Wheat Research, Karnal, India

10. References

[1] Amin, I., Patil, B., Briddon, R. W., Mansoor, S., and Fauquet, C. M. (2011). A common set of developmental miRNAs are upregulated in *Nicotiana benthamiana* by diverse begomoviruses. *Virology Journal* 8(1), 143.

[2] Argüello-Astorga, G. R., and Ruiz, M. R. (2001). An iteron-related domain is associated to Motif 1 in the replication proteins of geminiviruses: identification of potential interacting amino acid-base pairs by a comparative approach. *Arch Virol* 146, 1465–1485.

[3] Azzam, O., Frazer, J., de-la-Rosa, D., Beaver, J. S., Ahlquist, P., and Maxwell, D. P. (1994). Whitefly transmission and efficient ssDNA accumulation of *bean golden mosaic geminivirus* require functional coat protein. *Virology* 204(1), 289-96.

[4] Bisaro, D. (2006). Silencing suppression by geminivirus proteins. *Virology* 344, 158 - 168.

[5] Boulton, M. I., Steinkellner, H., Donson, J., Markham, P. G., King, D. I., and Davies, J. W. (1989). Mutational analysis of the virion-sense genes of *maize streak virus*. *J Gen Virol* 70(Pt 9), 2309-23.

[6] Briddon, R. W. (2003). Cotton leaf curl disease, a multicomponent begomovirus complex. *Molecular Plant Pathol*4, 427 - 434.

[7] Briddon, R. W., Bull, S., Amin, I., Idris, A., Mansoor, S., Bedford, I., Dhawan, P., Rishi, N., Siwatch, S., and Abdel-Salam, A. (2003a). Diversity of DNA β : a satellite molecule associated with some monopartite begomoviruses. *Virology* 312, 106 - 121.

---

* Corresponding Author

[8] Briddon, R. W., Bull, S. E., Amin, I., Mansoor, S., Bedford, I. D., Dhawan, P., Rishi, N., Siwatch, S. S., Zafar, Y., Abdel-Salam, A. M., and Markham, P.G. (2004). Diversity of DNA 1: a satellite molecule associated with some monopartite begomoviruses-DNA β complex. *Virology* 324(2),462-474.

[9] Briddon, R. W., Mansoor, S., Bedford, I. D., Pinner, M. S., Saunders, K., Stanley, J., Zafar, Y., Malik, K. A., and Markham, P. G. (2001). Identification of DNA components required for induction of cotton leaf curl disease. *Virology* 285, 234 - 243.

[10] Briddon, R. W., Pinner, M., Stanley, J., and Markham, P. (1990). Geminivirus coat protein replacement alters insect specificity. *Virology* 177, 85 - 94.

[11] Briddon, R. W., and Stanley, J. (2006). Sub-viral agents associated with plant-infecting single-stranded DNA viruses. *Virology* 344, 198 - 210.

[12] Briddon, R. W., Watts, J., Markham, P. G., and Stanley, J. (1989). The coat protein of *beet curly top virus* is essential for infectivity. *Virology* 172(2), 628-633.

[13] Cheng, X., Wang, X., Wu, J., Briddon, R. W., and Zhou, X. (2011). βC1 encoded by tomato yellow leaf curl China betasatellite forms multimeric complexes in vitro and in vivo. *Virology* 409(2), 156-162.

[14] Clerot, D., and Bernardi, F. (2006). DNA helicase activity is associated with the replication initiator protein rep of *tomato yellow leaf curl geminivirus*. *J Virol* 80(22), 11322-30.

[15] Cui, X., Li, G., Wang, D., Hu, D., and Zhou, X. (2005a). A begomovirus DNA betaencoded protein binds DNA, functions as a suppressor of RNA silencing, and targets the cell nucleus. *J Virol* 79, 10764 - 10775.

[16] Cui, X., Li, G., Wang, D., Hu, D., and Zhou, X. (2005b). A begomovirus DNA β-encoded protein binds DNA, functions as a suppressor of RNA silencing, and targets the cell nucleus. *J Virol* 79, 10764 - 10775.

[17] Cui, X., Tao, X., Xie, Y., Fauquet, C. M., and Zhou, X. (2004). A DNAbeta associated with *tomato yellow leaf curl China* virus is required for symptom induction. *J Virol* 78, 13966 - 13974.

[18] Dry, I. B., Leslie, R., Krake, Justin, E., Rigden, and Rezaian, M. A. (1997). A novel subviral agent associated with a geminivirus: The first report of a DNA satellite. *Proceedings of the National Academy of Sciences* 94(13), 7088-7093.

[19] Eagle, P. A., Orozco, B. M., and Hanley-Bowdoin, L. (1994). A DNA sequence required for geminivirus replication also mediates transcriptional regulation. *Plant Cell* 6(8), 1157-1170.

[20] Eini, O., Akbar Behjatnia, S. A., Satish, D., Dry, I. B., Randles, J. W., and Rezaian, M. A. (2009a). Identification of sequence elements regulating promoter activity and replication of a monopartite begomovirus-associated DNA beta satellite. *J Gen Virol* 90, 253–260.

[21] Eini, O., Dogra, S., Selth, L. A., Dry, I. B., Randles, J. W., and Rezaian, M. A. (2009b).Interaction with a host ubiquitin-conjugating enzyme is required for the pathogenicity of a geminiviral DNA β satellite. *Molecular Plant Microbe Interaction* 22(6), 737-46.

[22] Elmer, J. S., Brand, L., Sunter, G., Gardiner, W. E., Bisaro, D. M., and Rogers, S. G. (1988). Genetic analysis of the *tomato golden mosaic virus*. II. The product of the AL1 coding sequence is required for replication. *Nucleic Acids Research* 16(14B), 7043-60.

[23] Etessami, P., Watts, J., and Stanley, J. (1989). Size reversion of *African cassava mosaic virus* coat protein gene deletion mutants during infection of *Nicotiana benthamiana*. *J Gen Virol* 70(Pt 2), 277-89.

[24] Frischmuth, T., and Stanley, J. (1998). Recombination between viral DNA and the transgenic coat protein gene of *African cassava mosaic geminivirus*. *J Gen Virol* 79(Pt 5), 1265-71.

[25] Gilbertson, R. L., Sudarshana, M., Jiang, H., Rojas, M. R., and Lucas, W. J. (2003). Limitations on geminivirus genome size imposed by plasmodesmata and virusencoded movement protein: insights into DNA trafficking. *Plant Cell* 15(11), 2578-91.

[26] Gopal, P., Kumar, P., Sinilal, B., Jose, J., Kasin Yadunandam, A., and Usha, R. (2007). Differential roles of C4 and βC1 in mediating suppression of post-transcriptional gene silencing: Evidence for transactivation by the C2 of *bhendi yellow vein mosaic virus*, a monopartite begomovirus. *Virus Res*123, 9 - 18.

[27] Gorlich, D., and Kutay, U. (1999). Transport between the cell nucleus and the cytoplasm. *Annual Review of Cell and Development Biology* 15, 607-60.

[28] Hanley-Bowdoin, L., Settlage, S., and Robertson, D. (2004). Reprogramming plant gene expression: a prerequisite to geminivirus DNA replication. *Molcular Plant Pathol*5, 149 - 156.

[29] Hanley-Bowdoin, L., Settlage, S. B., Orozco, B. M., Nagar, S., and Robertson, D. (1999). Geminiviruses: Models for Plant DNA Replication, Transcription, and Cell Cycle Regulation. *Critical Reviews in Plant Sciences* 18(1), 71-106.

[30] Heyraud-Nitschke, F., Schumacher, S., Laufs, J., Schaefer, S., Schell, J., and Gronenborn, B. (1995). Determination of the origin cleavage and joining domain of geminivirus Rep proteins. *Nucleic Acids Research* 23, 910-916.

[31] Ho, S. N., Hunt, H. D., Horton, R. M., Pullen, J. K., and Pease, L. R. (1989). Sitedirected mutagenesis by overlap extension using the polymerase chain reaction. *Gene* 77(1), 51-9.

[32] Hofer, P., Engel, M., Jeske, H., and Frischmuth, T. (1997). Nucleotide sequence of a new bipartite geminivirus isolated from the common weed Sida rhombifolia in Costa Rica. *J Gen Virol* 78, 1785 - 1790.

[33] Höhnle, M., Höfer, P., Bedford, I. D., Briddon, R. W., Markham, P. G., and Frischmuth, T. (2001). Exchange of three amino acids in the coat protein results in efficient whitefly transmission of a non transmissible *abutilon mosaic virus* isolate. *Virology* 290, 164-171.

[34] Hong, Y., Stanley, J., and van Wezel, R. (2003). Novel system for the simultaneous analysis of geminivirus DNA replication and plant interactions in *Nicotiana benthamiana*. *J Virology* 77(24), 13315-22.

[35] Idris, M. A., Shahid, M. S., Briddon, R. W., Khan, A. J., Zhu, J. K., and Brown, J. K. (2011). An unusual alphasatellite associated with monopartite begomoviruses attenuates symptoms and reduces betasatellite accumulation. *J Gen Virol* 92, 706–717.

[36] Karpova, O. V., Ivanov, K. I., Rodionova, N. P., Dorokhov Yu, L., and Atabekov, J. G. (1997). Nontranslatability and dissimilar behavior in plants and protoplasts of viral RNA and movement protein complexes formed in vitro. *Virology* 230(1), 11-21.

[37] Karpova, O. V., Rodionova, N. P., Ivanov, K. I., Kozlovsky, S. V., Dorokhov, Y. L., and Atabekov, J. G. (1999). Phosphorylation of *Tobacco Mosaic Virus* Movement Protein Abolishes Its Translation Repressing Ability. *Virology* 261(1), 20-24.

[38] Kharazmi, S., Behjatnia, S. A., Hamzehzarghani, H., and Niazi, A. (2012). Cotton leaf curl Multan betasatellite as a plant gene delivery vector trans-activated by taxonomically diverse geminiviruses. *Arch Virol* 5, 5.

[39] Kheyr-Pour, A. B., M. Matzeit, V. Accotto, G. P. Crespi, S. Gronenborn, B. (1991). *tomato yellow leaf curl virus* from Sardinia is a whitefly-transmitted monopartite geminivirus. *Nucl Acids Res* 19(24), 6763–6769.

[40] Kon, T., Rojas, M., Abdourhamane, I., and Gilbertson, R. (2009). Roles and interactions of begomoviruses and satellite DNAs associated with Okra leaf curl disease in Mali, West Africa. *J Gen Virol* 90, 1001 - 1013.

[41] Kon, T., Sharma, P., and Ikegami, M. (2007). Suppressor of RNA silencing encoded by the monopartite *tomato leaf curl Java begomovirus*. *Arch Virol* 152(7), 1273-1282.

[42] Kumar P, P., Usha, R., Zrachya, A., Levy, Y., Spanov, H., and Gafni, Y. (2006). Protein-protein interactions and nuclear trafficking of coat protein and βC1 protein associated with Bhendi yellow vein mosaic disease. *Virus Res* 122(1–2), 127-136.

[43] Kunik, T., Palanichelvam, K., Czosnek, H., Citovsky, V., and Gafni, Y. (1998). Nuclear import of the capsid protein of *tomato yellow leaf curl virus* (TYLCV) in plant and insect cells. *Plant J* 13(3), 393-9.

[44] Lacatus, G., and Sunter, G. (2008). Functional analysis of bipartite begomovirus coat protein promoter sequences. *Virology* 376(1), 79-89.

[45] Laufs, J., Traut, W., Heyraud, F., Matzeit, V., Rogers, S. G., Schell, J., and Gronenborn, B. (1995). *In vitro* cleavage and ligation at the viral origin of replication by the replication protein of *tomato yellow leaf curl virus*. *PNAS* 92, 3879-3883.

[46] Lazarowitz, S. G., Pinder, A. J., Damsteegt, V. D., and Rogers, S. G. (1989). *Maize streak virus* genes essential for systemic spread and symptom development. *EMBO J* 8(4), 1023-32.

[47] Li, D., Akbar Behjatnia, S. A., Dry, I. B., Randles, J. W., Eini, O., and Rezaian, M. A. (2007). Genomic regions of *tomato leaf curl virus* DNA satellite required for replication and for satellite-mediated delivery of heterologous DNAs. *J Gen Virol* 88(7), 2073-2077.

[48] Lin, B., Akbar Behjatnia, S. A., Dry, I. B., Randles, J. W., and Rezaian, M. A. (2003). High-Affinity Rep-Binding Is not Required for the Replication of a Geminivirus DNA and Its Satellite. *Virology* 305(2), 353-363.

[49] Liu, H., Boulton, M. I., Oparka, K. J., and Davies, J. W. (2001). Interaction of the movement and coat proteins of *maize streak virus*: implications for the transport of viral DNA. *J Gen Virol* 82(1), 35-44.

[50] Liu, H., Boulton, M. I., Thomas, C. L., Prior, D. A., Oparka, K. J., and Davies, J. W. (1999). *Maize streak virus* coat protein is karyophyllic and facilitates nuclear transport of viral DNA. *Mol Plant Microbe Interact* 12(10), 894-900.

[51] Liu, Y., Robinson, D. J., and Harrison, B. D. (1998). Defective forms of *cotton leaf curl virus* DNA-A that have different combinations of sequence deletion, duplication, inversion and rearrangement. *J Gen Virol* 79(P6), 1501-8.

[52] Llave, C., Kasschau, K. D., Rector, M. A., and Carrington, J. C. (2002). Endogenous and Silencing-Associated Small RNAs in Plants. *The Plant Cell Online* 14(7), 1605-1619.

[53] Mansoor, S., Briddon, R. W., Bull, S. E., Bedford, I. D., Bashir, A., Hussain, M., Saeed, M., Zafar, Y., Malik, K. A., Fauquet, C. M., and Markham, P. G. (2003a). Cotton leaf curl

disease is associated with multiple monopartite begomoviruses supported by single DNA β. *Arch Virol* 148(10), 1969-1986.

[54] Mansoor, S., Briddon, R. W., Zafar, Y., and Stanley, J. (2003b). Geminivirus disease complexes: an emerging threat. *Trends Plant Sci* 8, 128 - 134.

[55] Mansoor, S., Khan, S. H., Bashir, A., Saeed, M., Zafar, Y., Malik, K., Briddon, R. W., Stanley, J., and Markham, P. G. (1999a). Identification of a novel circular single-stranded DNA associated with cotton leaf curl disease in Pakistan. *Virology* 259, 190 -199.

[56] Mansoor, S., Khan, S. H., Bashir, A., Saeed, M., Zafar, Y., Malik, K. A., Briddon, R. W., Stanley, J., and Markham, P. G. (1999b). Identification of a novel circular singlestranded DNA associated with cotton leaf curl disease in Pakistan. *Virology* 259, 190 -199.

[57] Meister, G., and Tuschl, T. (2004). Mechanisms of gene silencing by double-stranded RNA. *Nature* 431(7006), 343-349.

[58] Mubin, M., Amin, I., Amrao, L., Briddon, R. W., and Mansoor, S. (2010). The hypersensitive response induced by the V2 protein of a monopartite begomovirus is countered by the C2 protein. *Mol Plant Pathol* 11(2), 245–254.

[59] Nawaz-ul-Rehman, M. S., Mansoor, S., Briddon, R. W., and Fauquet, C. M. (2009). Maintenance of an Old World betasatellite by a New World helper begomovirus and possible rapid adaptation of the betasatellite. *J Virol* 83, 9347 - 9355.

[60] Nawaz-ul-Rehman, M. S., Nahid, N., Mansoor, S., Briddon, R. W., and Fauquet, C. M. (2010). Post-transcriptional gene silencing suppressor activity of the alpha-Rep of non-pathogenic alphasatellites associated with begomoviruses. *Virology* 405, 300 -308.

[61] Noris, E., Jupin, I., Accotto, G. P., and Gronenborn, B. (1996). DNA-binding activity of the C2 protein of *tomato yellow leaf curl geminivirus*. *Virology* 217(2), 607-612.

[62] Noris, E. V., A. M. Caciagli, P. Masenga, V. Gronenborn, B. Accotto,G. P. (1998). Amino acids in the capsid protein of *tomato yellow leaf curl virus* that are crucial for systemic infection, particle formation, and insect transmission. *J Virol*72 (12), 10050-10057.

[63] Orozco, B. M., Kong, L.-J., Batts, L. A., Elledge, S., and Hanley-Bowdoin, L. (2000). The multifunctional character of a geminivirus replication protein is reflected by its complex oligomerization properties. *J Biol Chem* 275, 6114-6122.

[64] Padidam, M., Beachy, R. N., and Fauquet, C. M. (1996). The role of AV2 ("precoat") and coat protein in viral replication and movement in *tomato leaf curl geminivirus*. *Virology* 224, 390 - 404.

[65] Pascal, E., Goodlove, P. E., Wu, L. C., and Lazarowitz, S. G. (1993). Transgenic tobacco plants expressing the geminivirus BL1 protein exhibit symptoms of viral disease. *Plant Cell* 5(7), 795-807.

[66] Qazi, J., Amin, I., Mansoor, S., Iqbal, M. J., and Briddon, R. W. (2007). Contribution of the satellite encoded gene β C1 to cotton leaf curl disease symptoms. *Virus Res* 128(1-2), 135-139.

[67] Qian, Y., and Zhou, X. (2005). Pathogenicity and stability of a truncated DNA β associated with *tomato yellow leaf curl China* virus. *Virus Res* 109(2), 159-163.

[68] Rhee, Y., Gurel, F., Gafni, Y., Dingwall, C., and Citovsky, V. (2000). A genetic system for detection of protein nuclear import and export. *Nature Biotechlogy* 18(4), 433-437.

[69] Rigden, J. E., Dry, I. B., Mullineaux, P. M., and Rezaian, M. A. (1993). Mutagenesis of the virion-sense open reading frames of *tomato leaf curl geminivirus*. *Virology* 193(2),1001-1005.

[70] Rojas, M. R., Jiang, H., Salati, R., Xoconostle-Cazares, B., Sudarshana, M.R., Lucas, W.J., Gilbertson, R.L. (2001). Functional analysis of proteins involved in movement of the monopartite begomovirus, *tomato yellow leaf curl virus*. *Virology* 291, 110–125.

[71] Ruiz, M. T., Voinnet, O., and Baulcombe, D. C. (1998). Initiation and Maintenance of Virus-Induced Gene Silencing. *Plant Cell* 10(6), 937-946.

[72] Saeed, M., Behjatnia, S., Mansoor, S., Zafar, Y., Hasnain, S., and Rezaian, M. (2005). A single complementary-sense transcript of a geminiviral DNA β satellite is determinant of pathogenicity. *Mol Plant-Microbe Interact* 18, 7 - 14.

[73] Saeed, M., Mansoor, S., Rezaian, M. A., Briddon, R. W., and Randles, J. W. (2008). Satellite DNA beta overrides the pathogenicity phenotype of the C4 gene of *tomato leaf curl virus*, but does not compensate for loss of function of the coat protein and V2 genes. *Arch Virol* 153, 1367 - 1372.

[74] Saeed, S., Zafar, Y., Randles, J. W., and Rezaian, M. A. (2007). A monopartite begomovirus-associated DNA beta satellite substitutes for the DNA B of a bipartite begomovirus to permit systemic infection. *J Gen Virol* 88, 2881 - 2889.

[75] Sanderfoot, A. A., and Lazarowitz, S. G. (1995). Cooperation in Viral Movement: The Geminivirus BL1 Movement Protein Interacts with BR1 and Redirects It from the Nucleus to the Cell Periphery. *Plant Cell* 7(8), 1185–1194.

[76] Saunders, K., Bedford, I., Briddon, R., Markham, P., Wong, S., and Stanley, J. (2000). A unique virus complex causes Ageratum yellow vein disease. *PNAS* 97, 6890 - 6895.

[77] Saunders, K., Bedford, I. D., and Stanley, J. (2001). Pathogenicity of a natural recombinant associated with ageratum yellow vein disease: implications for geminivirus evolution and disease aetiology. *Virology* 282(1), 38-47.

[78] Saunders, K., Briddon, R. W., and Stanley, J. (2008). Replication promiscuity of DNAbetasatellites associated with monopartite begomoviruses; deletion mutagenesis of the *ageratum yellow vein virus* DNA β satellite localises sequences involved in replication. *J Gen Virol* 89, 3165 - 3172.

[79] Saunders, K., Norman, A., Gucciardo, S., and Stanley, J. (2004). The DNA beta satellite component associated with ageratum yellow vein disease encodes an essential pathogenicity protein (β C1). *Virology* 324, 37 - 47.

[80] Saunders, K., and Stanley, J. (1999). A nanovirus-like DNA component associated with yellow vein disease of Ageratum conyzoides: evidence for interfamilial recombination between plant DNA viruses. *Virology* 264, 142 - 152.

[81] Settlage, S. B., See, R. G., and Hanley-Bowdoin, L. (2005). Geminivirus C3 protein replication enhancement and protein interactions. *J Virol* 79, 9885 - 9895.

[82] Shahid, M. S., Ali, L., Andleeb, S. (2009). The function of the a-rich region of the alphasatellite associated with the cotton leaf curl disease in Pakistan. *EurAsia J BioSci* 3, 152-156.

[83] Sharma, A. M., A. Osaki, T. Ikegami, M. (1998). Characterization of virus-specific DNA forms from tomato tissues infected by *tobacco leaf curl virus*: evidence for a single genomic component producing defective DNA molecules. *Plant Pathol* 47, 787-793.

[84] Sharma, P., Ikegami, M., and Kon, T. (2010). Identification of the virulence factors and suppressors of posttranscriptional gene silencing encoded by *ageratum yellow vein virus*, a monopartite begomovirus. *Virus Res* 149(1), 19-27.

[85] Stanley, J., Bisaro, D. M., Briddon, R. W., Brown, J. K., Fauquet, C. M., Harrison, B. D., Rybicki, E. P., and Stenger, D. C. (2005a). Geminiviridae. *Virus Taxonomy, VIIIth Report of the ICTV*, 301 - 326.

[86] Stanley, J., Bisaro, D. M., Briddon, R. W., Brown, J. K., Fauquet, C. M., Harrison, B. D., Rybicki, E. P., and Stenger, D. C. (2005b). Geminiviridaea. *Virus Taxonomy, VIIIth Report of the ICTV*, 301 - 326.

[87] Stanley, J., Bisaro, D. M., Briddon, R. W., Brown, J. K., Fauquet, C. M., Harrison, B. D., Rybicki, E. P., Stenger, D. C., by, E., Fauquet, C. M., Mayo, M. A., Maniloff, J., Desselberger, J., and Ball, L. A. (2005c). Geminiviridae. In Virus Taxonomy, 5th report of the ICTV. *London: Elsevier/Academic Press*, pp 301–326.

[88] Stanley, J., and Latham, J. R. (1992). A symptom variant of *beet curly top geminivirus* produced by mutation of open reading frame C4. *Virology* 190(1), 506-509.

[89] Stanley, J., Saunders, K., Pinner, M. S., and Wong, S. M. (1997). Novel defective interfering dnas associated with *ageratum yellow vein geminivirus* infection of ageratum conyzoides. *Virology* 239(1), 87-96.

[90] Sunter, G., and Bisaro, D. (1997). Regulation of a geminivirus coat protein promoter by AL2 protein (TrAP): evidence for activation and derepression mechanisms. *Virology* 232, 269 - 280.

[91] Sunter, G., and Bisaro, D. M. (1991). Transactivation in a geminivirus: AL2 gene product is needed for coat protein expression. *Virology* 180(1), 416-9.

[92] Sunter, G., Hartitz, M. D., Hormuzdi, S. G., Brough, C. L., and Bisaro, D. M. (1990). Genetic analysis of *tomato golden mosaic virus*: ORF AL2 is required for coat protein accumulation while ORF AL3 is necessary for efficient DNA replication. *Virology* 179(1), 69-77.

[93] Sunter, G., Stenger, D. C., and Bisaro, D. M. (1994). Heterologous complementation by geminivirus AL2 and AL3 genes. *Virology* 203(2), 203-10.

[94] Talya Kunik, L. M., Vitaly Citovsky and Yedidya Gafni (1999). Characterization of a tomato karyopherin α that interacts with the *tomato yellow leaf curl virus* (TYLCV) capsid protein. *Journal of Experimental Botany* 50(334), 731-732.

[95] Tao, X., Qing, L., and Zhou, X. (2004). Function of A-Rich region in DNA β associated with *tomato yellow leaf curl China* virus. *Chinese Sci Bull* 49(14), 1490-1493.

[96] Tao, X., and Zhou, X. (2004). A modified viral satellite DNA that suppresses gene expression in plants. *Plant J* 38(5), 850-60.

[97] Trinks, D. R., R. Shivaprasad, P. V. Akbergenov, R. Edward J. Oakeley, K. Veluthambi, Thomas Hohn, Pooggin, M. M. (2005). Suppression of RNA silencing by a *geminivirus* nuclear protein, ac2, correlates with transactivation of host genes. *J Virol* 79(4), 2517-2527.

[98] Van Wezel, R., Dong, X., Liu, H., Tien, P., Stanley, J., and Hong, Y. (2002). Mutation of three cysteine residues in *tomato yellow leaf curl China virus* C2 protein causes dysfunction in pathogenesis and posttranscriptional gene silencing suppression. *Mol Plant Microbe Interaction* 15, 203 - 208.

[99] Vanitharani, R., Chellappan, P., Pita, J. S., and Fauquet, C. M. (2004). Differential roles of AC2 and AC4 of cassava geminiviruses in mediating synergism and suppression of posttranscriptional gene silencing. *J Virol* 78, 9487 - 9498.

[100] Voinnet, O. (2005). Induction and suppression of RNA silencing: insights from viral infections. *Nature Rev Genet* 6(3), 206-220.

[101] Von Arnim, A., Frischmuth, T., and Stanley, J. (1993). Detection and possible functions of *African cassava mosaic virus* DNA B gene products. *Virology* 192(1), 264-72.

[102] Wang, M. B., X. Wu, L. Liu, L. Smith, N. A. Isenegger, D. Wu, R. Masuta, C. Vance, V. B. Watson, J. Rezaian M.A. (2004). On the role of RNA silencing in the pathogenicity and evolution of viroids and viral satellites. *PNAS* 101(9), 3275-3280.

[103] Wartig, L., Kheyr-Pour, A., Noris, E., De Kouchkovsky, F., Jouanneau, F., Gronenborn, B., and Jupin, I. (1997). Genetic analysis of the monopartite *tomato yellow leaf curl geminivirus*: roles of V1, V2, and C2 orfs in viral pathogenesis. *Virology* 228(2), 132-140.

[104] Woolston, C. J., Reynolds, H. V., Stacey, N. J., and Mullineaux, P. M. (1989). Replication of wheat dwarf virus DNA in protoplasts and analysis of coat protein mutants in protoplasts and plants. *Nucl Acids Res* 17(15), 6029-6041.

[105] Wu, P., and Zhou, X. (2005). Interaction between a nanovirus-like component and the *tobacco curly shoot virus*/satellite complex. *Acta Biochim Biophys Sin* 37, 25 - 31.

[106] Yaakov, N., Levy, Y., Belausov, E., Gaba, V., Lapidot, M., and Gafni, Y. (2011). Effect of a single amino acid substitution in the NLS domain of *tomato yellow leaf curl virus*-Israel (TYLCV-IL) capsid protein (CP) on its activity and on the virus life cycle. *Virus Res* 158(1–2), 8-11.

[107] Zhang, S. C., Wege, C., and Jeske, H. (2001). Movement proteins (BC1 and BV1) of *abutilon mosaic geminivirus* are cotransported in and between cells of sink but not of source leaves as detected by green fluorescent protein tagging. *Virology* 290(2), 249-60.

[108] Zhou, X., Xie, Y., Tao, X., Zhang, Z., Li, Z., and Fauquet, C. M. (2003). Characterization of DNAbeta associated with begomoviruses in China and evidence for co-evolution with their cognate viral DNA-A. *J Gen Virol* 84, 237 - 247.

[109] Zrachya, A., Glick, E., Levy, Y., Arazi, T., Citovsky, V., and Gafni, Y. (2003). Suppressor of RNA silencing encoded by *tomato yellow leaf curl virus*-Israel. *Virology* 358, 159 - 165.

[110] Kon, T., Hidayat, S.H., Hase, S., Takahashi, H., and Ikegami, M. (2006). The natural occurrence of two distinct begomoviruses associated with DNA β and a recombinant DNA in a tomato plant from Indonesia. *Phytopathology* 96(5), 517-525.

[111] Ito, T., Kimbara,J., Sharma,P., and Ikegami, M. (2009). Interaction of tomato yellow leaf curl virus with diverse betasatellites enhances symptom severity. *Arch Virol* 154, 1233-1239.

[112] Ogawa, T., Sharma, P. and Ikegami, M. (2008). The begomoviruses Honeysuckle yellow vein mosaic virus and Tobacco leaf curl Japan virus with DNA β satellites cause yellow dwarf disease of tomato. *Virus Res* 137(2), 235-244.

[113] Sharma, P., and Ikegami, M. (2009). Characterization of signals that dicate nuclear/nucleolar and cytoplasmic shuttling of the capsid protein of Tomato leaf curl Java virus associated with DNA β satellite. *Virus Res* 144, 145-153.

[114] Sharma, P., and Ikegami, M. (2010). Tomato leaf curl virus V2 protein is a determinant of virulence, hypersensitive response and suppression of posttranscriptional gene silencing. *Virology* 396, 85-93.

[115] Sharma, P., Gaur, R.K. and Ikegami, M. (2011). Subcellular localization of V2 protein of Tomato leaf curl Java virus by using green fluorescent protein and yeast hybrid system. *Protoplasma* 248. 281-288.

Chapter 11

In vitro Regeneration, Acclimatization and Antimicrobial Studies of Selected Ornamental Plants

A. Bakrudeen Ali Ahmed, S. Mohajer, E.M. Elnaiem and R.M. Taha

Additional information is available at the end of the chapter

1. Introduction

Tissue culture has been applied to diverse research techniques such as viral elimination, clonal propagation, gene conservation, *in vitro* fertilization, mutation, induction for genetic diversity, genetic transformation, protoplast isolation and somatic hybridization, secondary metabolite production and other related techniques. The commercial production of ornamental plants is growing worldwide. Its monetary value has significantly increased over the last two decades and there is a great potential for continued further growth in both domestic and international markets. About 156 ornamental genera are propagated through tissue culture in different commercial laboratories worldwide. About 212.5 million plants including 157 million ornamental plants amounting to 78% of the total production were reported [1]. These plants are over exploited due to their high medicinal value and hence, propagation of the plants by tissue culture may be mandatory, which offers a greater potential to deliver large quantities of disease-free, true-to-type healthy stock within a short span of time. Biotechnological interventions for *in vitro* regeneration, mass micropropagation and gene transfer methods in forest tree species have been practiced with success, especially in the last decade. Against the background of the limitations of long juvenile phases and lifespan, developments of plant regeneration protocols of ornamental species are gaining importance. Ornamental industry has applied immensely *in vitro* propagation approach for large-scale plant multiplication of elite superior varieties. During *in vitro* condition, plantlets are grown under fixed and controlled environment in sterile formulated medium which contained macronutrients, micronutrients, vitamins and plant growth regulators. After the plantlets reached optimum growth in the culture containers after a certain growth period, it can be transferred to *ex vitro* condition to allow continuous growth of the plantlets. As a result, hundreds of plant tissue culture laboratories have been set up worldwide, especially in the developing countries due to cheap labour costs.

Plant tissue culture media is normally rich in sucrose and other organic nutrients that can support organogenesis in plants but also the growth of many microorganisms (like bacteria and fungi). To overcome and prevent contamination in media preparation, sterilization should be done thoroughly. Sterilization of nutrient media can be done in an autoclave (large pressure cooker), less often by filtration and seldom by irradiation [2]. The container with the medium should be properly closed and autoclaved at 121ºC, 105 kPa, for 20 minutes. It also identified that good sterilization relies on time, pressure, temperature and volume of the object to be sterilized [2]. The sterilized nutrient media should be stored in a sterile box that has previously been disinfected with 70% alcohol [2]. Some of the plant growth regulators such as giberellic acid (GA3), zeatin, abscisic acid (ABA), urea, certain vitamins, pantothenic acid, antibodies, colchicines, plant extracts and enzymes used in tissue culture is thermolabile. These compounds should not be autoclaved and filter-sterilization is often used if a thermolabile substance is needed in a nutrient medium.

2. Problem statement

In vitro plant tissue culture needs the formulation of a complete nutritional medium and for exploration of plant physiological processes; it needs the addition of effective plant growth regulators. These two aspects can be considered to plant tissue culture the wings to take off. With the starting of common or specific media and the selection of appropriate plant tissue culture, enable induction of cell division, callus growth, differentiation of shoots, roots and embryos. Commonly, synthetic analogues are used, mainly; 1-naphthalene acetic acid (NAA), indole butyric acid (IBA), 2,4-Dichlorophenoxyacetic acid (2,4-D) in plant tissue culture for the induction of plant cells and the regeneration of the root, bud, embryo-like body which has a high organizational structure of specialized cells. The main physiological role of cytokinins; promotion of cell division and expansion, buds induction, differentiation, promote lateral bud sprouting and inhibition of senescence. *In vitro* tissue or organ will be soon aging. Such as cytokinin treatment can slow the aging process. Another class of hormones used in tissue culture is gibberellins. The main physiological effects of gibberellin (GA) are buds induction. It is commonly used in plant tissue culture to promote the growth of the seedling stem elongation. It is also affect cell differentiation of cambium and it often works in synergy with auxins. In addition, the other hormone such as abscisic acid (ABA) is one of the plant growth substances; it can be biosynthesized from mevalonic acid. It can control water and ion uptake by roots and to promote the adventitious shoots and absorb and prevent the phenolic production. Whereas, the ethylene one of the gases plant hormone, it is moved by diffusion around the plant rather than translocation. It has stimulates the final stage of fruit development and flower fall. The main function of ethylene in plant tissue culture, it can stimulate the respiration, seed germination, peroxidase enzymes and regulates the level of auxins. The low concentration of ethylene induces the proper resistance to the developed plant.

3. Application area

The physiological role of auxins promote cell growth and cell division, induction of the injured tissue of one to several layers of cells recovering the ability to divide to form callus; promote rooting, in the conventional cutting propagation and in organized tissue culture. Auxins can also promote plant sex differentiation, the formation of flowers and to promote the formation of seedless fruit. Auxins main use in tissue culture to induce callus and, also they are important use is in line with a certain amount of cytokinin for root differentiation, lateral bud germination and growth, as well as in certain plants induce embryogenesis. Since, many research articles have dealt at great length about the status, applications, potentials and needs in tissue culture of ornamental plants. Since, the major effects have been made to develop basic back ground technologies for consistent production and regeneration of calluses from diverse group of medicinal plants [3]. The techniques which have been so far described for propagation of ornamental plants through tissue culture have been tested on the laboratory scale and have not been validated for their suitability in commercial scale production. However, the following aspects have to be critically studied if the economic prospects associated with *in vitro* culture technology are to be realized. Generally, the application of plant cell and tissue culture technique which is commonly known as *in vitro* cloning can also be divided into several procedures, including meristematic cultures, vegetative explant cultures, callus induction, suspension cultures, direct and indirect somatic embryogenesis, synthetic seed production, *in vitro* flowering, *in vitro* mutation breeding, protoplast and also somatic hybridization process. Some of these techniques would selectively applied to selected plants overcome generation incapability of the plants. This *in vitro* technique is very useful in ensuring sustainable, optimized sources of plant-derived natural products. However, *ex situ* cultivation should be preceded by proper evaluation of the plants for their ability to produce the required bioactive constituents before commencing cultivation or introducing the technology to potential growers. The ability of plants to produce certain bioactive substances is largely influenced by physical and chemical environments in which they grow. Plants also produce certain chemicals to overcome abiotic stresses. In this aspects plant tissue culture developed callus influenced by medium, explants, plant growth regulators, color lights, temperature, photoperiod and carbon sources are helpful to produce valuable secondary metabolites compounds in many studies [3,4]. Growing a plant outside its natural environment under ideal conditions may therefore, result in being unable to produce the desired bioactive substances, hence the need for prior evaluation.

4. Research course

Ornamental plants are used especially as decorative houseplants and for landscaping. Ornamental plants are unique for their sheer beauty and variety of leaves. For example Begonias, which have a medicinal value. It is a temperate plant, which is commercially used as a flowering pot plant. It does not produced seeds. It is a winter flowering plant and was developed from cross between other species of the same family. Begonia plant normally has thick, shiny, dark green leaves. Although Begonias can be readily vegetative propagated, they are susceptible to many pathogenic bacteria, fungi, and nematodes [5]. Nowadays, the seaweed market has grown as predicted with prospects to go even further. Therefore potential improvements introduced through the application on *in vitro* techniques are expected to be even higher.

Modern techniques of propagation through tissue culture technique have been developed to meet the demand of the horticultural industries including nursery industries. For pot plant production, the priority is to obtain early, synchronized and profuse flowering, together with a compact and homogenous plant size, rather than continuous flowering. For ornamental pot plant production to be successful, an efficient method for flower induction in small plantlets is thus required. Thus, the application of plant tissue culture technique is always required. The technology is widely applied in both research and development of improved crops [6]. Rout *et al.* reported that about 156 ornamental genera were propagated through tissue culture in different commercial laboratories worldwide [7].

5. Method used

5.1. Surface sterilization

Surface of plant parts carry a wide range of microbial contaminants. The presence of any contaminant will interfere with the growth of explant or cultures and fungal or bacterial explant contamination in plant cultures is usually detectable 1-14 days after culturing. Therefore, sterilization or disinfection of tissues is necessary in order to eradicate surface microorganism. In order to disinfect plant tissues, 5-50% (v/v) commercial bleach Clorox (Sodium hypochlorite), 70% (v/v) alcohol and a few drops of Tween 20 can be used in sterilization technique. Pierik *et al.* had suggested that sterilization plant for a few seconds in alcohol is not sufficient to kill all microorganisms and after this they are usually treated with sodium hypochlorite [2]. Diluted solution of sodium hypochlorite (0.25-2.63%) (v/v) is used as a disinfectant and tween 20 is an emulsifier which is added at the rate of 1 drop per 100 ml of solution. The development of techniques for the culture of isolated plant organs, tissues and cells have led to several exciting opportunities in the area of plant biotechnology, and allowed widespread use of cell culture for *in vitro* genetic manipulation, plant propagation and production of commercially useful products. The techniques of cultivating cells and tissues have been referred to sometimes as "aseptic culture of plants". Therefore, the absence of contaminants is assumed to be a fundamental requisite *in vitro*. Surface sterilization of ornamental plants is difficult as they lack a thick protective surface, and therefore sodium hypochlorite and similar agents can easily damage the delicate tissues. The scope of these techniques has been extended for use in bioprocess technology for production of high value chemicals of immense commercial value in the pharmaceutical and nutraceutical sectors [8]. Finally, the instruments including forceps and scalpels were sterilized by dipping them into hot bead sterilizer at 250ºC and allowed to cool. Glassware, empty test tubes, empty flasks, petri dishes, filter paper and distilled water can be sterilized using an autoclave at 121ºC, 105 kPa, for 20 minutes. The bottles and glassware should not be too tightly packed and their tops should be loosen during autoclaving.

5.1.1. Explants sterilization procedure

To initiate cultures, various explants of African violet (*Saintpaulia ionantha* H. Wendl) were excised from 2-month-old intact plants which were grown in greenhouse. The flower buds (3-5 mm), leaf (10 x 10 mm), petiole (10 mm) and peduncle (10 mm) were used as explants for *in vitro* studies [9]. In some cases aseptic seedlings need to be used. Seeds of *Dianthus caryophyllus* were washed with chlorox concentration of 70%, 50%, 30%, 20% and 10% (v/v). First, the seeds were shaken in 70% chlorox with three drops of tween-20 for 15 minutes. They were then washed three times with sterile distilled water. These steps were repeated with other concentrations of chlorox. The seeds were cultured into test tubes containing MS media without hormone under aseptic condition. The work has to be done under sterile conditions. The seeds were germinated in a culture room at the temperature of 25±1 °C with 16 h light and 8 h dark. The growth was evaluated after 30 days. For tissue culture studies of *Gerbera jamesonii* Bolus ex. Hook f. various explants were obtained from 8-week-old aseptic seedlings. *Gerbera* seeds were first soaked in distilled water for 30 min with addition of 1-2 drops of Tween-20, followed by 40% (v/v) Sodium hypochlorite solution and gently agitated. The seeds were then rinsed 3 times with distilled water and then soaked in 70% (v/v) alcohol for 1 min. Finally the seeds were rinsed 3 times with sterile distilled water. Sterilized seeds were cultured in MS basal medium [10]. Surface sterilization process for seeds of cauliflower (*Brassica oleracea* var. *botrytis*) was slightly different; the seeds were soaked in distilled water with 1 or 2 drops of Tween-20 for 20 min, followed by 60% (v/v) Sodium hypochlorite solution, gently agitated for 15 min. The seeds were then rinsed 3 times in distilled water, soaked in 70% ethanol (v/v) for 30 sec and rinsed again in 3 changes of sterile distilled water prior to culturing in MS basal medium [11]. The conventional methods of propagation are problematic due to rapid occurrence of diseases. The production of large numbers of genetically homogenous plants is also very difficult. Plant cell culture technique is an alternative method for mass cloning of Begonia plants and also to overcome the problems occurring in the conventional propagation. The regeneration frequency and average number of shoots per explant varied among the cultivars. Shoot tip size also plays an important role in shoot regeneration efficiency [7]. Wang and Ma reported that shoot tip between 0.2 and 0.5 mm and shoot meristems between 0.1 and 0.2 mm diameter produced only a single shoot [12]. The techniques of stimulating axillary branching or culturing nodal sections *in vitro* are probably most commonly used in micropropagation [13]. Size of the meristem (both shoot tip and nodal explants) of *Floribunda* and miniature roses had significant effect on shoot multiplication; on an average 2.5-5.0 shoots were obtained per culture cycle, dependent on cultivars [14]. Recently, Teixeira de Silva and Fukai published a detailed review on tissue culture of chrysanthemum, which highlights organogenesis, thin cell layer, and somatic embryogenesis for plant regeneration [15].

5.1.2. Media and plant growth regulators roles

MS basal medium [16] was used for these experiments. The constituent of the media was adjusted to 1 liter after the sucrose addition and the pH was adjusted to 5.7 prior to the adding of 7.0 g agar and 0.1 g charcoal to the media. Media was autoclaved at 121 °C 1.5 kpa for 20 minutes. Media were then dispensed into sterile plastic vials containing 20-25 ml of aliquots, inside the laminar air flow cabinet. For culturing of *Pereskia grandifolia* the axillary bud explants were sliced and cultured on MS medium consisted of 0.1-10 mg/l BAP and 30 mg/l adenine as well as BAP and NAA. Petioles obtained from aseptically grown young plantlet of *Gerbera jasmesonii* were used as source of explants. Leaves and petioles were cultured for shoot induction on MS media containing BAP and NAA at various concentrations. Plant regeneration from leaf disk callus of *Begonia elatior* was achieved on MS medium supplemented with 1.0 mg/l Kn and 0.1 mg/l zeatin [17]. Liquid medium seems to be more effective for shoot regeneration and root induction, which is due to better aeration. Simmonds and Werry used liquid medium for enhancing the micropropagation profile of *Begonia hiemalis* [18]. Liquid media have been used for plant cells, somatic embryos and cell suspension cultures in either agitated flasks or various types of bioreactors [19-24]. Wated et al. compared performance of agar-solidified medium and interfacial membrane drafts floating on liquid medium for shoot multiplication and root induction [25]. The regenerated shoots were rooted on half-strength MS medium supplemented with 0.1 mg/l NAA and 0.2 mg/l Kn. Nearly 300 plantlets of each cultivar were transferred to soil with 95% survival rate [26].

5.1.3. Active chemical roles

During *in vitro* conditions, plantlets are grown under fixed and controlled environment in sterile formulated medium which contained macronutrients, micronutrients, vitamins and plant growth regulators. When explants are first placed onto a nutrient medium, there is often an initial leakage of ions from damaged cells, especially metallic cations (Na^+, Ca^{2+}, K^+, Mg^{2+}) for the 1-2 days, so that the concentration in the plant tissues actually decreases [27]. Cells then commence active absorption and the internal concentration slowly rises. Phosphate and nitrogen (particularly ammonium) are absorbed more rapidly than other ions. Both growth and morphogenesis in tissues cultures are markedly influenced by the availability of nitrogen and the form in which it is presented [28]. Chemical and substances are synthesized in particular cells and are transferred to other cells, which in extremely small quantities influence the development process. The plant growth regulators are implicated in many biological processes in ornamental plants, including cell division, root and floral initiation, fruit development, senescence and abiotic stress responses.

The rooting efficiency enhanced by addition of 0.05% Poly vinyl pyrrolidone (PVP) in the culture medium containing 0.5 mg/l IBA [29]. The addition of PVP helps in oxidizing polyphenols leached in the medium, and promotes high rate of organogenesis. Dijkshoorr-Dekker studied the influence of light and temperature on propagation profile of *Ficus benjamina* [30]. Propagation of different *Ficus* species by using shoot tips or axillary bud explants had been reported [29,31-34]. In the most cases, shoots were rooted in hormone free medium. Both orientation of the petiole explants and auxin transport system are crucial factors for the induction of somatic embryogenesis of *Saintpaulia* [35], and TDZ helped in the development of somatic embryos. Winkelmann *et al.* used cell suspension culture of *Cyclamen* for rapid development of somatic embryos [36], and later on followed by Hohe et al. [37], who developed a large scale propagation system of *Cyclamen* from embryogenic cell suspension cultures. Kumari *et al.* developed an efficient protocol for micropropagation of *Chrysanthemum* on MS medium supplemented with 1.0-2.5 μM TDZ [38]. Castillo and Smith induced direct somatic embryogenesis from petiole and leaf blade explants of *Begonia gracilis* on MS medium supplemented with 0.5 mg/l kinetin and 2% (v/v) coconut water [5]. Kim et al. established a large-scale propagation of *chrysanthemum* through bioreactor system, and obtained 5000 plantlets after 12 weeks of culture in 10±l column type bioreactor [39].

6. Results

6.1. Micropropagation

In vitro propagation through meristem culture is the best possible means of virus elimination and produces a large numbers of plants in a short span of time. It is a powerful tool for large-scale propagation of ornamental plants. The term 'meristem culture' specially means that a meristem with no leaf primordial or at most 1-2 leaf primordial which are excised and cultured. The pathway of regeneration undergoes several steps. Starting with an isolated explant, with de-differentiation followed by re-differentiation and organization into meristematic centers. More than 600 million micropropagated plants are produced every year in the world [40]. Micropropagation is one of the few areas of plant tissue culture in which the techniques have been applied commercially. To circumvent these impediments, clonal or vegetative propagation has been deployed for recovering dominant, additive and epistatic genetic effects to select superior genotypes. Plant tissue culture methods offer an important option for effective multiplication and improvement of ornamental plants within a limited time frame.

6.1.1. Gerbera jamesonii

Gerbera jamesonii is an ornamental flowering perennial belonging to the Asteraceae family. This plant is very well known to be planted as cut flowers, bedding plant and also pot crops. The *in vitro* shoots were successfully obtained in petiole explants of *Gerbera jamesonii* [10]. The adventitious shoots were observed on MS medium supplemented with BA (1.0 mg/l) and NAA (0.5 mg/l). The developed shoots were subcultured every two week interval for shoot elongation. The elongated shoots were transferred for rooting on MS media with IAA (2.0 mg/l). The developed plantlets were maintained in the following acclimatization conditions : the garden soil (black soil: red soil, 2:1) which gave best result with 86.0 ± 0.9% survival rates, followed by vermiculite with 73 ±1.3% survival rates. In this study, plantlets established from *in vitro* regeneration of *Gerbera jamesonii* were morphologically identical to the mother plant and developed normally and produced flowers after 6 months being transplanted to the greenhouse [41,42].

6.1.1.1. Multiple shoots induction

Leaf and petiole explants were cultured on MS medium for shoot induction containing BAP (0.5-2.0 mg/l) and NAA (0.5-1.0 mg/l). Normal adventitious shoots of *Gerbera jamesonii* were successively obtained from petiole explants cultured on MS medium supplemented with BAP (2.0 mg/l) and NAA (0.5 mg/l) with 94.3% regeneration rate and 9.3 shoots per explant, followed by BAP (1.5 mg/l) and NAA (1.0 mg/l) with 83.1% regeneration rates and 8.3 shoots per explant. Pierik et al. stated that the addition of strong auxin such as NAA in combination with BAP promoted shoots induction in plant tissue culture [2]. In this study, higher concentration of auxin, NAA (2.0 mg/l) in combination with lower concentration of BAP (1.0 mg/l) showed the lowest shoots regeneration rates (4.6) with the lowest number of shoots (1.6). Son et al. investigated the micropropagation of different plant varieties using the plant buds as explants [43]. They found that the best hormone combination for the *in vitro* initiation of *Gerbera jamesonii* shoots was 3 mg/l BAP + 0.1 mg/l IAA producing 11.29 number of shoots per explant on MS medium.

6.1.1.2. Rooting and hardening

The induced shoots were best rooted on MS media supplemented with BAP (0.1 mg/l) with frequency of 73.7% and 22.1 roots per explant, while Son et. al.[43] found that MS medium supplemented with 2.0 mg/l NAA was the best medium for *in vitro* rooting of the shoots (94.0%). The developed plantlets were maintained in the following acclimatization conditions: the garden soil (black soil: red soil, 2:1) which gave the best results with 86.0 % survival rates, followed by vermiculite with 73% survival rates. The regenerated plantlets failed to survive, when they were cultured in the autoclaved garden soil (black soil: red soil, 2:1). The plantlets established from *in vitro* regeneration of *Gerbera jamesonii* were morphologically identical to the mother plant and developed normally and also produced flowers after 6 months being transplanted to the greenhouse.

6.1.2. Pereskia grandifolia Haworth var. grandifolia

A protocol for an *in vitro* propagation was developed for the ornamental plant *Pereskia grandifolia Haworth var. grandifolia* from axillary bud explants. Optimum multiplication of shoots was achieved on MS [16] medium supplemented with 3.0 mg/l BAP and 30.0 mg/l adenine. Plants were maintained *in vitro* on MS medium while callus were induced on MS basal medium supplemented with the combination of 5.0 mg/l BAP and 5.0 mg/l NAA. The somatic embryogenic callus of the plant species was induced by Chuah and Chan on B5 medium supplemented with 6.0 mg/l 2, 4-D [44,45].

6.2. Organogenesis

Regeneration in plant tissue culture will be successful by maintaining various factors involved, including media factors and environmental factors. The media factors include media constituents, macronutrients, micronutrients, vitamins, amino acids, carbon source, complex nutritive mixtures, gelling agents, activated charcoal, plant growth regulators and pH of the medium. Environmental factors on the other hand are the culture conditions under which explants are maintained. The environmental factors involved include the temperature and illumination of the culture room, agitation process and incubation period of the cultures [4]. For the initiation of callus culture, the following factors are important-:

the origin of explants used for the establishment of callus culture, the cellular/tissue differentiation status, external plant growth regulators, culture media and culture conditions [46]. Cellular competence to plant hormones is understood as the status in which a cell must possess the ability to perceive a transducer and respond to a signal [47]. Organogenesis refers to the formation of shoots/roots. The callus may remain in a differentiated condition regardless of the hormones and nutrients to which it is exposed the secondary metabolites and these metabolites have biological activity [48-50]. Organ formation generally follows cessation of unlimited proliferation of callus. Individual cells or group of cells of smaller dimensions may form small nets of cells scattered throughout the callus tissue, the so-called meristemoids. These meristemoids become transformed into cyclic nodules from which shoot bud or root primordia may grow as shoots/roots. Shoot bud formation may decrease with age and subculture duration of the callus tissue but the capacity of rooting may persist for longer period. In some calli, rooting occurs more often than in other forms of organogenesis. During organogenesis, if the roots are first formed, then it is very difficult to induce adventitious shoot bud formation from the same callus tissue. If the shoots are first formed, it may form roots later on or may remain in rootless condition unless and until the shoots are transformed to another medium or hormone less medium or conditions that induce root formation. In certain cases root and shoot formation may occur simultaneously, but organ connection (vascular connection) between root and shoot primordial is essential for the regeneration of complete plantlet from the culture. Shoot formation followed by rooting is the general feature of organogenesis. The color of callus tissue may remain unchanged during rhizogenesis or may develop yellow pigmentation. During shoot bud formation, the callus tissue generally develops green or pale green pigmentation. Organogenesis is a process by which a cell or group of cells differentiate to form organs. It is the reflection of the intrinsic genetic constitution of a taxon [51]. Since recovery of plants is the usual objective, regeneration of shoots is of greater interest. Organogenesis is commonly induced by manipulation of exogenous phytohormone levels and occurs either directly from explant tissue or through callus.

6.2.1. Saintpaulia ionantha

Saintpaulia ionantha known worldwide as African violet belongs to the family Gesneriaceae. African violet is one of the most popular ornamental plants. The complete plant regeneration was obtained from leaf, petiole, peduncle and floral parts of the plant. For shoot regeneration MS medium supplemented with IAA (1.0 mg/l) with Zeatin (2.0 mg/l). The developed plantlets were transferred for acclimatization, whereby *in vitro* plantlets were transferred into planting pots containing mixed soil, which is a mixture of compost, sand and black soil with the ratio of 1:1:2. The morphological characters were compared to the mother plants [52].

6.2.2. Somatic embryogenesis

The commercial exploitation of the ornamental plants for the production and conventional propagation is hampered due to their poor seed viability, low rate of germination and poor rooting ability of the vegetative cuttings. Somatic embryogenesis is an alternative method.

However, there is lack of information for the embryo induction process [53]. The aim or work was to study the germination capability and development of somatic embryos (SEs) from ornamental plants. A sustainable plant regeneration system *in vitro* through somatic embryos from mature sexual embryos has been reported in *Clitoria ternatea*. Somatic embryos developed through callus from seedling roots on hormone-free MS medium (MS$_1$). Addition of growth hormones, Kn 0.5 mg/l (MS$_2$) or Kn +IAA 0.5 mg/l of each (MS$_3$) induced direct somatic embryos, in high frequency, on split root and hypocotyl systems. The embryogenic potential varied with the organ, roots or hypocotyls, and also with the medium. The morphogenetic capacity of the somatic embryos is retained for more than 2 years by sub-culturing at intervals of 4 weeks on MS$_3$ in complete darkness. Somatic embryos, under the appropriate subculture conditions (16 h light/8 h dark photoperiod at 24± 1 °C on media MS$_3$, MS$_4$ and MS$_5$), resulted in recurrent-somatic embryogenesis and was profuse at the shoot and root apices of the somatic embryos. Mature somatic embryos were transplanted to MS$_1$ to stimulate germination and plantlet regeneration. Plantlets, developed from primary and secondary embryos on MS1 were successfully hardened and grown in natural outdoor conditions. The morphology and histology of the somatic embryo and plantlet and the culture conditions for continuous production of plantlets through direct somatic embryogenesis are discussed. In our laboratory, plant regeneration of *Clitoria ternatea* was reported from leaf explants cultured on DKW (Driver and Kuniyuki) medium supplemented with various concentrations of NAA and BAP [54].

Embryogenic callus was induced from leaf explants of *Gerbera jamesonii* Bolus ex. Hook f. in cell suspension cultures. A cream friable embryogenic callus was formed within two weeks when leaf explants were cultured on MS medium containing 2,4-D. hormone (1.0 to 2.0 mg/l). A hundred percent (100%) induction frequency was obtained in 2,4-D concentration range of 1.7-2.0 mg/l. While maximum percentage response for somatic embryos induction (64.56%) from callus clumps was obtained on MS medium fortified with BAP (0.5 mg/1) and NAA (1.0 mg/1) by Ranjan and Gaurav [27]. For proliferation, embryogenic callus was transferred to MS liquid medium containing the same hormone; 2,4-D with a small amount of NAA and subcultured at 2 weeks interval. Induction of somatic embryos different stages (globular, heart and torpedo) were observed after 2 weeks of culture. Somatic embryos were developed in MS suspension medium containing 1.0 to 2.0 mg/l 2,4-D with 0.1 or 1.0 mg/l NAA and the globular embryos were further differentiated into the cotyledonary phase embryos. The addition of 5.0 mg/l amino acids (L-glutamine or L-proline) to the culture media, in the range of the tested medium condition, yielded higher enhancement of the embryo growth and development. Transfering of individual embryos onto a fresh basal MS medium without plant growth regulators enabled the achievement of complete maturation Relatively, only a small number of the induced embryos developed shoots and roots when they were transferred to MS medium supplemented with 2.0 mg/l BAP and 0.5 mg/l NAA in addition of 3% (w/v) sucrose and 0.8% (w/v) agar. Nearly, 11% of somatic embryos were able to be converted to fertile plants. This similar result was supported by other authors [3,55].

6.2.3. Dianthus caryophyllus

The in vitro regeneration of the carnation plant species was established from aseptic shoots produced from surface sterilized seeds grown in basal MS medium. High percentage of grown seeds was obtained (50-90%) after 3 weeks of culturing the seeds. All the plantlets were rooted efficiently on the multiplication MS medium without hormone supplementation. This may be due to the presence of endogenous auxins which causes rooting as it was stated by Mosleh et al. [56]. After 4 weeks, full grown plantlets were transferred to sterile soil at ratio of 1:1:1 for garden soil, sand and loam for acclimatization process. They were maintained in the culture room at 25±1°C with 16 hours light and 8 hours dark planted in pots and covered with polystyrene plastic bags. The growth was encouraging and after 4 weeks in the culture room, well grown and healthy plantlets were transferred to the greenhouse. Survival rate of the acclimatized plantlets was 78% and it showed that the most of the plantlets successfully survived after transferred to the greenhouse. Best response for hardening of the plant species was also obtained by Aamir et al. [57] (95%) in mixture containing sand, peat and soil (1:1:1) under natural light conditions [58].

6.2.4. Saintpaulia ionantha

The complete plant regeneration was obtained from leaf, petiole, peduncle and floral parts of the plant. For shoot regeneration, MS medium supplemented with IAA (1.0 mg/l) and Zeatin (2.0 mg/l) induced the highest number of shoots (15.0) in 8 weeks. While shoots regeneration and multiplication obtained from the same plant by Azura et al. [59] with maximum induction rate on MS medium supplemented with 3.0 mg/l BAP and 1.0 mg/l of NAA. It was reported by Hasbullah et al. [10] that a good combination of cytokinins and auxin in the plant culture medium enhanced good shoot formation and plantlet regeneration. Most of the developed plantlets were hardened (84%) and acclimatized in the greenhouse by transferring the plantlets into planting pots containing mixed soil, which is a mixture of compost, sand and black soil with the ratio of 1:1:2. Meanwhile, Khan et al. found that among the different potting mixes used for the acclimatization of rooted plantlets, 100% sand was found to be the best [59].

However, the obtained new plants failed to flower even after twelve months from planting. The morphological characters of these plantlets were compared to those of the mother plants. They were found to be different in some of their morphological characters such as plant height, leaf size and leaf texture and showed similarity in leaf arrangement and leaf margin and they showed unstable morphological characters. But variations in flowering period, number of flowers per plant and flower morphology were observed by Jain [17] in the plants directly regenerated from leaf disk explants. So, he concluded that the cytokinins, benzylaminopurine and zeatin tested in the culture medium did not affect the basic plant characteristics including flower colour which remained stable in both species. An attempt to induce in vitro flowering from african violet was also reported by Daud and Taha [9]. They found that the floral buds were formed in vitro with sepal (calyx) and petal (corolla) but did not show any formation of reproductive organs (stamens or pistils).

6.3. Acclimatization

Acclimatization process were carried out while the plants still under *in vitro* condition. A few days before the process was to be carried out, the cover of test tube was removed. With the relative humidity at 50-70% in the culture room, this will increase the epicuticular wax development on the upper leaf surfaces of the plantlets and their survival rate rose from 70 to 90% [60]. The plantlets were exposed to the normal environment in stages as they will wilt due to rapid changes of relative humidity and light intensity. *In vitro* plantlets that reached 3-5 cm height were taken out from culture tubes and the excess media were rinsed to avoid contamination. They were then put into plastic pots and planted out in soil at a ratio of 1:1:1 for garden soil, sand and loam. There were 3 types of treatments which were carried out:

1. Plantlets were planted in a pot and placed a beaker to cover it up.
2. Plantlets were planted in a pot and put in Mistifier device.
3. Plantlets were planted in a pot and covered with transparent plastic lid.

For acclimatization purpose of (*Saintpaulia ionantha*), various substrates were used, such as autoclaved mixed soil (compost, sand, and black soil in the ratio 1:1:2) and non-autoclaved mixed soil. The regenerated plants must reach 4-5 cm before transferring them into pots of mixed soil. After transplanting, the plantlets were watered regularly to prevent from drying. For the first 3 weeks the regenerated plants were maintained in the culture room at 25 ± 2 °C. Gamma irradiations of 10-60 gray were also tested on the regenerated plants to induce flowering and also to observe the effect of radiation on the plantlets. Successful micropropagation of plants which can survive under the natural environmental conditions depends on acclimatization process. Most species grown *in vitro* required an acclimatization process in order to ensure that sufficient number of plants can survive and grow vigorously after being transferred to ex *vivo* soil. The excess media was cleaned from the roots and the plants were transplanted in an adequate substrate such as peat or soil. Plantlets were maintained in a confined environment temporarily before they can be adapted progressively in typical environment within drier air, high light intensity and temperature variations.

6.4. Antimicrobial studies

To determine the antimicrobial activity, *Pereskia grandifolia* fresh leaves (300 g) were dried in oven (30-35 °C) for about 5-7 days. Dried leaves were crushed and ground using mortar and pestle in the laboratory. The final weights of the dried powdered materials were 35 g. The leaf powder was extracted with methanol as a solvent. Extracts were filtered and concentrated to dryness using a rotary evaporator. Extract was then ready for the antimicrobials tests. The same procedure was done with aseptic callus to obtain the extract of callus. The antimicrobial screening battery consisted of: gram positive bacterium, *Bacillus subtilis, Staphylococcus aureus*; gram negative bacteria, *Escherichia coli, Proteus mirabilis, Proteus vulgaris, Klebsiella pneumoniae* and *Pseudomonas aeruginosa*; and Fungi, *Candida albicans, Microsporium cants, Trichophyton rubrum, Trichophyton mentagrophyte* and *Aspergillus niger*. The antimicrobials tests were done using standard microbial test culture.

6.4.1. Media for microbial cultivation and maintenance

Meuller Hinton agar, Meuller Hinton Broth, Sabouraud Dextrose Agar (Difco) and Sabouraud Dextrose Broth (Difco) culture media were used. Each medium was prepared to manufacturers` specification and adjusted to the appropriate pH before sterilized by autoclaving at 121 °C for 15 minutes. About 20 ml sterile agar media were poured into petri dishes and let to solidify at a slanted position in Universal bottles. Broth or liquid media were distributed into final containers before autoclaving. Antioxidant and antibacterial activities of ethanolic extracts of *Asparagus officinalis* cv. Mary Washington grown *in vivo* and *in vitro* were compared in our laboratory [61]. Although no antibacterial activity was detected from both *in vivo* and *in vitro* grown plant extracts in the disc diffusion antimicrobial assay, ethanolic extract of *A. officinalis* offered antibacterial activity against *Bacillus cereus* (Table 2; Figure 1).

Plant name	Inhibition Zone of Tested Bactria (mm)			
	E.c.	P.a.	S.a.	B.c.
Zizyphus jujube	-	0.67±0.29	0.5±0.0	-
Thymus vulgaris L.	0.93±0.12	0.5±0.0	2.33±0.58	1.17±0.29
Carum carvi L.	-	0.5±0.0	0.83±0.29	-
Teucrium polium L.	-	0.5±0.0	1.0±0.0	-
Althaea officinalis L.	-	0.5±0.0	1.5±0.5	-
Borage officinalis L	-	-	-	-
Tetracycline (30μg)	13±0.1	2.97±0.06	12.97±0.06	8.07±0.12

Table 1. Inhibition effect of 100 mg/ml of ethanolic extracts of some ornamental plants against the growth of four pathogenic bacteria.

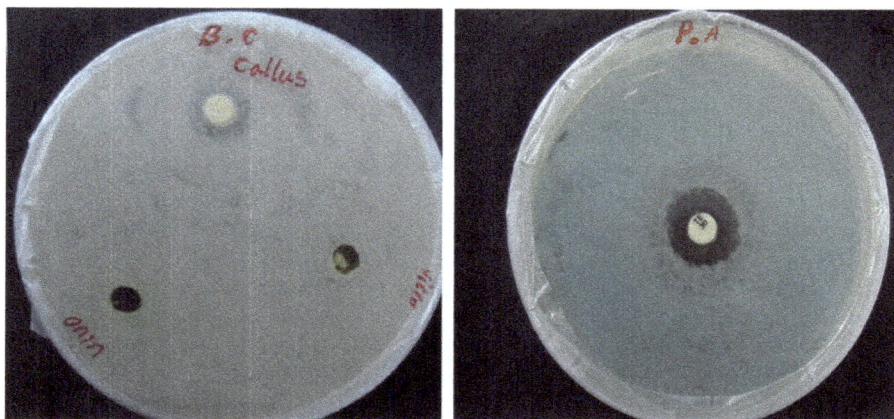

B.c: Bacillus cereus; P.a: Pseudomonas aeruginosa

Figure 1. Comparison of antibacterial activity of *in vitro, in vivo* and callus extracts of *Asparagus officinalis* against *Bacillus cerus* and *Pseudomonas aeruginosa*, using agar diffusion method

Bacteria	Inhibition Zone (mm)			
	In Vivo Plant	*In Vitro* Plant	Callus	Tetracycline (30 μg)
Escherichia coli	_b	_b	_b	42 ± 3.00[a]
Staphylococcus aureus	_b	_b	_b	20 ± 2.64[a]
Pseudomonas aeruginosa	_b	_b	_b	11 ± 2.00[a]
Bacillus cereus	_c	_c	12 ± 1.00[b]	40 ± 3.60[a]

-: No inhibition; The data were analyzed by one-way ANOVA and the inhibition means of samples were compared using Duncan's Multiple Comparison Test (DMCT). Mean of different samples labeled with different letters are significantly different in each row (p < 0.05).

Table 2. Inhibition effect of 100 mg/ml of *Asparagus officinalis* ethanolic extracts (*in vivo* plant, *in vitro* plant and callus) against the growth of four pathogenic bacteria

6.4.2. Inoculums for antimicrobial testing

Cultures of fungi and bacteria grown in Mueller Hinton Broth and Sabouraud Dextrose Broth for 18 hours, respectively, were standardized to an optical density of 1 at 600 nm (OD_{600} = 1) using NOVASPEC II Visible Spectrophotometer. The density was adjusted by adding sterile broth to the cultures. The concentrations of the resultant suspensions of fungi, bacteria and yeasts were approximately 10^8 cells/ml and 10^7 cells/ml, respectively [62]. The fungi and bacteria suspensions were prepared immediately before carrying out the antimicrobial assay. Conidial suspensions of the test fungi were prepared by pouring 20 ml broth containing 1 drop Tween 80 (Sigma P-8074), into 3-days-old cultures of *Aspergillus* species or 2-week-old culture of *Trichophyton mentagraphytes*. The culture of *Aspergillus* species and *Trichophyton mentagraphytes* were grown on Sabouraud Dextrose Agar slants at 37 °C and 27 °C, respectively. After homogenizing with glass beads, the concentrations of the resultant conidial suspensions of *Aspergillus* species and *Trichophyton mentagraphytes* were adjusted to 10^6 conidia/ml and 10^8 conidia/ml, respectively by using haemocytometer.

6.4.3. Semi-quantitative antimicrobial activity test

According to paper-disk Diffusion assay method, the suspension culture of fungi and bacteria were diluted to the final concentrations of approximately 10^6 cells/ml and 10^5 cells/ml, respectively. The bacterial suspensions were evenly spread on the surface of 4 mm thickness of Meuller Hinton Agar (MHA) plate and the fungi suspensions on Dextrose Sabouraud Agar (DSA) plates. Sterile cotton swabs were used to produce uniform growth of organism. Methanol (MeOH) extracts of the leaves (*in vivo*) and callus (*in vitro*) of *Pereskia grandifolia* were used as test extracts. These extracts were dissolved in methanol and applied to filter disks (Whatman No 1, 6 mm in diameter), at the concentrations of 100, 200, 400 and 800 mg/ml for fungi, yeasts and antibacterial screening. After evaporation of the solvent, the disks were placed in a good contact on the seeded agar plates. Chloramphenicol and 5-fluorocytosine (SIGMA F-7129) at the concentrations of the 1.0 mg/ml were used as positive controls for antifungal and antibacterial, respectively. Saturated filter paper disk of methanol (MeOH) as blank disks were used as negative controls. Incubation of bacteria and fungi were done at 37°C for 24 hrs. Each extract and control was employed in triplicate for each organism. Diameters of clear zones produced around the disks (if present) were measured after the incubation time. This study showed that the antioxidant and antimicrobial activities showed that these bioactivities differ between *in vitro* and *in vivo* grown plants. Total antioxidant capacity of *in vivo* grown plant was higher than *in vitro* grown plant, while the only antimicrobial activity was obtained from *in vitro* callus tissue against ornamental plants. Some phytochemical studies are required to investigate the production of antioxidant and antimicrobial compounds in differentiated and undifferentiated callus cells of ornamental plants.

6.4.4. Agar diffusion assay

Arnone et al. and Drouhet et al. reported the conidial suspensions of fungi test were diluted 10 times with molten Sabouraud Dextrose Agar at 40 °C and 20 ml was poured into each petri dish [63, 64]. Filter paper disks (Whatman No. 1, 6 mm in diameter) were impregnated with the test extract solutions in methanol (MeOH) at the concentration of 100, 200, 400 and 800 mg/ml. The disks were transferred into the surface of solidified agar after evaporation of solvent. 5-Fluorocytosine at the concentration of 1.0 mg/ml, saturated filter paper disks of methanol as blank disks were applied as controls. Three series of determination were run for each extract and species of fungi. Zones of growth inhibition were measured after 3 days incubation at 37°C for *Aspergillus niger,* whereas a week at room temperature (27 °C) for *Trichophyton mentagraphyte* and *Trichophyton rubrum.* The results of well diffusion antimicrobial activity carried out and showed that the ethanolic extract of *A. graveolens* callus inhibited the growth of *Bacillus subtitles* more than ethanolic extract of the plant roots. Antimicrobial activity of the callus extract might be either related to the production of a compound in only undifferentiated callus cells or may be produced in higher amounts in these cells when compared to differentiated cells. Several quantitative estimations and studies showed that the production of biocompounds can vary between differentiated and undifferentiated plant cells [50,65].

6.5. Future research

- The development of culture methods, particularly those far highly sealing plants, is expected to have a significant environmental benefit by controlling to retention of biodiversity.
- The ornamental and medicinal plants demand for raw materials raises questions surrounding the sustainability of the new industry.
- Development of *in vitro* culture technology is of fundamental importance if ornamental biotechnology is to play a central role in the growth of global ornamental plant production industry in future.
- Ornamental plant tissue culture techniques are expected to be developed enough in the near future when combined with molecular genetics. This may give support to be same

biotechnological applications as in ornamental plant in genomic age, a field in which ornamental plant is also far behind other higher plants.

• Thus intensive work on new strain selection and improvement of an efficient mass culture system is clearly needed. For the exploitation of ornamental plants are developed at the cellular level, plant tissue culture constituents a basic powerful tool.

• This efficient somatic embryogenesis protocols could be useful for conservation and agronomy and in the improvement of ornamental plants using gene transfer biotechnologies.

7. Conclusion

Successful *in vitro* propagation, organogenesis and somatic embryogenesis with acclimatization of ornamental plants is now being used for commercialization. Many commercial laboratories and national research institutes worldwide use *in vitro* culture system for rapid plant multiplication, germplasm conservation, elimination of pathogens, genetic manipulations and secondary metabolite production. This somatic embryos protocol could be useful for conservation and agronomy and in the improvement of ornamental plants using gene transfer biotechnologies. The results of present investigation clearly indicate that antimicrobial activity vary with the plant species of the ornamental plants and plant material used. Thus, the study ascertains the value of ornamental plants used in Ayurveda, which could be of considerable interest to the development of new drugs.

Author details

A. Bakrudeen Ali Ahmed, S. Mohajer, E.M. Elnaiem and R.M. Taha
Institute of Biological Sciences, Faculty of Science, University of Malaya, Kuala Lumpur, Malaysia

Acknowledgement

The authors thank University of Malaya, Malaysia for the financial support and facilities provided.

8. References

[1] Pierik R.L.M. Commercial aspects of micropropagation. In; Horticulture – New technologies and Applications. Kluwer, Dordrecht, The Netherlands 1991, Pp. 141-153.

[2] Pierik R.L.M, Steegmans H.H.M, Marelis J.J. Gerbera plantlets from *in vitro* cultivated capitulum explants. Scientia Horticulturae 1973, 1, 107-119.

[3] Ahmed A.B.A, Pallela R, Rao A.S, Rao M.V, Taha R.M. Optimized conditions for callus induction, plant regeneration and alkaloids accumulation in stem and shoot tip explants of *Phyla nodiflora*, Spanish Journal of Agricultural Research 2011a, 9 (4), 1262-1270.

[4] Ahmed A.B.A, Rao A.S, Rao M.V, Taha R.M. Effect picloram, additives and plant growth regulators on somatic embryogenesis of *Phyla nodiflora*, Brazilan Archives of Biology and Technology 2011b, 54 (1), 7-13.

[5]   Castillo B, Smith M.A.L. Direct somatic embryogenesis from *Begonia gracilis* explants. Plant Cell report 1997, 16, 385-388.

[6]   Maliro M.F.A, Lameck G. Potential of cassava flour as a gelling agent in media for plant tissue cultures. African Journal of Biotechnology 2004, 3(4), 244-247.

[7]   Rout G.R, Mohapatra A, Mohan J.S. Tissue culture of ornamental pot plant: A critical review on present scenario and future prospects. Biotechnology Advances 2006, 24, 531-560.

[8]   Munoz J.L, Garcia-Molina F, Varon R, Rodriguez-Lopez J.N, Garcia-Canovas F, Tudela J. Calculating molar absorptivities for quinones: Application to the measurement of tyrosinase activity. Analysis Biochemistry 2006, 351, 128-138.

[9]   Daud N, Taha R.M. Plant regeneration and floral bud formation from intact floral parts of African violet (*Saintpaulia ionantha* H. Wendl.) cultured *in vitro*. Pakistan Journal of Biological Sciences 2008, 11(7), 1055-1058.

[10] Hasbullah N.A, Taha R.M, Awal A. Growth optimization and organogenesis of *Gerbera jamesonii* Bolus ex. Hook f. *In vitro*. Pakistan Journal of Biological Science 2008, 11, 1449-1454.

[11] Siong P.K, Taha R.M, Rahiman F.A. Somatic embryogenesis and plant regeneration from hypocotyls and leaf explants of *Brassica oleracea* var. *botrytis* (Cauliflower). Acta Biologica Cracoviensia Series Botanica 2011, 53(1), 26-31.

[12] Wang S.O, Ma S.S. Clonal multiplication of *Chrysanthemum in vitro*. Journal Agriculture Associate China 1978, 32, 64–73.

[13] Lawrence Jr R.H. *In vitro* plant cloning system. Environment Experiment Biotechnology 1981, 21, 289-300.

[14] Douglas G.C, Rutledge C.B, Casey A.D, Richardson D.H.S. Micropropagation of *Floribunda* ground cover and miniature roses. Plant Cell Tissue Organ Culture 1989, 19, 55-64.

[15] Teixeira de Silva J.A, Fukai S. *Chrysanthemum* organogenesis through thin cell layer technology and plant growth regulator control. Asian Journal Plant Science 2003, 2, 505-514.

[16] Murashige T, Skoog T. A revised medium for rapid growth and bioassays with tobacco tissue cultures. Plant Physiology 1962, 15, 473-497.

[17] Jain S.M. Micropropagation of selected somaclones of *Begonia and Saintpaulia*. Journal of Bioscience 1997, 22, 585-592.

[18] Simmonds J, Werry T. Liquid shake cultures for improved micropropagation of *Begonia×hiemalis*. Horticulture Science 1987, 22, 122–124.

[19] Smart N.J, Fowler M.W. An airlift column bioreactor suitable for large scale cultivation of plant cell suspensions. Journal Experiment Botany 1984, 35, 531-537.

[20] Tautorus T.E, Dunstan D.I. Scale-up of embryogenic plant suspension cultures in bioreactors. In: Jain, S.M., Gupta, P.K. and Newton, R.J. editors. Somatic embryogenesis in woody plants, Vol. 1. The Netherlands: Kluwer Academy Publication 1995. Pp. 265-269.

[21] Takayama S. Bioreactor, airlift. In: Spier, R.E., Griffiths, B. and Scragg, A.H. editors. The encyclopedia of cell technology, vol. I & II. John Wiley & Son Inc.; 2000. p. 201-218.

[22] Ziv M. *In vitro* hardening and acclimatization of tissue culture plants. In: Withers, L.A. and Alderson, P.G. (eds.), Plant Tissue Culture and its Agricultural Applications 1986. Pp. 187-196.

[23] Peak D.E, Cumming B.G. *In vitro* propagation of *Begonia×tuberhybrida* from leaf sections. Horticulture Science 1984, 19, 395-397.

[24] Eide A.K, Munster C, Heyerdahl P.H, Lyngved R, Olsen O.A.S. Liquid culture systems for plant propagation. Acta Horticulturae 2003, 625, 173-85.

[25] Wated A.A, Raghothama K.G, Kochba M, Nissim A, Gaba V. Micropropagation of *Spathiphyllum* and *Syngonium* is facilitated by use of Interfacial membrane rafts. Horticulture Science 1997, 32, 307-308.

[26] Joseph D, Martin K.P, Madassery J, Philip V.J. *In vitro* propagation of three commercial cut flower cultivars of *Anthurium andraeanum*. Horticulture Indian Journal Experiment Biology 2003, 41,154-159.

[27] Ranjan S, Gaurav S. Somatic embryogenesis in Gerbera (*Gerbera jamesonii Bolus ex Hooker f.*) as influenced by explants. Journal of Ornamental Horticulture 2005, 8, 128-130.

[28] Ahmed A.B.A, Rao A.S, Rao M.V, Taha R.M. Production of gymnemic acid depends on medium, explants, PGRs, color lights, temperature, photoperiod and sucroses in batch culture of *Gymnema sylvestre*. The Scientific World Journal 2012, 897867, 1-11.

[29] Nobre J, Romano A. *In vitro* cloning of *Ficus carica* L. adult trees. Acta Horticulturae 1998, 480, 161-164.

[30] Dijkshoorn-Dekker M.W.C. The influence of light and temperature on the dynamic behaviour of *Ficus benjamina* 'Exotica'. Acta Horticulturae 1996, 417, 65-67.

[31] Deshpande S.R, Josekutty P.C, Prathapasenan G. Plant regeneration from axillary buds of a mature tree of *Ficus religiosa*. Plant Cell Report 1998, 17, 571-573.

[32] Demiralay A, Yalcin-Mendi Y, Aka-kacar Y, Cetiner S. *In vitro* propagation of *Ficus carica* L. var. *Bursa siyahi* through meristem culture. Acta Horticulturae 1998, 480, 165-167.

[33] Kumar V, Radha A, Kumar C.S. *In vitro* plant regeneration of fig (*Ficus carica* L. cv. Gular) using apical buds from mature trees. Plant Cell Report 1998, 17, 717-720.

[34] Nagaraju S, Reddy S.K, Farook S.A. Propagation of *Ficus reliosa* L. from maxillary buds and shoot tips. Advance Plant Science 1998, 11, 287-290.

[35] Murch S.J, Victor J.M.R, Saxena P.K. Auxin, calcium and sodium in somatic embryogenesis of african violet (*Saintpaulia ionantha* Wendl.) cv. Benjamin. Acta Horticulturae 2003, 625, 201–209.

[36] Winkelmann T, Hohe A, Schwenkel H.G. Establishing embryogenic suspension cultures in *cyclamen persicum* "Purple Flamed". Advance Plant Science 1998, 12, 25-30.

[37] Hohe A, Winkelmann T, Schwenkel H.G. Development of somatic embryos of *Cyclamen persicum* Mill. in liquid culture. Gartenbauwissenschaft 2001, 66, 219-224.

[38] Kumari M, Varghese T.M, Mehta P.K. Micropropagation of *Chrysanthemum* through shoot apex culture in two named varieties viz. Miss Universe and Snow Ball. Annals of Agriculture Research 2001, 11, 371-376.

[39] Kim S.J, Hahn E.J, Paek K.Y, Murthy H.N. Application of bioreactor culture for large scale production of *Chrysanthemum* transplants. Acta Horticulturae 2003, 625, 187-191.

[40] Maes M, Crepel C, Werbrouck S, Debergh P. Perspectives for a DNA-based detection of bacterial contamination in micropropagated plant tissue. Plant Tissue Culture Biotechnology 1998, 4, 49-56.

[41] Hasbullah N.A, Saleh A, Taha R.M. Establishment of somatic embryogenesis from *Gerbera jamesonii Bolus ex. Hook F.* Through Suspension Culture. African Journal of Biotechnology 2011, 10, 13762-13768.

[42] Taha R.M, Hasbullah N.A, Aziz A.H, Awal A. Establishment of *In vitro* plantlets and acclimatization of *Gerbera jamesonii Bolus ex. Hook.F.* Acta Horticulturae 2010a, 865, 401-404.

[43] Son N.V, Mokashi A.N, Hegde R.V, Patil V.S, Lingaraju T. Response of gerbera *(Gerbera jamesonii* Bolus) varieties to micropropagation. Karnataka Journal Agriculture Science 2011, 24, 354-357.

[44] Chuah E.L, Chan L.K. Induction of Somatic Embryogenic Callus From the Leaves of *Pereskia grandifolia*. Biotechnology 2007, 6, 45-48.

[45] Taha R.M, Latif F.A. *In vitro* studies and antimicrobial activities of *Pereskia grandifolia Haworth var. grandifolia*. *CATRINA*-International Journal of Environmental Sciences 2007, 2, 61-66.

[46] Yeoman M.M, Yeoman C.L. Manipulating secondary metabolism in cultured plant cells. New Phytologist 1996, 134, 553-569.

[47] Osborne D.J, McManus M.T. Hormones, Signals and target cells in plant development. Cambridge University Press 2005, Pp. 252.

[48] Arumugam S, Kavimani S, Ahmed A.B.A, Kadalmani B, Akbarsha M.A, Rao M.V. Antidiabetic activity of leaf and callus extracts of *Aegle marmelos* in rabbit. Science Asia 2008, 34, 317-321.

[49] Ahmed A.B.A, Rao A.S, Rao M.V. Role of *in vivo* and *in vitro* callus of *Gymnema sylvestre* (Retz) R.Br. Ex Roemer & Schultes in maintaining the normal levels of blood glucose and lipid profile in diabetic wistar rats. Biomedicine 2008, 28, 134-138.

[50] Ahmed A.B.A, Rao A.S, Rao M.V. *In vitro* callus and *in vivo* leaf extract of *Gymnema sylvestre* stimulate β-cells regeneration and anti-diabetic activity in Wistar rats. Phytomedicine 2010, 17 (13), 1033-1039.

[51] Tomar U.K, Gupta S.C. *In vitro* plant regeneration of leguminous trees *(Albizia* spp.). Plant Cell Report 1998, 15, 441-444.

[52] Taha R.M, Daud N, Hasbullah N.A. Establishment of efficient Regeneration System, Acclimatization and Somaclonal variation in *Saintpaulia ionantha H. Wendl.* Acta Horticulturae 2010b, 865, 115-122.

[53] Dodeman V.L, Ducreux G, Kreis M. Zygotic embryogenesis versus somatic embryogenesis. Journal of Experimental Botany 1997, 48, 1493-1509.

[54] Mohamed N, Taha R.M. Plant regeneration of *Clitoria ternatea* from leaf explants cultured *in vitro*. Journal of Food, Agriculture & Environment 2011, Vol.9 (3&4), 268-270.

[55] Ahmed A.B.A, Rao A.S, Rao M.V. Somatic embryogenesis and plant regeneration from cell suspension culture of *Gymnema sylvestre* (Retz). R.Br. Ex. Roemer & Schultes. KMITL Science and Technology Journal 2009b, 9(1), 18-26.

[56] Mosleh M.S, Duhoky M, Salman A, Media E.M.A. *Micropropagation of carnation* (*Dianthus caryophyllus L.*). Journal of Duhok University 2009, 12, Pp. 61-66.

[57] Aamir A, Humera A, Shhagufta N, Mamoona R, Javed I. An efficient protocol for *in vitro* propogation of carnation (*Dianthus caryophyllus*). Pakestanian Journal of Botany 2008, 40, 111-121.

[58] Azura H.B.A, Taha R.M, Hasbullah N.A. Acclimatization of *Dianthus caryophyllus Linn.* Acta Horticulturae 2010, 865, 397-400.

[59] Khan S, Naseeb S, Ali K. Callus induction, plant regeneration and acclimatization of African violet (*Saintpaulia ionantha*) using leaves as explants. Pakistan Journal of Botany 2007, 39, 1263-1268.

[60] Ziv M. Bioreactor technology for plant micropropagation. In: Janick. J. editor. Horticulture reviews, Vol. 24. New York: John Wiley & Sons Inc. 2000. Pp. 1–30.

[61] Khorasani A, Sani W, Philip K, Taha R.M, Rafat A. Antioxidant and antibacterial activities of ethanolic extracts of *Asparagus officinalis* cv. Mary Washington: Comparison of *in vivo* and *in vitro* grown plant bioactivities. African Journal of Biotechnology 2010, 9 (49), 8460-8466.

[62] Rahalison L, Hamburger M, Hostettman K. A bioautographic agar overlay method for the detection of antifungal compounds from higher plants. Phytochemical Analysis 1991, 2, 199-203.

[63] Arnone A, Assante G, Montorsi M, Nisim G. Asteromine - A bioactive secondary metabolite from a strain of *Mycospharella asteroma*. Phytochemistry 1994, 38, 595-597.

[64] Drouhet E, Dupont B, Improvisi L, Viviani M.A, Tortorano A.M. Disk agar diffusion and microplate automatized technics for *in vitro* evaluation of antifungal agents on yeast and sporulated pathogenic fungi. In: Iwata, K. and Vanden Bossche, H. (eds.), proceeding of the international symposium on *in vitro* and *in vivo* evaluation of antifungal agents held in Tokyo (Japan), 1985. Pp.19-22.

[65] Ahmed A.B.A, Rao A.S, Rao M.V. *In vitro* production of gymnemic acid from *Gymnema sylvestre* (Retz) R.Br. Roemer and Schultes through callus culture under stress conditions. Methods Molecular Biology 2009a. 547, 93-105.

Chapter 12

Microalgal Biotechnology: Prospects and Applications

Soha S.M. Mostafa

Additional information is available at the end of the chapter

1. Introduction

There is a current worldwide interest in finding new and safe antioxidants from natural sources such as plant material to prevent oxidative deterioration of food and to minimize oxidative damage to living cells [1]. Microalgae are photosynthetic microorganisms that are able to rapidly generate biomass from solar energy, CO_2 and nutrients in bodies of water. This biomass consists of important primary metabolites such as sugars, oils and lipids, for which process path-ways exist for the production of high-value products including human and animal feed supplements, transport fuels, industrial chemicals and pharmaceuticals. Algal biomass and algae-derived compounds have a very wide range of potential applications, from animal feed and aquaculture to human nutrition and health products. Some algae are considered as rich sources of natural antioxidants. Although macroalgae have received much attention as potential natural antioxidants [2]. Furthermore, the qualities of the microalgal cells can be controlled, so that they contain no herbicides and pesticides, or any other toxic substances, by using clean nutrient media for growing the microalgae. The value of microalgae as a source of natural antioxidants is further enhanced by the relative ease of purification of target compounds. Reports on the antioxidant activity of microalgae are limited. Because cyanobacteria are largely unexplored, they represent a rich opportunity for discovery; the expected rate of rediscovery is far lower than for other better- studied groups of organisms Li et al. 2007 [3]. In this chapter, we focus on many desirable chemicals are the products of secondary metabolism triggered under conditions not conducive to fast growth. For those chemicals to be produced by microalgae, one needs to develop new strains (faster growth, higher substrate tolerance, etc.) by classical selection or genetic manipulation so microalgal biomass can be produced consistently. Highlight the role of dietary antioxidants and their potential benefits in health and disease directly or indirectly by the plant nutrition and animal feed to produce healthy organic food. Investigate the different biological activities of algae and the relations with its biochemical composition, pigments and different constituents which may vary with salt stressed culture conditions and describe the antioxidant characteristics of algae.

2. What are microalgae?

Microalgae are prokaryotic or eukaryotic photosynthetic microorganisms that produce carbohydrates, proteins and lipids as a result of photosynthesis. They can grow rapidly and live in harsh conditions due to their unicellular or simple multicellular structure. Examples of prokaryotic microorganisms are Cyanobacteria (Cyanophyceae) and eukaryotic microalgae are for example green algae (Chlorophyta) and diatoms (Bacillariophyta). Microalgae are present in all existing earth ecosystems, not just aquatic but also terrestrial, representing a big variety of species living in a wide range of environmental conditions. It is estimated that more than 50,000 species exist, but only a limited number, of around 30,000, have been studied and analyzed [4]. Sunlight, water, nutrients and arable land are the major requirements for growing algae. Micro algae have the ability to fix CO_2 using solar energy with efficiency 10 times greater than that of the terrestrial plants with numerous additional technological advantages. Algae are more efficient at utilizing sunlight than terrestrial plants, consume harmful pollutants, have minimal resource requirements and do not compete with food or agriculture for precious resources [5].

3. Algal metabolites

Metabolites are the intermediates and products of metabolism. The term metabolite is usually restricted to small molecules. A primary metabolite is directly involved in the normal growth, development, and reproduction. A secondary metabolite is not directly involved in those processes, but usually has important ecological function. The induction of secondary metabolism is linked to particular environmental conditions or developmental stages. Secondary metabolites are those chemical compounds in organisms that are not directly involved in the normal growth, development or reproduction of organisms.The exploration of these organisms for pharmaceutical purposes has revealed important chemical prototypes for the discovery of new agents, stimulating the use of sophisticated physical techniques and new syntheses of compounds with biomedical application. In this regard, both secondary and primary metabolisms have been studied as a prelude to future rational economic exploitation (Figure 1). The secondary metabolism is of restricted distribution, while the primary metabolism furnishes intermediates for the synthesis of essential macromolecules [6].

4. What are phytochemicals?

"Phyto" is the Greek word for plant. The term "phytochemicals" refers to a wide variety of compounds produced by plants. Phytochemicals are chemical compounds formed during the plants normal metabolic processes. There are many "families" of phytochemicals and they help the human body in a variety of ways. Phytochemicals may protect human from a host of diseases. These chemicals are often referred to as "secondary metabolities" of which

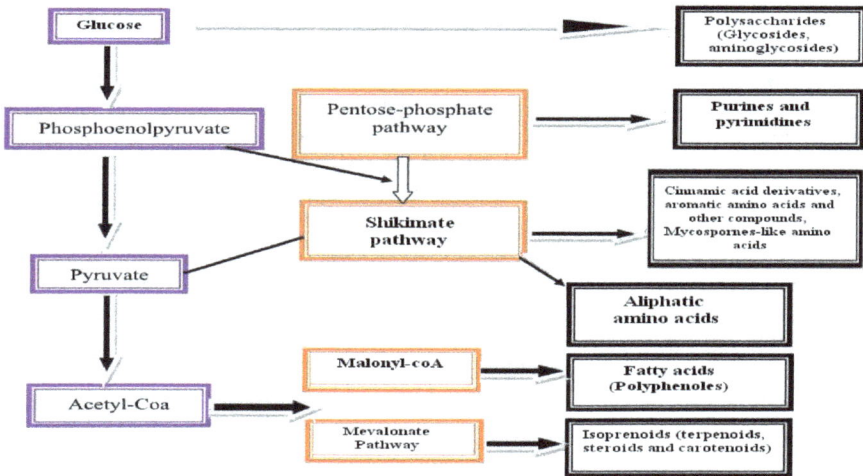

Figure 1. Main pathways of some secondary and primary metabolites biosynthesis modified from Burja et al. [7]

there are several classes including alkaloids, flavonoids, coumarins, glycosides, gums, polysaccharides, phenols, tannins, terpenes and terpenoids Phytochemicals are naturally occurring, nonnutritive chemicals. They appear to work alone and in combination, and perhaps in conjunction, with vitamins [8].

5. Microalgal bioactive compounds

Microalgae are significant resource for bioactive metabolites, particularly cytotoxic agents with applications in cancer chemotherapy. From the marine microalgae such as from the blooms of *Phaeocystis* sp., antibiotic substances were listed. *Phaeocystis pouchetii* is reported to produce chemicals such as acrylic acid, which constitutes about 7.0% of the dry weight. The antibiotic substances thus produced are transferred throughout the food chain and found in the digestive tract of *Antartic penguins*. Production of ß carotene and vitamins by the halotolerant alga *Dunaliella* sp. is documented. These compounds have much importance for the Mariculture activities [9]. Cyanobacteria have been identified as one of the most promising group of organisms from which novel and biochemically active natural products are isolated. Cyanobacteria such as *Spirulina, Anabaena, Nostoc* and *Oscillatoria* produce a great variety of secondary metabolites. Cyanobacteria produce a wide variety of bioactive compounds, which include 40% lipopeptides, 5.6% amino acids, 4.2% fatty acids, 4.2% macrolides and 9% amides. Cyanobacterial lipopeptides include different compounds like cytotoxic (41%), antitumor (13%), antiviral (4%), antibiotics (12%) and the remaining 18% activities include antimalarial, antimycotics, multi-drug resistance reversers, antifeedant, herbicides and immunosuppressive agents [7]; besides the immune effect, blue green algae improves metabolism. Cyanobacteria are also known to produce antitumor, antiviral, antifungal compounds and have a cholesterol-lowering effect in animals and humans [10]. Many of the pharmaceutically interesting compounds in cyanobacteria are peptides, including cyanobacterial toxins and important candidates for anti-cancer drugs. Peptide synthetases are common in cyanobacteria and responsible for the production of cyanobacterial hepatotoxins and other peptides. Polyketide synthetases are also involved in the biosynthesis of certain cyanobacterial bioactive compounds (e.g. microcystins). A number of extracts were found to be remarkably active in protecting human lymphoblastoid T-cells from the cytopathic effects of HIV infection. Active agents consisting of sulfolipids with different fatty acid esters were isolated from *Lyngbya lagerheimii* and *Phormidium tenue*. Cyanovirin is a protein isolated from an aqueous cellular extract of *Nostoc elipsosporum* prevents the in vitro replication and citopathicity of primate retroviruses.Cryptophycin 1, an active compound isolated from *Nostoc* strain, exerts antiproliferative and antimitotic activities by binding to the ends of the microtubules, thus blocking the cell cycle at the metaphase of mitosis. Research has been focused on its potent antitumor activity and a synthetic analogue, cryptophycin-52, is at present in Phase II clinical trials. Sulfated polysaccharide, calcium spirulan. A novel water soluble extracts of cyanobacteria have found to be an antiviral agent. This compound appears to be selectively inhibiting the penetration of enveloped viruses into host cells, thereby preventing the replication. The effect was described for many different viruses like herpex simplex, measles, and even HIV-1. Among eukaryotic microalgae, a glycoprotein prepared from *Chlorella vulgaris* culture supernatant exhibited protective activity against tumor metastasis and chemotherapy-induced immunosuppression in mice [11]. Hereafter, a brief discuss of the commercial application of the most explored compounds from algae and the biosynthetic pathways of fatty acids, steroids and carotenoids.

5.1. Fatty Acids (FA)

Microalgae include essential fatty acids (EFAs) such as linoleic, arachidonic, linolenic, ?-linolenic acids etc. that must be in diet for healthy growth. These acids cannot be synthesized fast enough by body to meet needs [12]. Fatty acids are structural components of many lipids, and the types and amounts of fatty acids vary considerably among algae. In recent years, fatty acids compositions in large scale production of microalgae including marine algae have created considerable interest among researchers. This is mainly because of the health benefit of mono and polyunsaturated fatty acids (MUFA and PUFA) that can be found in plants including microalgae. Moreover, polyunsaturated fatty acids (PUFAs) play key roles in cellular and tissue metabolism, including the regulation of membrane fluidity, electron and oxygen transport, as well as thermal adaptation [13]. The biosynthesis of EPA occurs through a series of reactions that can be divided into two distinct steps. First is the de novo synthesis of oleic acid (18:1 ω9) from acetate, followed by conversion to linoleic acid (18:2 ω-6) and α-linolenic acid (18:3 ω-3). The subsequent stepwise desaturation and elongation steps form an ω-3 PUFA (Fig. 2). Inside the cell, EPA is normally esterified (by cyclooxygenase and lipooxygenase activities) to form complex lipid molecules and plays an important role in higher animals and humans as the precursor of a group of eicosanoids, hormone-like substances such as prostaglandins, thromboxanes and leucotrienes that are crucial in regulating developmental and regulatory physiology (Figure 2) [14]. Consumption of n-3 PUFAs from both seafood and plant sources may reduce coronary heart disease (CHD) risk as reported by Mozaffarian et al. [15] in a cohort study of 45,722 men. Thus, many health supplement stores now sell preparation of microalgae such as *Spirulina* and *Chlorella* packed in capsule or caplets, or even in food and beverages known to have therapeutic values in treating hypercholesterolemia, hyperlipidaemia and atherosclerosis [16]. The fatty acid contents of microalgae are influenced by the environmental and cultural condition selected for its growth [17]. Some of the environmental conditions include heterotrophic, photoautotrophic and nitrogen deprivation or stimulation. Although some microalgae species are cultivated as sources of these fatty acids, transgenic algae engineered to produce EPA, like transgenic oilseed crops, could provide an alternative sustainable source of oil for human consumption [18]. However, the possibility for deploying transgenic organisms nutritionally enhancedwith EPA is currently limited by continued consumer antipathy to transgenic food products. One alternative would be to use EPA from transgenic algae as a high potential food source in aquaculture. In this way, the significant health benefits of these fatty acids could be delivered into the human diet, without the requirement of direct ingestion of genetically modified food.

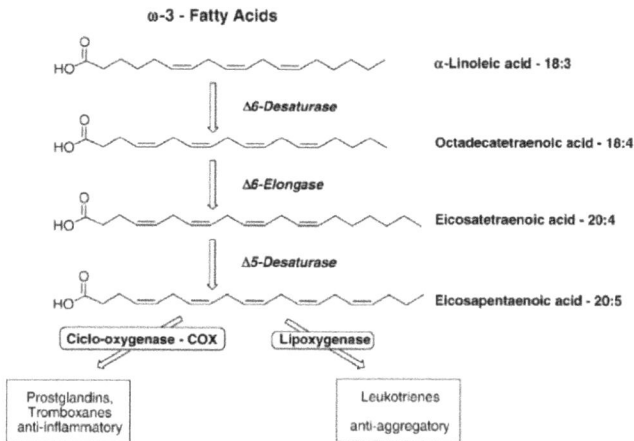

Figure 2. A simplified biosynthesis scheme of eicosapentaenoic acid and eicosanoid (prostaglandins, thromboxanes, leukotrienes) modified from Sayanova and Napier [19].

5.2. Sterols

Sterols are one of the most important chemical constituents of microalgae [20]. Sterols are the main component of eukaryote organisms and different classes of organisms have divergent sterols patterns. It is because of this that sterols act as a fingerprint for organic matter input into an aquatic environment. Furthermore, sterols have a relatively high resistance to degradation when settled in anoxic sediments and persist in the environment for a longer period of time. Of all the sterol compounds, cholesterol is the most abundant and ubiquitous one in the environment, which is due to it having a variety of sources [21]. Most biologically produced sterols are planar 3β-hydroxy tetracyclic structures commonly containing a methyl- or ethyl- substituted C7-C11 hydrocarbon side chain, and exhibiting a range of methyl-substitution (C4, C14) patterns on the polycyclic nucleus with varying degrees and positions of unsaturation (C5, C7, C8). The rigid structure of the sterols (Figure 3), caused by the fused ring system, provides the cell membrane integrity and stability thus, holds the membrane together. In general, there is not a specific sterol that can be uniquely linked to one algal source. Many of the sterols previously discussed are also found in other groups of algae [22].

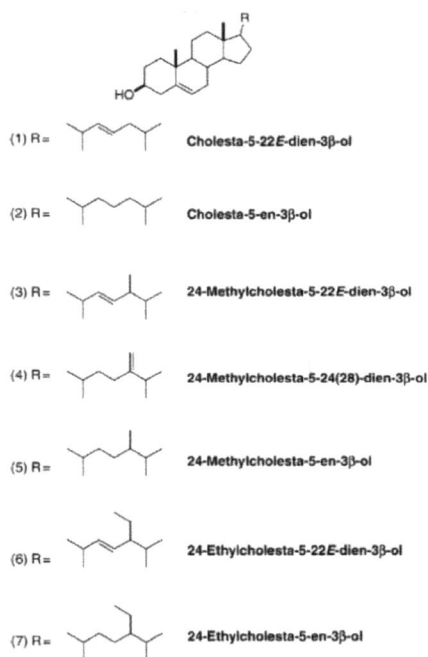

(1) R =   Cholesta-5-22E-dien-3β-ol

(2) R =   Cholesta-5-en-3β-ol

(3) R =   24-Methylcholesta-5-22E-dien-3β-ol

(4) R =   24-Methylcholesta-5-24(28)-dien-3β-ol

(5) R =   24-Methylcholesta-5-en-3β-ol

(6) R =   24-Ethylcholesta-5-22E-dien-3β-ol

(7) R =   24-Ethylcholesta-5-en-3β-ol

Figure 3. Some sterols found in marine and freshwater microalgae (modified from Ponomarenko et al. [34]

5.3. Pigmentation in aquacultures

Astaxanthin (Figure 4) is a red pigment common to several aquatic organisms including microalgae, seagrasses, shrimp, lobsters and fish such as salmon and trout. Crustaceans are unable to synthesize carotenoids de novo and require astaxanthin (or appropriate precursors) in their diet in order to acquire the adequate color for seafoodmarket acceptance [23]. Several natural sources–such as the algae *Dunaliella salina* and *Spirulina maxima*–or synthetic β-carotene, canthaxanthin and astaxanthin have been used for this purpose. Astaxanthin is, in fact, one of the most expensive components of salmon farming, accounting for about 15%of total production costs [48]. Among the several natural sources of astaxanthin applied in aquaculture, the green unicellular freshwater alga *Haematococcus pluvialis* has been explored by biotechnology companies [24].

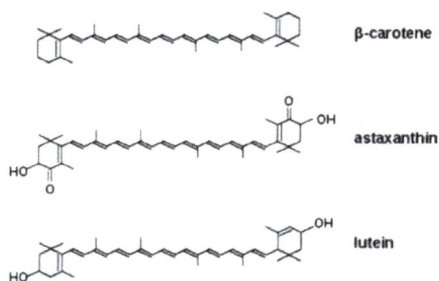

Figure 4. Figure 4. Chemical structure of β-carotene and of the xanthophylls astaxanthin and lutein, main carotenoids from microalgae with commercial interest

β-Carotene is one of the important members of the family of carotenoids; a group of natural fat-soluble stereoisometric pigments. β-Carotene shows pro-vitamin A activity and as such it plays an important role in the human body[25]. β-Carotene can be also used as a coloring agent. Therefore, β-carotene has several applications in food, pharmaceuticals and cosmetics. The great demand of β-carotene has been met by industry, mainly by synthetic production. Increasing demand for natural carotenoids has resulted in growing interest in extracting β-carotene from different natural sources. *Dunaliella salina* is the main source for the natural β-carotene in the market. The estimated market size for natural β-carotene is 10-100 tonnes.year-1 and its price is >750 €.Kg-1 [26]. In addition, β-carotene (like other carotenoids) is a strong antioxidant, scavenging potentially harmful oxy radicals, which are commonly associated with the induction of certain cancers (Leach et al., 1998) and there is an inverse relation between the consumption of certain carotenoids and the risk of cancer [25]. The demonstrated antioxidant activity of carotenoids is the basis of the protective action of these compounds against oxidative stress in many organisms and situations. Effects of carotenoids on human health are, in general, associated with their antioxidant properties. Notwithstanding, not all of the biological activities ascribed to carotenoids must be necessarily linked to their ability to prevent accumulation of free radicals and reactive oxygen species. The halophilic green biflagellate microalga *Dunaliella salina* has since long been recognized as an efficient biological source of this carotenoid. Many epidemiological studies suggest that humans fed on a diet high in β-carotene from *Dunaliella*, which maintains higher than average levels of serum carotenoids, have a lower incidence of several types of cancer and degenerative diseases [27]. The xanthophyll astaxanthin has many applications in nutraceuticals, cosmetics, and food and feed industries. Recently, a variety of additional potential applications of this carotenoid, mainly related to human health and nutrition properties, have been claimed [28]. Lutein is among the most important carotenoids in foods and human serum and, together with zeaxanthin, is the essential component of the pigment present in the macula lutea (or yellow spot) in the eye retina and in the eye lens. Lutein is used as food dyes and especially as feed additives in aquaculture and poultry farming. During the last few years, additional applications for lutein have received considerable interest, especially those related to human health. Mainly on the basis of epidemiological studies, lutein is currently considered as effective agent for the prevention of a variety of human diseases. The microalga *Muriellopsis* sp. and other chlorophycean species are able to accumulate lutein as a part of their biomass. An established commercial system for the production of lutein from microalgae does not exist yet, although the basis for outdoor production of lutein-rich cells of strains of Muriellopsis and *Scenedesmus* at a pilot scale has alreadybeen set up [29].

5.4. Mycosporine-like amino acids

A remarkable group of marine natural products are the mycosporine-like amino acids (MAAs). An outstanding characteristic of these compounds is their high UV absorption with molar absorptivities (ε) of around 40 000 l mol^{-1} cm^{-1} (e.g. Takano et al. [30]). MAAs are water-soluble, low molecular-weight (generally <400) compounds composed of either an aminocyclohexenone or an aminocyclohexenimine ring, carrying nitrogen or amino alcohol substituents [31]. They are found in a wide variety of marine, freshwater and to a smaller degree in terrestrial organisms. There is limited evidence that MAAs are derived from early steps of the shikimate pathway. However, the biochemical pathway of MAA synthesis is still largely unknown, as well as its genetic base. The most primitive organisms capable of MAA synthesis are cyanobacteria [32].

6. Biological activity of microalgae

Many of the microalgal metabolites have chemical structure and possess interesting biological activity. Microalgae are a unique source of therapeutic substances, particularly from cyanobacteria. Among cyanobacteria *Spirulina* sp. has undergone numerous and rigorous toxicological studies that have highlighted its potential therapeutic applications in the area of immunomodulation, anticancer, antiviral, and cholesterol reduction effects.

6.1. Antioxidant activity

Hydrogen peroxide is a product of microalgae and plants through of photosynthesis, photorespiration, respiration and other metabolic processes, as result from the enzymatic activity of glycolate oxidase, urate oxidase and amino acid oxidase. However, major pathway for production of H_2O_2 is conversion from superoxide (O_2^{-}) produced through the transfer of an electron from ferredoxin of photosystem I (PSI) to O_2 (Mehler reaction) by the action of Superoxide Dismutase (SOD). However, it is suspected that those antioxidants are responsible for some side effects such as liver damage and carcinogenesis. Antioxidants can involve with the oxidation process by scavenging free radicals, chelating catalytic metals and by acting as oxygen scavengers [33]. Recently many researchers are interested in finding any natural antioxidants having safety and effectiveness, which can be substituted for current and commercial synthetic antioxidants, BHA and BHT. Microalgae have become good candidates for sources of natural antioxidants, as revealed by a number of recent studies [34-35]. Algae contain several enzymatic and nonenzymatic antioxidant defense systems to maintain the concentration of ROS (O_2^- and H_2O_2) to protect cells from damage [36]. The main cellular components susceptible to damage by these ROS are lipids (peroxidation of poly-unsaturated fatty acids in membranes), proteins (denaturation), carbohydrates and nucleic acids. The essential for ROS detoxification during normal metabolism and particularly during stress, are antioxidant defenses system [37]. The primary scavenging enzymatic defenses system include SOD, calalase (CAT) and glutathione peroxidase, (GPX) and peroxiredoxin (PrxR) [38]. These enzymic detoxification system involving the action of SOD and reductase, either quench toxic compounds or regenerate antioxidants with the help of reducing power provided by photosynthesis [39]. However, at low levels, H_2O_2 resulted in induction of defense genes such as glutathione S-transferase and glutathione peroxidase. The hydrophilic antioxidants AA and GSH effectively scavenge oxygen radicals. Carotenoids and TOH remove ROS directly from the pigment bed [40]. Also, Foyer and Noctor [41] reported that the changes in ROS, fluctuations in the antioxidants concentrations in photosynthetic cells might have important consequences not only for defense metabolism but also for the regulation of genes associated with adaptive responses. Several bioactive metabolites produced by cyanobacteria and algae have been discovered by screening programs, employing target organisms quite unrelated to those for which the metabolites evolved [42]. Shanab et al. [43] studied the antioxidant activity of aqueous extracts of nine microalgal species namely, *Nostoc muscorum, Anabaena flos aquae, Anabaena oryzae, Nostoc humifusum, Oscillatoria* sp., *Spirulina platensis, Phormedium fragile, Wollea saccata* and *Chlorella vulgaris*. Antioxidant activity of the algal extracts was performed using 2,2 diphenyl-1-picrylhydrazyl (DPPH) test and 2,2'-azino-bis(ethylbenzthiazoline-6-sulfonic acid (ABTS) radical action assay which revealed higher antioxidant activity than DPPH mehod. Concerning DPPH, the antioxidant activity of nine tested algal species ranged between 30.1 and 72.4% comparing with the standarad antioxidant BHT (80.2%). Using ABTS method, which was more sensitvie than the DPPH method (Figure 5), the antioxidant activity ranged between 31.2 and 75.9% (Standarad BHT showed 85.6%); *Spirulina platensis, Oscillatoria* sp, *Anabaena flous-aqua* and *Nostoc muscorum* recorded the highest (75.9, 75.6, 73.6 and 72.8%, respectively) antioxidant activity which is could be attributed to the extracellular and intracellular secondary metabolits content (Total phenolic content, terpenoids and alkaloids) of these microalgae (Tables 1,2). The extractracellular phytochemicals metabolites (%) released in the algal cultures show large variability. *Anabaena oryza, Phomidium fragile* and *Wallea saccate* (Table 1) recorded the highest extracellular total phenolic compounds (0.0085, 0.0078 and 0.0074% respectively). The highest terpenoids contents were achieved by *Phormidium fragile, Spirulina platensis* and *Wollea saccata* (0.0055. 0.0050 and 0.0049%, respectively). The maximum values of the extracellular alkaloids were recorded by *Anabaena oryza, Phomidium fragile, Anabaena oryza, Spirulina platensis* and *Phomidium fragile* (0.075, 0.068 and 0.068%, respectively). While, the highest percentages of these metabolities complaining with the released extracellularly metabolities, *Spirulina platensis*, *Nostoc muscourum* and *Oscillatoria* sp. recorded the greatest intracellular total phenolic compounds (0.71, 0.6 and 0.55% respectively), while *Wollea saccata* showed the least content (0.1%) as shown in Table (2).

Concerning terpenoids, *Anabaena flos aquae*, *Spirulina platensis* and *Wollea saccata*, recorded the highest contents (0.15, 0.14 and 0.14% respectively). Alkaloids determination in algal cultures showed that *S. platensis*, *Oscillatoria* sp. and *Chlorella vulgaris* showed the highest contents (3.02, 2.6 and 2.45% respectively). Phycobiliprotein pigments (Table 3) were determined in water extracts of the tested algal species. Normally, phycobilin pigments in cyanobacteria comprised phycocyanin, allophycocyanin and phycoerytherin (the blue, gray and red colors respectively). *Phormidium fragile* recorded the higher C-phycocyanin (CPC) content (0.13 mg/ml) while *Anabaena oryzae* and *A. flos aquae* recorded the least and absence of phycocyanin (0.0089 and 0.0 mg/l respectively).

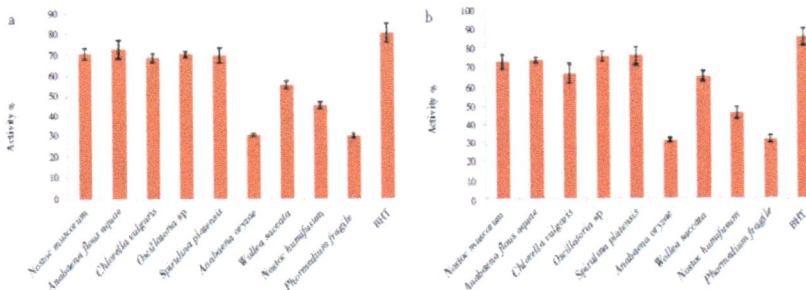

Figure 5. Antioxidant activity of algal wter extracts at 100 g/ml against DPPH (a) and ABTS (b) radicals [43]

Metabolites	Nostoc muscorum	Anabaena flous aquae	Chlorella vulgaris	Oscillatoria sp	Spirulina platensis	Anabaena oryzae	Wollea saccata	Nostoc humifusum	Phormedium fragile
T. phenolic	1.91±0..02	4.10±0.51	2.90±0.05	4.40±0.65	4.80±0.74	8.50±1.87	7.40±0.84	3.10±0.055	7.80±0.96
Terpenoids	1.80±0.01	2.10±0.16	2.10±0.27	2.50±0.68	5.00±0.52	8.10±0.65	4.90±0.64	2.10±0.0	5.50±0.32
Alkaloids	35.00±1.65	39.00±2.98	36.00±1.64	49.00±2.65	68.00±3.61	75.00±4.95	59.00±3.94	42.00±1.66	68.00±3.74
LSD 0.05	2.60	2.61	2.61	2.64	2.60	1.423	1.36	2.60	3.93

Table 1. Secondary metabolites (as mg/100 g) in algal filterates (extracellular) [43]

Metabolites	Nostoc muscorum	Anabaena flous aquae	Chlorella vulgaris	Oscillatoria sp	Spirulina platensis	Anabaena oryzae	Wollea saccata	Nostoc humifusum	Phormedium fragile
T. phenolic	0.61±0.06	0.32±0.05	0.20±0.00	0.55±0.0	0.71±0.14	0.40±0.00	0.10±0.00	0.34±0.0	0.36±0.00
Terpenoids	0.10±0.01	0.15±0.02	0.09±0.00	0.20±0.0	0.14±0.00	0.12±0.00	0.14±0.00	0.10±0.0	0.08±0.30
Alkaloids	2.30±0.20	1.9±0.03	2.45±0.05	2.62±0.15	3.02±0.06	1.60±0.14	1.50±0.40	1.65±0.16	1.80±0.25
LSD 0.05	0.066	0.105	0.0036	0.018	0.029	0.067	0.031	0.031	0.066

Table 2. Secondary metabolites (as %) in algal cells (inracellular) [43]

Algal species	CPC	APC	CPE	Total phycobilins
Nostoc muscorum	0.089±0.010	0.140±0.020	0.0±0.0	0.229±0.020
Anabaena flous aquae	0.000±0.000	0.017±0.000	0.0±0.0	0.017±0.000
Chlorella vulgaris	ND	ND	ND	ND
Oscillatoria sp	0.021±0.000	0.001±0.000	0.0±0.0	0.022±0.000
Spirulina platensis	0.030±0.000	0.113±0.000	0.0±0.0	0.143 ±0.000
Anabaena oryzae	0.009±0.000	0.002±0.000	0.0±0.0	0.011±0.000
Wollea saccata	0.002±0.000	0.033±0.000	0.0±0.0	0.035±0.000
Nostoc humifusum	0.012±0.000	0.019±0.000	0.0±0.0	0.031±0.000
Phormedium fragile	0.130±0.010	0.030±0.000	0.0±0.0	0.160±0.030
LSD 0.05	0.005	0.0057	---	0.003

Table 3. Phycobilins pigments (mg/ml) in different aqueous filtrate of some microalgae [43]

Nostoc muscorum contained the greatest allophycocyanin (APC) pigment (0.14 mg/l), while, *Oscillatoria* sp have the least content (0.001 mg/l). Concerning total phycobilin pigments (phycocyanin and allophycocyanin), *Nostoc muscrum*, *Phormedium fragile* recorded the highest contents (0.229 and 0.16 mg/l) followed by that of S. platensis (0.143 mg/ml), while *Anabaena oryzae* showed the least total phycobilin pigment content (0.011 mg/l).*Chlorella vulgaris* is green alga which have other pigments (Chlorophyll and carotenoids) than the phycobilins (absent).

Aqueous extracts of the tested algal species showed wide range of colours (green, blue, violet, pink, ligh-blue) in spite of the fact that eight of the tested algae were of cyanobacteria and only one species was a green alga, their water extracts showed highly variable colors (Figure 6) which may be attributed in part to their phycobiliprotein constituents (ratios of phycocyanin to allophycocyanin and approximately absence of phycoerytherin pigments), and in part to the produced major polar secondary metabolites. All these substances may not only caused the alteration of the pH values of the algal aqueous extracts, but also the induced biological activities which may be attributed to the synergistic effects of these compounds. The aqueous extract of the tested algal species (8 cyanobacteria and one green alga) have variable colors (Figure 6) ranging from green, violet, blue, light blue and pink color, which can be used as an additive coloring agents to different food products (natural, non toxic) instead of the synthetic coloring substances which may be carcinogenic [43].

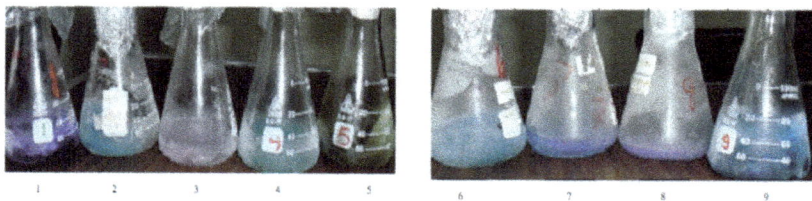

Figure 6. Color of aqueous extracts of the different microalgae (1. Nostoc muscorum, 2. Anabaena flos aquae, 3.Chlorella vulgari, 4. Oscillatoria sp., 5. Spirulina platensis, 6. Anabaena oryzae, 7. Wollea saccat, 8. Nostoc humifusum and 9. Phormedium fragile [43]

Phycobiliprotein pigments were known by its antioxidant activity [44], increasing of these pigments production as a result of doubling nitrate concentration in the growth culture media, led to a progressive increase in the antioxidant activity recorded by both DPPH and ABTS assays in the two cyanobacteria under investigation.Keeping in mind that, synergetic effect occurred between the polar secondry metabolites especially the phenolic compounds and the polysaccharides in antioxidant activity. Increasing nitrate concentrations in the culture media of both cyanobacteria species (*N. muscorum* and *Oscillatoria* sp) led to a marked enhancement in phycobiliprotein production (Table 4) which was translated in an obvious increase in antioxidant activity (by DPPH and ABTS) in both species under study while decreasing the nitrate content, phycobilin pigments production were consequently decreased in both species and its complete absence was recorded on nitrogen starvation especially in case of *Oscillatoria* sp., and no allophycocyanin pigments were produced as found by Shanab et al.[43]. They also investigated that increasing nitrate conc. and the consequent increase in phycobilin pigments production, have the major role in enhancing the antioxidant activity may be attributed. The decrease in nitrate conc. was followed by an obvious decrease in phycobilin pigment and even an absence of one of its constituents on nitrate starvation. The antioxidant activity in both species (by both assays) was apparently not affected comparing with the control (1.5 g/l nitrate). Under stress conditions, it was known that, deviation in metabolic pathways may occur. In presence of nitrate, nitrogenous compounds, including the phycobilin pigments were increasingly produced leading, together with other antioxidant active secondary metabolites (as phenolics), to a marked increase in biological activity.

Treatment	*Oscillatoria* sp.		*Nostoc muscorum*	
	DPPH	ABTS	DPPH	ABTS
Control (1.5 g/L NaNO₃)	59.80±0.95	69.80±1.45	69.80±1.22	72.60
3 g/L	60.20±1.58	70.00±0.95	70.60±1.00	75.30±2.30
6	61.50±0.88	71.60±0.63	70.20±0.98	76.10±3.00
9	68.00±3.60	73.60±2.80	72.90±0.51	80.30±1.65
0.75	60.30±1.50	69.80±4.31	71.50±0.64	70.00±0.58
0.37	62.50±2.45	73.60±2.55	70.00±1.60	74.30±0.47
0.0	66.80±1.67	73.00±1.60	74.00±2.78	76.80±0.61
LSD 0.05	0.968	0.956	0.987	1.001

Table 4. Antioxidant activity of the nitrogen stressed promising algal species using DPPH and ABTS radicals [43]

The decrease in nitrate content induced a stress condition and not only a decrease in nitrogen skeleton compounds as phycobilin pigment production, but an increase in the carbon skeleton compounds (as phenolics) as a result of metabolic alterations under these stress conditions. So on decreasing nitrate content, the antioxidant activity remain at a level comparable or even higher than the control due to the synergistic effect of the phycobilin pigment and the phenolic compounds produced in excess under stress nitrate condition which have high redox potentials. On nitrogen starvation the recorded antioxidant activity

(Comparable to those in presence of high nitrate content (6-9 g/l) was largely due to the high production of the carbon skeleton compounds (phenolic compounds) which show potent antioxidant activity [45]. Shalaby et al. [46] stated that cultivation of *Spirulina platensis* under salt stress conditions (0.02 M as control), 0.04 and 0.08 M NaCl led to a remarkable alteration of algal metabolism as well as an enhancement or induction of biologically active compounds. Biochemical analysis of salt stressed algal revealed that lipid content was slightly increased together with certain saturated and unsaturated fatty acids especially the polyunsaturated ones (γ-inolenic acid, omega 3 fatty acid).

6.2. Anticancer activity

Today cancer is the largest single cause of death in men and women, and chemoprevention has been a promising anticancer approach aimed at reducing themorbidity andmortality of cancer by delaying the process of carcinogenesis. A variety of compounds fromnature sources have been shown to be beneficial for the inhibition of cancer, such as flavonoids, phenolic acids, carotenoids, etc.; the mechanisms which suppress tumorgenesis often involve inhibition of tumor cell mediated protease activity, attenuation of tumor angiogenesis, promotion of cell cycle arrest, induction of apoptosis and immunostimulation, etc. In addition, Chinery et al. [47] also reported their use with the chemotherapy agents 5-fluorouracil and antioxidants could cause complete remissions in colorectal cancer, where only partial remission is possiblewith chemotherapy agents only; therefore, antioxidants have been proposed to have potential for the prevention and treatment of diseases associated with active oxygen species, especially in cancer diseases. Moreover, experimental and epidemiological evidence suggests that anti-inflammatory drugs may also decrease the incidence of mammary cancer, tumor burden, and tumor volume [48].The medicinal value of cyanobacteria was appreciated as early as 1500 Bc, when strains of Nostoc were used to treat gout, fistula and several forms of cancer. Cyanobacteria are a rich source of potentially useful natural products. Over 40 different Nostocales species, the majority of which are Anabaena and Nostoc spp. Produce over 120 natural products (Secondry metabolities) having activities such as anti-HIV anticancer, antifungal, antimalarial and antimicrobial. Cyanovirin (CV-N, cyanoviorin-N), a 101 amino acid protein extracted from *Nostoc ellipsosporum* was found to have potent activity against all human immunodeficiency viruses such as HIV-1, M and T tropic strains of HIV-1, HIV-2, SIV (Simian), and FIV (Feline) [7]. The cosmopolitan distribution of cyanobacteria indicates that they can cope with a wide spectrum of global environmental stress, such as heat, cold, desiccation, salinity, nitrogen starvation, photooxidation, anaerobiosis and osmotic stress. They have developed a number of mechanisms by which cyanobacteria defend themselves against environmental stressors. Important among them are the production of photoprotective compounds such as mycosporine-like amino acids (MAAs) and Scytonemin enzymes such as superoxide dismutase, catalase and peroxidases repaire of DNA damage and synthesis of shock proteins [49]. Shanab et al. [43] investigated anticancer efficiency of the algal water extracts against Ehrlich Ascites Carcinoma cell (EACC) and Human hepatocellular cancer cell line (HepG2). Anticancer efficiency of the algal water extracts was investigated against Ehrlich

Ascites Carcinoma cell (EACC) and Human hepatocellular cancer cell line (HepG2). The anticancer efficiency of the algal aqueous extracts illustrated in Figure (7) and using EACC and HepG2 cell lines, recorded that the anticancer activity ranged between 15.68 to 87.25 % in case of EACC cell line and from 9.5 to 89.4% using HepG2 cell line *Nostoc muscorum* aqueous extracts recorded the highest anticancer activity in both cell lines (87.25% in case of EACC and 89.4% in case of HepG2), followed by *Oscillatoria* sp. (67.40 and 77.8% in EACC and HepG2 respectively). In case of *N. muscorum*, the anticancer activity against EACC cell line ranged between 83.0 and 90.4% at all nitrate concentrations (increase and decrease) compared to the control (85.9%). Comparable anticancer activity was recorded at both the highest nitrate conc and starvation (90.4 and 89.9% respectively). The anticancer activity against HepG2 cell line recorded more or less comparable activities were recorded at most nitrate conc compared to the control (85.6, 86.9, 88.7 and 88.6 % at 3, 6, 9 and 1.5 g/l). At nitrate starvation the highest anticancer activity against HepG2 cell line was recorded (92.3%). In case of *Oscillatoria* sp., the anticancer activity against EACC and HepG2 recorded an increase in activity on both increasing and decreasing nitrate conc comparing with the control. Higher activity was recorded against both cell lines at higher nitrate conc (82.6 and 75.9% in case of EACC and HepG2 respectively) and at nitrate starvation (82.9 and 82.0% respectively) compared to the control (68.3 and 70.4 % against EACC and HepG2 respectively). Water extracts of the tested promising algal species demonstrated higher anticancer efficiencies against both EACC and HepG2 cell lines (87.25 and 89.4% respectively) in case of *N. muscorum* and 67.40 and 77.8% in case of *Oscillatoria* sp.). Under stress nitrogen conditions, these two cyanobacteria species recorded higher anticancer activities on exess limitation or starvation of nitrate comparing with its normal content in growth media.

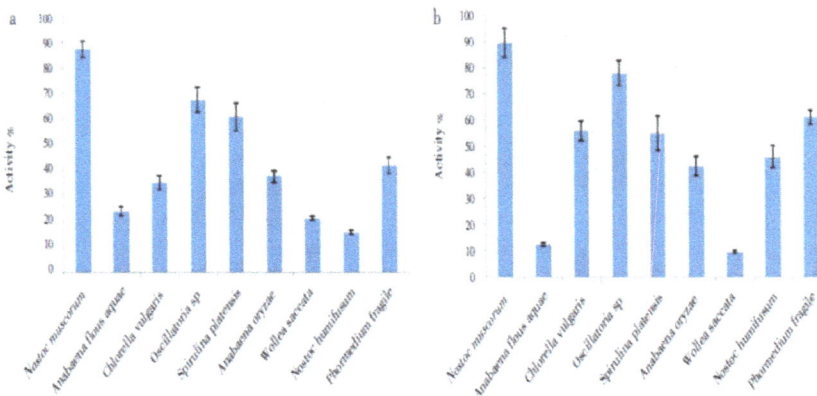

Figure 7. (a, b). Anticancer activity of algal water extracts at 100 µg/ ml against EACC (a) and HepG2 (b) cell lines [43]

The recorded maximum activity in both species against both cell lines at the highest nitrate content (9 g/L) may be attributed mainly to the higher content of the phycobiliprotein pigments produced under excess nitrate contents. Nitrate limitation and starvation, in spite of the caused decrease in phycobilin pigment production due to metabolic alteration expected under stress conditions, the carbon skeleton compounds as phenolic may replace phycobilin shortage in inducing similar anticancer activity of or even higher efficiency caused by great phycobilin contents at higher nitrate supplementation [43]. These results demonstrated that the compounds responsible for anticarcinogenic activity was highly polar as the phycobilins, phenolic compounds and polysaccharides which induced apoptosis of the cancer cells as reported by Aboul-Enein et al [50], which go parallel with these results coincides with the results obtained by Wang et al [51] who reported that the aqueous extract of red algae mainly contain c-phycocyanin, exhibited higher antipraleferation inducing apoptosis body formation. The authors explained that phycocyanin interact with membrane associated B-tubulin and glyceraldehydes-3-phosphate dehydrogenase (GAPDH), caused polymerization of microtubules and actins filaments leading to arrested the cell cycle at G0/G1 phase. As these aqueous extracts exhibited antioxidant and anticancer activities, its effect as coloring agent is amplified by these biological efficiencies which are very important for human health. Also, it can be used for the manufacture of pharmaceutical drugs (antioxidant and anticancer).

Table (5) recorded the anticancer efficiency of nitrate stressed *N. muscorum* and *Oscillatoria* sp. against EACC and HepG2 cell lines. The promising cyanobacterial species *N. muscorum* and *Oscillatoria* sp. induced both the highest antioxidant (by DPPH and ABTS methods) and anticancer activities (using EACC and HepG2 cell lines) which may be attributed to their large content in total phycobiliprotein pigments together with the higher secondary metabolites content (phenolic compounds, terpenoids, alkaloids [43].

6.3. Antimicrobial activity

The antimicrobial activity of microalgae has been attributed to compounds belonging to several chemical classes –including indoles, terpenes, acetogenins, phenols, fatty acids and volatile halogenated hydrocarbons [52] for instance, the antimicrobial activity of supercritical extracts obtained from the microalga *Chaetoceros muelleri* were related to its lipid composition [53]. However, the antimicrobial activity detected in several pressurized extracts from *Dunaliella salina* may be explained not only by several fatty acids, but also by such compounds as - and -ionone, -cyclocitral, neophytadiene and phytol. Efforts to identify the compounds directly responsible for those antimicrobial features –e.g. chlorellin [54] have been on the run, but are still relatively incipient owing the some new classes of compounds found. Microalgal cell-free extracts are already being tested as additives for food and feed formulation, in attempts to replace antimicrobial compounds of synthetic origin currently in use – including subtherapeutical doses of antibiotics employed as prophilatic measure in animal breeding [55].

Recall, in this regard, the growing resistance of some bacterial strains arising from the widespread and essentially unrestricted use of antibiotics in cattle handling, and by domestic consumers use via self-prescription [56]. However, a key factor for their eventual economic feasibility is the possibility of operating large photobioreactors under aseptic conditions, which are able to produce biomass and metabolites to sufficiently high levels [57].

Treatment	Oscillatoria sp.		Nostoc muscorum	
	HepG2	EACC	HepG2	EACC
Control (1.5 g/L NaNO₃)	70.40±1.87	68.30±0.58	88.60±2.96	85.90±3.60
3 g/L	72.10±2.60	69.70±1.64	85.60±2.30	84.60±4.00
6	70.60±0.80	72.60±2.34	86.90±1.80	85.60±2.98
9	75.90±2.61	82.60±4.65	88.70±0.95	90.40±4.82
0.75	72.60±3.05	68.30±0.72	86.70±1.00	83.00±1.64
0.37	74.80±2.50	78.30±5.96	88.00±2.70	84.00±0.50
0.0	82.00±4.85	82.90±3.40	92.30±1.65	89.90±1.96
LSD 0.05	0.575	0.907	1.63	0.505

Table 5. Anticancer activity of the nitrate stressed promising algal species using EACC and HepG2 cell lines [43]

6.3.1. Antiviral activity

A number of infectious diseases caused by viruses have emerged (and re-emerged) in recent years. Although several antiviral drugs have been specifically developed, drug-resistant mutations are constantly occuring – so new antiviral active principles are necessary, especially those from sources that do not constitute (or are exposed to) viral pools. This is why microalgae have received a strong attention as potential suppliers of antiviral agents [58]; Viral growth is generally divided into three stages, and antiviral action may take place at a single or more stages: Stage I, which consists on adsorption and invasion of cells; Stage II, or eclipse phase, during which the cell is forced to synthesize multiple copies of said virus; and Stage III, or maturity and release of virus particles. For instance, the anti-HSV activity of the antiviral compound acyclovir® is expressed at stage II, but the anti-HSV factor from *Dunaliella* sp. inactivates the viral function at stage I. Sulphated exopolysaccharides from marine microalgae have been claimed to interfere with Stage I of some enveloped viruses they offer competitive advantages because of their broad antiviral spectrum against e.g. HSV and HIV-1 [59]. Apparently, their inhibitory effect arises from interaction with the positive charges on the virus or on the cell surface – which prevents penetration of the former into the host cells; they may also selectively inhibit reverse transcriptase in the case of HIV, thus hampering production of new viral particles after infection yet the exact step during viral replication when they act remains to be elucidated. Antiviral highly sulfated polysaccharides from several species of red microalgae consist mainly of xylose, glucose and galactose [60]; they are unusually stable when exposed to extreme pH and temperature [61]. Despite their successful antiviral performance, the metabolic pathways leading to sulfated polysaccharides are still poorly known. Their secretion by unicellular red algae was originally characterized via radiolabeling – which howed biosynthesis of the carbon chain, and sulfation of the resulting polysaccharide to occur in the Golgi apparatus [62]; these findings were confirmed in *Porphyridium* sp. [63] and other red microalgae [64]. More recently, [65] used 14C pulse-chase experiments and ultrastructural microscopy to conclude that brefeldin A–a membrane-traffic inhibitor of the Golgi apparatus, decreases the contents of the bound and the soluble forms of polysaccharides, while inhibiting cell-wall binding of polysaccharides to a greater extent than its soluble counterpart (in both actively growing and resting cells). Discovery of small molecules that can specifically disrupt a particular protein-protein interface remains a challenge-but is of a particular interest in virology, since the antiviral drugs currently available target only viral proteins.

6.3.2. Antibacterial activity

Most efforts were devoted to the study of antibiotic resistance in bacteria for several reasons: (i) bacterial infections are responsible for most community-acquired and nosocomial infections; (ii) the large and expanding number of antibacterial classes offers a more diverse range of resistance mechanisms; and (iii) the ability to move bacterial resistance determinants into standard, well-characterized bacterial strains facilitates more detailed studies of the underlying molecular mechanisms [66]. Pratt et al. [67] isolated the first antibacterial compound from a microalga, *Chlorella*; a mixture of fatty acids, viz. chlorellin, was found to be responsible for that inhibitory activity against both Gram+ and Gram- bacteria. Research aimed at identifying antibacterial active principles produced by microalgae has meanwhile boomed [68]. This realisation arose e.g. from the risk associated with several multidrug-resistant *Staphylococcus aureus* (MRSA) strains, which have been causing an increased concern in healthcare institutions worldwide – since they are not susceptible to most conventional antibiotics. Hence, discovery of novel antibacterial compounds following distinct biochemical mechanisms of action is urged. Antibiotics are typically less effective against Gram- bacteria because of their complex, multilayered cell wall structure – which makes it more difficult for the active compound to penetrate them [69]; this justifies why the antibacterial activity of the supernatant (and methanolic extracts) is more potent against Gram+ than Gram- bacteria [68,70]. The exact mechanism of action of fatty acids remains unknown: they may act upon multiple cellular targets, even though cell membranes are the most probable ones – as membrane damage will likely lead to cell leakage and reduction of nutrient uptake, besides inhibiting cellular respiration; conversely, Desbois [71] claimed a peroxidative process. Furthermore, compounds synthesized by *Scenedesmus costatum*, and partially purified from its organic extract, exhibited activity against aquaculture bacteria because of their fatty acids longer than 10 carbon atoms in chain length –which apparently induce lysis of bacterial protoplasts. The ability of fatty acids at large to interfere with bacterial growth and survival has been known for quite some time, but recent structure-function relationship studies suggest that said ability depends on both their chain length and degree of unsaturation. Such compounds as cholesterol can antagonize antimicrobial features [53] so both composition and concentration of free lipids should be taken into account [72]. Among microalgal-derived oxylipins, the antibacterial activities of polyunsaturated aldehydes deserve a special ention. Such compounds are synthesized by diatoms, e.g. *S. costatum* and *Thalassiosira rotula*. One illustrative example is decadienal–probably derived from (the polyunsaturated) arachidonic acid (C20:4 n-3), which exhibits a strong activity against such important human pathogens as MRSA and *Haemophilus influenza* – with MIC values of 7.8 and 1.9 µg/mL, respectively, and well as against *E. coli* and *Pseudomonas aeruginosa,* and *S. aureus and Staphylococcus epidermidis* (Gram- and Gram+ bacteria, respectively). Furthermore, it impairs growth of diverse marine bacteria, such as (the Gram-) *Aeromonas hydrophila, L. anguillarum, Alteromonas haloplankti, Photobacterium phosphoreum* and *Psychrobacter immobilis,* and the (Gram+) *Planococcus citreus* and *Micrococcus luteus* [73].

6.3.3. Antifungal activity

Algae are one of the chief biological agents that have been studied for the control of fungi plant pathogens [74]. Various strains of cyanobacteria are known to produce intracellular and extracellular metabolites with diverse biological activities such as antibacterial, antifungal and antiviral activity [75]. These biologically active compounds include antibiotics and toxins [76]. Hussien *et al.* [77] screened the effect of culture filtrates of nine algal strains (*Anabaena flosaquae, Anabaena oryzae, Chlorella vulgaris, Nostoc muscorum, Nostoc humifusu, Oscillatoria* sp., *Phormedium fragile, Spirulina platensis* and *Wollea saccata*) at concentrations of 10, 20, 30 and 40% on mycelium growth of the plant pathogenic fungi *Cercospora beticola* causing leaf spot disease in sugar beet comparing with different concentrations of the synthetic fungicide Topsin M70 (100, 200, 300 and 400 ppm). They found that generally, all the algal culture filtrates reduced the fungal mycelium growth but the highest fungal mycelium growth inhibition percentages were achieved by *Spirulina platensis, Oscillatoria* sp. and *Nostoc muscorum* (Figure 8) at the concentrations of 30% (100, 100 and 82%, respectively) and at 40% (100, 100 and 100%, respectively). While, the chemical synthetic fungicide Topsin M70 completely inhibited the fungal mycelium growth at the concentrations of 200, 300 and 400 ppm. Fungal spores production (number of spores) was completely inhibited by the previous three algal culture filtrates at 300 and 400 ppm particularly at the concentration of 40%. Same results were obtained by Topsin M70. The antifungal activity of the algal culture filtrates has been attributed to the presence of bioactive compounds i.e. total phenolic compounds, total saponins and alkaloids in the algal culture filtrates.

6.4. Nematicidal activity

Root-knot nematodes, *Meloidogyne* spp., are among the most damaging nematodes in agriculture, causing an estimated US$ 100 billion loss/year worldwide [78]. The application of chemical nematicides has been found as an effective measure for controlling nematodes but they have toxic residual effect on the environment particularly on non-target organisms and human health. In addition, the use of chemical nematicides is prohibited in organic farming. Nematicidal potential of cyanobacteria has remained unexplored except for a few reports, which suggest that endospores of Microcoleus and *Oscillatoria* spp. killed nematodes [79].

(A):   *Nostoc muscorum*   Control

(B):   *Oscillatoria* sp   Control

(C):   *Spirulina platensis*   Control

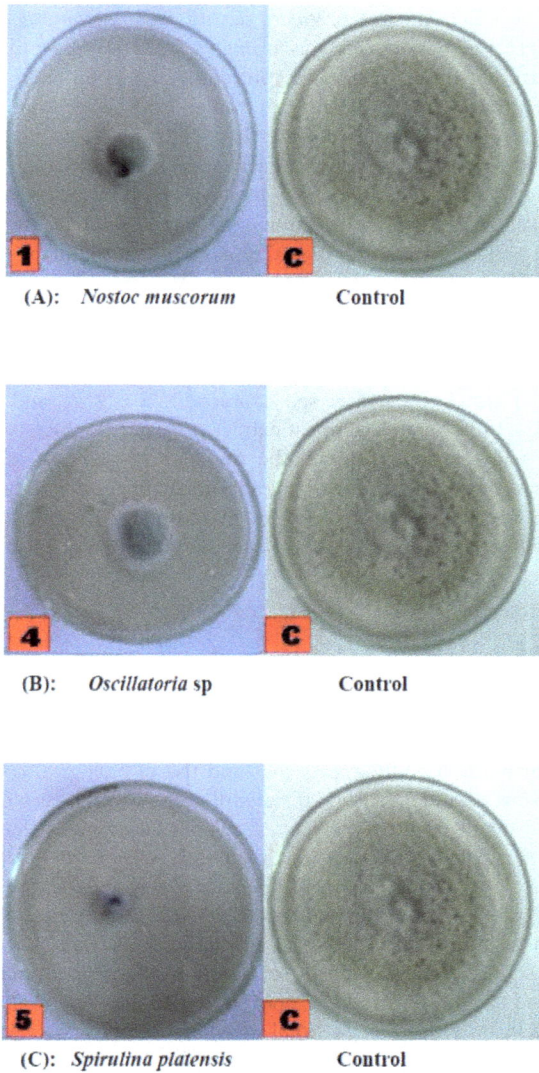

Figure 8. The efficacy of (A) *Nostoc muscorum,* (B) *Oscillatoria* sp and (C) *Spirulina platensis* culture filtrates (40 ml.L⁻¹) in suppressing the fungal mycelia growth diameter of *Cercospora beticola* [77]

Culture filtrates of Microcoleus vaginatus inhibited hatching of *Meloidogyne incognita* eggs and killed second stage juveniles. Microalgal metabolites have attracted attention, because they are a resource for toxins, and potential new drugs [80]. Shawky et al. [81] studied the nematicidal effect of nine culture filtrates of algal strains (*Nostoc muscorum, Anabaena flos aquae, Anabaena oryzae, Chlorella vulgaris, Wollea saccata, Phormedium fragile, Oscillatoria* sp., *Nostoc humifusum* and *Spirulina platensis*), *Azolla pinnata* aqueous extract filtrate (1:2w/v) and compost watery extract filtrate (1:5w/v) in controlling the population of the root knot nematode, *Meloidogyne incognita* in cucumber under both laboratory and greenhouse conditions. Laboratory experiment revealed that high juvenile mortality percentage occurred during all the exposure periods of all treatments, the best results were after 72 hr exposure. Only five cyanobacterial strains, namely, *Spirulina platensis, Oscillatoria* sp., *Anabaena oryzae, Nostoc muscorum* and *Phormedium fragile*, followed by compost watery extract, significantly increased juveniles mortality over 70% at the highest concentration of 1:10 (84.3, 80.4, 78.9, 75.4, 72.5 and 70.1%, respectively). *Azolla pinnata* aqueous extract filtrate achieved 69.8% at the same concentration while, *Anabaena flos aquae* and *Chlorella vulgaris* recorded the lowest effect on mortality percentage (52.1 and 40.1%, respectively) at the concentration of 1:10. In the greenhouse experiment, the combination of mixing five algal culture filtrates of *S. platensis, Oscillatoria* sp., *A. oryzae, N. muscorum* and *P. fragile* with *A. pinnata* aqueous extract filtrate and compost extract achieved the highest reduction in the number of the 2nd stage juveniles in soil, the numbers of galls, developmental stages, females, egg masses, egg numbers/egg mass in roots of cucumber plants comparing with the individual treatment and the non treated control. In addition, all combinations significantly improved fresh weight of roots and shoots and increased the yield of cucumber plants.

6.5. Mollscicidal activity

The snail intermediate hosts of schistosomiasis are the sites of intense multiplication of this parasite, thus their control strategies are considered a priority of the reduction of schistosomiasis transmission [82]. Although chemical molluscicides are to certain extent quite successful in curbing the disease concerned. However, in view of their side effects, interest in environmentally friendly approaches and use of biological control agents have been revived [83]. Mostafa and Gawish [84] stated that the algal culture filtrate of *Spirulina platensis* was proved for its mollscicidal activity against *Biomphalaria alexandrina* snails the intermediate host of *Schistosoma mansoni* in Egypt that accommodates their survival and fecundity and also for its effect on hatchability of snail's eggs and viability of the free living larval stages of the parasite (miracidia and cercariae). The culture filtrate proved to have a lethal effect on snails with 90 LC 0.23% of the filtrate. *B. alexandrina* snails stopped egg laying after one week of continuous exposure to the sublethal concentrations 0.19 and 0.17%, while those exposed to 0.07% laid few ones. It could be due to the phytochemical constituents of the culture filtrate i.e. total phenolic compounds, alkaloids and total saponins. This was confirmed by histological examination that showed a severe damage in the hermaphrodite gland cells of snails (Figure 9) exposed to these concentrations. *B. alexandrina* eggs of 3 and 7 days old failed to hatch post exposure to 0.1% of algal filtrate. while 22% and 10% only hatched after exposure to 0.05% of this filtrate. Free cell culture filtrate shows also marked miracidicidal and cercaricidal activities as 2% of this filtrate killed most of these organisms within 15 minutes of exposure. It is concluded that the byproducts of the blue-green alga *Spirulina platensis* has a lethal effect against adult *B. alexandria* snails, reduced or stopped their oviposition, hence minimize the snail populations available for the parasite transmission. Therefore it may be a potential source of effective compounds for control of *Schistosoma mansoni*.

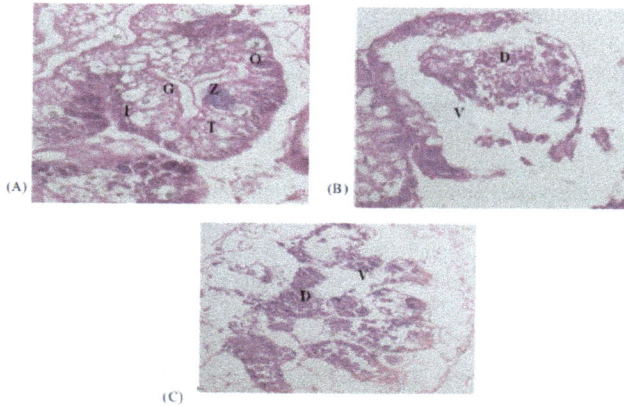

Figure 9. Section in the hermaphrodite gland of *Biomphalaria alexandrina* snail: control (A), exposed to 0.07% (B) and to 0.17% (C) concentrations of *Spirulina platensis* culture filtrate. G=Spermatogonia, T=Spermatid, Z=Spermatozoa, I= Oogonia, O=Oocyte V= vacules D=degeneration [84]

7. Other applications and products from microalgae

Microalgae have found commercial applications as natural sources of valuable macromolecules, including carotenoids, long-chain polyunsaturated fatty acids, and phycocolloids. As photoautotrophs, their simple growth requirements make them attractive for bioprocesses aimed at producing high added-value compounds that are in large demand by the pharmaceutical market. The productivity and biochemical composition of microalgae depend strongly on the mode of cultivation, medium composition, and nutrient profile. Consequently, numerous efforts aimed at elucidating the practical impacts of the aforementioned parameters have been developed [56]. Thus, there is a growing interest in the area of research on the positive effect of algae on human health and other benefits.

7.1. Food

The first use of microalgae by humans dates back 2000 years to the Chinese, who used *Nostoc* to survive during famine. Microalgae for human nutrition are nowadays marketed in different forms such as tablets, capsules and liquids. They can also be incorporated into pastas, snack foods, candy bars or gums, and beverages [85-86]. Owing to their diverse chemical properties, they can act as a nutritional supplement or represent a source of natural food colorants. The commercial applications are dominated by four strains: *Arthrospira*, *Chlorella*, *Dunaliella salina* and *Aphanizomenon flos-aquae*. *Arthrospira* is used in human nutrition because of its high protein content and its excellent nutritive value [87-88]). In addition, this microalga has various possible healthpromoting effects: the alleviation of hyperlipidemia, suppression of hypertension, protection against renal failure, growth promotion of intestinal *Lactobacillus*, and suppression of elevated serum glucose level [85-86]. A significant amount of *Arthrospira* production is realized in China and India. *Chlorella* can also be used as a food additive owing to the taste- and flavour-adjusting actions of its coloring agent [85]. *D. salina* is exploited for its β-carotene content that can reach 14% of dry weight [89]. For human consumption, Cognis Nutrition and Health, the world's largest producer of this strain, offers *Dunaliella* powder as an ingredient of dietary supplements and functional foods. The last major commercial strain application is *A. flos-aquae*. According to many research studies, used alone or in combination with other nutraceuticals and natural food products, *A. flos-aquae* promotes good overall health [85,90].The consumption of *Arthrospira (Spirulina)* by the Kanembu was also reported by Delisle et al. [91] in a survey of household food consumption and nutritional adequacy in Wadi zones of Chad. Table (6) presents a comparison of the general compositions of human food sources with that of different microalgae.

Commodity	Protein	Carbohydrate	Lipid
Bakers' yeast	39	38	1
Meat	43	1	34
Milk	26	38	28
Rice	8	77	2
Soybean	37	30	20
Anabaena cylindrica	43-56	25-30	4-7
Chlamydomonas rheinhardii	48	17	21
Chlorella vulgaris	51-58	12-17	14-22
Dunaliella salina	57	32	6
Porphyridium cruentum	28-39	40-57	9-14
Scenedesmus obliquus	50-56	10-17	12-14
Spirulina maxima	60-71	13-16	6-7
Synechococcus sp.	63	15	11

Table 6. General composition (%) of dry matter of different human food sources and algae [92].

7.2. Feed

Microalgae can be incorporated into the feed for a wide variety of animals ranging from fish (aquaculture) to pets and farm animals. In fact, 30% of the current world algal production is sold for animal feed applications [92] and over 50% of the current world production of *Arthrospira* is used as feed supplement [85]. In 1999, the production of microalgae for aquaculture reached 1000 t (62% for molluscs, 21% for shrimps, and 16% for fish) for a global world aquaculture production of 43×10^6 t of plants and animals [93]. The importance of algae in this domain is not surprising as they are the natural food source of these animals. The main applications of microalgae for aquaculture are associated with nutrition, being used fresh (as sole component or as food additive to basic nutrients) for coloring the flesh of salmonids and for inducing other biological activities. The most frequently used species are *Chlorella, Tetraselmis, Isochrysis, Pavlova, Phaeodactylum, Chaetoceros, Nannochloropsis,*

Skeletonema and *Thalassiosira* [85,94]. Many nutritional and toxicological evaluations have proved the suitability of algal biomass as feed supplement [92]. *Arthrospira* is largely used in this domain and concerns many types of animal: cats, dogs, aquarium fish, ornamental birds, horses, cows and breeding bulls. Algae positively affect the physiology (by providing a large profile of natural vitamins, minerals, and essential fatty acids; improved immune response and fertility; and better weight control) and their external appearance (resulting in healthy skin and a lustrous coat) of animals. In poultry rations, algae up to a level of 5-10% can be used safely as partial replacement for conventional proteins. Prolonged feeding of algae at higher concentrations produces adverse effects. The yellow color of broiler skin and shanks as well as of egg yolk is the most important characteristic that can be influenced by feeding algae [92].

7.3. Agricultural purposes

Humans have practiced agriculture for more than 10,000 years, but only in the past 50 years or so have farmers become heavily dependent on synthetic chemical fertilizers and pesticides. It contributes to numerous forms of environmental degradation, including air and water pollution, soil depletion and diminishing biodiversity. Synthetic chemical pesticides and fertilizers are polluting soil, water, and air, harming both the environment and human health. Soil is eroding much faster than it can be replenished—taking with it the land's fertility and nutrients that nourish both plants and those who eat them. Chemical fertilizers can gradually increase the acidity of the soil until it begins to impede plant growth. Chemically fertilized plots also show less biologic activity in the soil food web (the microscopic organisms that make up the soil ecosystem) than do plots fertilized organically with manure or other biologic sources of fertility [95]. The best way, however, is to use as much as possible microbial products, functional bio-fertilizers and bio-controllers and reduce the amount of the use of chemical fertilizers or pesticides. Heterocystous cyanobacteria and several nonheterocystous cyanobacteria are known for their ability to fix atmospheric nitrogen. The fertility of many tropical rice field soils has been mainly attributed to the activity of nitrogen-fixing cyanobacteria. An estimation showed that more than 18 kg N ha^{-1} year^{-1} was added to the soils by cyanobacteria. Inoculation of cyanobacteria to increase the fertility of soils has been successfully attempted. Recently, nitrogen-fixing cyanobacteria have been reported to dominate desert crusts worldwide. This is believed to contribute significantly to the fertility of desert soils and may eventually facilitate vegetation of deserts [96]. Algae as biofertilizers are a promising alternative to avoid soil pollution caused by agrochemicals. Also, they recover the nutrients content to soil as they secrete exo-polysaccharides that improve soil structure and bio-active substances that enhance the plant growth. Algae are known to be one of the most promising sources as bio-control agents of any residues, thereby having positive impact on human health [97]. Microorganisms play an important role in various chemical transformations of soils and thus, influence the availability of major nutrients like nitrogen, phosphorus, potassium and sulphur to the plants. Cyanobacteria and phosphate-solubilizing bacteria were used as biofertilizers to increase crop production [98].The cyanobacterial ability to mobilize insoluble forms of inorganic phosphates is evident from the finding of kleiner and Harper

[99] who reported more extractable phosphates in soils with cyanobacterial cover than in nearby soils without cover. Cyanobacteria can fix about 25 kg N/ha/season. Apart from nitrogen fixation, inoculation with cyanobacteria is also reported to reduce considerably the total sulphides and ferrous iron content of the soil. Blue-green algae constitute an important group of microorganism capable of nitrogen fixation. Most of the species possess nitrogen fixation ability to the order Nostocales and Stigonematales. Over 100 species of blue-green algae are known to fix atmospheric nitrogen. These have been found to be very effective on the rice and banana plantation. In field condition, overall increase in the gram yield of rice is amounted to about 586 kg/ha. In case of crops other than rice, algalization increased nearly 34 per cent yield. India is one of the countries where agro-chemical conditions appear to be favourable where blue-green algae technology has been put forward. In some parts of the country, production of BGA inoculants has been commercialised. Producing inoculum in artificially controlled conditions is well defined, but relatively expensive. On the other hand open-air soil culture is simpler, less expensive and easily adaptable by the farmers. Field scale production of algae biofertilizer is also possible. 20-25 kg dry algae can be obtained on 40 m field. Adopting this method, 15 t/ha of wet BGA can be obtained by the farmers. Farmers can also produce algae for countryard of the house [96]. Blue-green algal extracts comprise a great number of bioactive compounds that influence plant growth and development. They mostly contain growth phyto-regulators like gibberellins, auxin, cytokinin, ethylene and abscisic acid [100]. This group of microorganisms have been reported to benefit plants by producing growth promoting regulators resemble gibberellin and auxin, vitamins, amino acids, polypeptides, antibacterial and antifungal substances that exert phytopathogen biocontrol and polymers especially exopolysaccharides that were reported to enhance growth and productivity of plants like *Daucus carota* [101], *Santalum album* [102], *Oryzae sativa* [103], *Lilium alexandrae* hort [104] and *Beta vulgaris* L.[100]. Non-nitrogen fixing cyanobacteria can enrich phosphorus and potassium contents in soils, laying indirect major role in plant growth promotion [105]. Cyanobacteria also enhance the soil biological activity in terms of increasing the total bacterial, total cyanobacterial counts, CO_2 evolution, dehydrogenase and nitrogenase activities. Many researches suggested that up to 50% of the recommended dose of the mineral nitrogen fertilizers could be saved by using some species of nitrogen fixing cyanobacteria. The obtained results emphasized the prospects and potentials of using cyanobacteria biofertilizers as renewable natural nitrogen resources for many crops. They are none polluting, inexpensive, utilize renewable resources (inorganic nutrients and atmospheric CO_2) in addition to their ability in using free available solar energy, atmospheric nitrogen and water [106-108].

7.4. Mitigation of CO_2: Why algae for CO_2 sequestration?

Many options that have been proposed and that are in use for capturing CO_2 can be seen as economically, socially and environmentally short-sighted. A common approach is taking measures to offset any immediate effects, often by simple relocation of the emissions. Injection of flue gases into oceanic or geological sinks is examples of such "end-of-pipe" solutions [109]. Algae cultivation can yield a broad range of useful end products, apart from biofuels. The sequestration of CO_2 into algal biomass can become profitable also through the production of high value products such as pigments and high-grade lipids, which are extractable from several species of algae. Brennan and Owende [110] also mention high value products such as animal feed supplements being extractable from the microalgae species *Chlorella, Scenedesmus* and *Spirulina*. The urgent need for substantive net reductions in CO_2 emissions to the atmosphere can be addressed via biological CO_2 mitigation, coupled with transition to more extensive uses of biofuel, nuclear and renewable energy sources. Microalgae have attracted a great deal of attention for CO_2 fixation and biofuel production because they can convert CO_2 (and supplementary nutrients) into biomass via photosynthesis at much higher rates than conventional biofuel crops can. This biomass may then be transformed into methane or hydrogen, using processes mediated by anaerobic bacteria; an integrated process for hydrothermal production of methane via microalgae has been discussed recently [111-112]. Of particular interest is the production of oils by microalgae because of the ease of their synthesis (a lack of a nitrogen source usually suffices to trigger this form of secondary metabolism). Lipid extraction and re-esterification are accomplished with short-chain alcohols and other by-products of secondary metabolism (i.e. polyunsaturated fatty acids, bcarotenes or polymers [111]. Upon extraction, such oils can be hydrolyzed and then re-esterified with methyl- or ethyl alcohol moieties to obtain biodiesel. Microalga-mediated CO_2 fixation and biofuel production can be rendered more sustainable by coupling microalgal biomass production with existing power generation and wastewater treatment infrastructures (Figure 10). Microalgae can utilize low-quality water, such as agricultural runoff or municipal, industrial or agricultural wastewaters, as a source of water for the growth medium as well as a source of nitrogen, phosphorus and minor nutrients [113].

7.5. Wastewater treatments

Wastewater nitrogen and phosphorous as microalgae nutrients aquaculture systems involving microalgae production and wastewater treatment (e.g. of amino acids, enzyme, or food industries wastewaters) seems to be quite promising for microalgae growth combined with biological cleaning. This allows nutrition of microalgae by using organic compounds (nitrogen and phosphorous) available in some manufactures wastewater, not containing heavy metals and radioisotopes. Additionally, microalgae can mitigate the effects of sewage effluent and industrial sources of nitrogenous waste such as those originating from water treatment or fish aquaculture and at the same time contributing to biodiversity. Moreover, removing nitrogen and carbon from water, microalgae can help reduce the eutrophication in the aquatic environment. Aslan and Kapdan [114] used *C. vulgaris* for nitrogen and phosphorus removal from wastewater with an average removal efficiency of 72% for nitrogen and 28% for phosphorus (from 3 to 8 mg/L NH_4^+ and 1.5–3.5 mg/L PO_4^{-3}). Other widely used microalgae cultures for nutrient removal are *Chlorella* [115] and *Spirulina* species [116]. Nutrient removal capacities of *Nannochloris* [117]), *Botryococcus brauinii* [118] and cyanobacterium *Phormidium bohneri* have also been investigated [119-120]. Environmental applications production of biodiesel and other bio-products from microalgae can be more environmentally sustainable, cost-effective and profitable, if combined with processes such as wastewater and flue gas treatments. In fact various studies demonstrated the use of microalgae for production of valuable products combined with environmental applications [113,121].

7.6. Biofuel production

Microalgae can potentially be employed for the production of biofuels in an economically effective and environmentally sustainable manner. The production of these biofuels can be coupled with fuel gas CO_2 mitigation, wastewater treatment and the production of high-value chemicals. The efficiency is low but there is much room for improvement. The use of microalgae is seen as, at least, a partial solution to climate change and energy problem [122]. Many microalgae are exceedingly rich in oil which can be converted to biodiesel using existing technology. More than 50% of their biomass as lipids, sometimes even up to 80%, and oil levels of 20-50% are quite common [123].

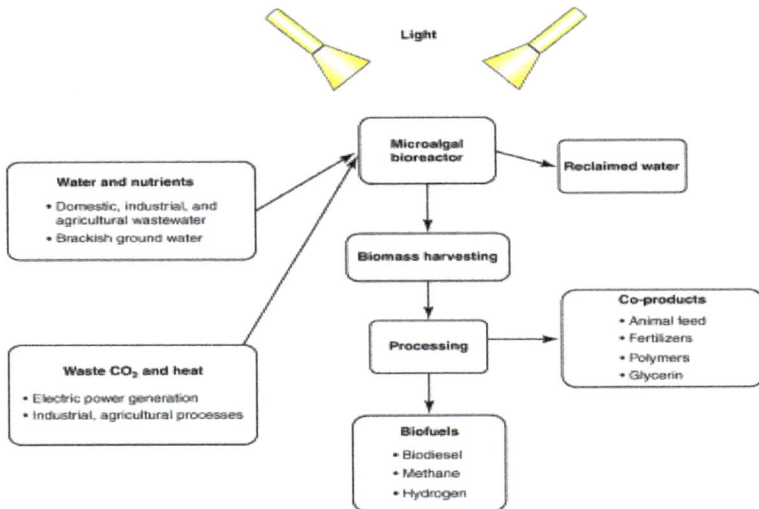

Figure 10. Integration of microalgal bioreactors into existing wastewater and power generation infrastructures. The overall process uses microalgae to capture industrially produced waste CO_2 in photobioreactors, coupled with treatment of nutrients in wastewater. CO_2 is converted into algal biomass by photosynthesis in the presence of light. After processing (biological, physical or thermochemical), the biomass generated can be used for production of biodiesel, methane or other fuels and co-products (e.g. animal feeds and polymers)

Lipids production and biodiesel extraction from algae depend on algal species and extraction solvent system [124].There is a unique opportunity to both treat wastewater and provide nutrients to algae using nutrient-rich effluent streams. By cultivating microalgae, which consume polluting nutrients in municipal wastewater, and abstracting and processing this resource, then the goals of sustainable fuel production and wastewater treatment can be combined [174,125].The efforts span over many areas of "algae to fuels" technologies including production system development, algae harvest, algae strain development and genetic modification, algae products development, etc. Screening and genetic modification of algae strains will play an increasingly important role. Genetic engineering has the potential to improve the overall algal biomass yield and lipid yield. Discovery of new strains and genetically modified strains capable of secreting hydrocarbons to extracellular spaces will open some new opportunities; however, challenges with recovering the secreted liquids or volatiles remain. There is a need to develop high throughput screening and analysis methods. Current harvest and dewatering are still too energy intensive. New techniques and strategies must be devised to lower the costs. Direct conversions such as in situ transesterification and hydrothermal liquefaction offer the possibility to process wet algae. Fractionation of algal biomass, before or after oil extraction, deserves a closer look because it may play an important role in offsetting the costs. New techniques to disrupt algae cellular structures to improve oil extraction efficiency are needed [126].

7.7. Heavy metals and phycoremediation

Metals are directly or indirectly involved in all phases of microbial growth. Many metals such as sodium, potassium, iron, copper, magnesium, calcium, manganese, zinc, nickel and cobalt are vital for biological functions, while others such as aluminum, cadmium, silver, gold, mercury and lead are not known to have necessary biological functions. All these elements can interact with microbial cells and be accumulated as a result of different mechanisms [127]. Some of these mechanisms have biotechnological importance and can be applied for the bioremediation of metals from industrial effluents. The capability of some microbial species to adsorb some heavy metals on their surface [128-129] or accumulate them within their structure is a chief route for the removal of heavy metals from contaminated environment [130-132]. Another fashion for the detoxification of heavy metals by microorganisms is the chelation of these metals inside or outside their cells after converting them into other forms to reduce their toxicity. In 2007, Lefebvre et al. [133] working with some cyanobacterial strains (*Limnothrix planctonica*, *Synechococcus leopoldiensis* and *Phormidium limnetica*) demonstrated their ability to convert Hg^{2+} into elemental mercury $Hg°$ and meta-cinnabar (β-HgS) under pH controlled and aerated conditions. The transformation of mercury into β-HgS was attributed to the interaction with metal binding sulfhydryl protein as an intermediate step in metal sulfide synthesis. Moreover, some of the freshwater algae *Limnothrix planctonica* and *Selenastrum minutum* were recorded for their ability to bio-transform Hg^{2+} into a form with the analytical properties of β-HgS under aerobic conditions due to the presence of some protein and non-protein thiol chelators [134]. Furthermore, Lengke et al. [135] investigated the gold bioaccumulation by cyanobacterium *Plectonema boryanum* from gold (III)-chloride solutions. They confirmed that the reduction mechanism of gold (III) to metallic gold by this organism involves the formation of an intermediate gold (I)-sulfide due to a chelation process via some thiol compounds. Recntly Essa and Mostafa [136] studied the effeicincy of three cyanobacterial isolates (*Spirulina platensis*, *Nostoc muscorum*, and *Anabaena oryzae*) individually or as a mixed culture to precipitate some heavy metals (Hg^{2+}, Cd^{2+}, Cu^{2+} and Pb^{2+}) out of their solutions through using the culture biogas produced during their aerobic growth in a batch bioreactor. Variable capabilities of metal bioprecipitation were recorded by the three algal isolates. FT-IR studies showed the existence of –OH groups in the metal precipitate produced by the algal isolates while –NH groups were identified only in the metal precipitates produced by *N. muscorum*, and *A. oryzae*. This study highlighted a novel approach for heavy metals bioremediation through the transformation of these metals into nitrogen complexes and/or hydroxide complexes via using the culture biogas produced by some cyanobacterial species.

8. Microalgal production

Microalgae for human nutrition are nowadays marketed in different forms such as tablets, capsules and liquids. They can also be incorporated into pastas, snack foods, candy bars or gums, and beverages. In addition, this microalga has various possible healthpromoting effects: the alleviation of hyperlipidemia, suppression of hypertension, protection against renal failure, growth promotion of intestinal Lactobacillus, and suppression of elevated serum glucose level [86-87]. Owing to their diverse chemical properties, they can act as a nutritional supplement or represent a source of natural food colorants. The commercial applications are dominated by four strains: *Arthrospira, Chlorella, D. salina* and *Aphanizomenon flos-aquae*. *Arthrospira* is used in human nutrition because of its high protein content and its excellent nutritive value [87,88,137]. A significant amount of *Arthrospira* production is realized in China and India. The world's largest producer Hainan Simai Enterprising Ltd. is located in the Hainan province of China. This company has an annual production of 200 t of algal powder, which accounts for 25% of the total national output and almost 10% of the world output. The largest plant in the world is owned by Earthrise Farms and streches over an area of 440,000 m^2 (located at Calipatria, CA, USA; Figure 11). Their production process is presented in Figure 12. Their Arthrospira-based products (tablets and powder) are distributed in over 20 countries around the world. Many other companies sell a wide variety of nutraceuticals made from this microalga. For example, the Myanmar *Spirulina* Factory (Yangon, Myanmar) sells tablets, chips, pasta and liquid extract, and Cyanotech Corp. (a plant in Kona, Hawaii, USA) produces products ranging from pure powder to packaged bottles under the name *Spirulina pacifica*. Cyanotech Corp. has developed an orginal process for drying the biomass in order to avoid the oxidation of carotenes and fatty acids that occurs with the use of standard dryers. The patented process employs a closed drying system that is kept at low oxygen concentrations by flushing with nitrogen and carbon dioxide. The process relies on a very cold ocean water crown from a depht of 600 m just offshore to provide dehumidification and actually dries microalgal products in less than 6 s (Figure 13). *Chlorella* is produced by more than 70 companies; Taiwan Chlorella Manufacturing and Co. (Taipei, Taiwan) is the largest producer with 400 t of dried biomass produced per year. Significant production is also achieved in Klötze, Germany (130 – 150 t dry biomass per year) with a tubular photobioreactor. This reactor consists of compact and vertically arranged horizontal running glass tubes with a total length of 500,000 m and a total volume of 700 m3 (Figure 14). The world annual sales of

Chlorella are in excess of US$ 38 billion [85]. The most important substance in *Chlorella* is β-1,3-glucan, which is an active immunostimulator, a free-radical scavenger and a reducer of blood lipids (9, Ryll et al., Abstr. Europ. Workshop Microalgal Biotechnol., Germany, p. 56, 2003). However, various other health-promoting effects have been clarified (efficacy on gastric ulcers, wounds, and constipation; preventive action against atherosclerosis and hypercholesterolemia; and antitumor action) [85,122,138].

Figure 11. Earthrise Farms *Arthrospira* production plant (Calipatria, CA, USA)

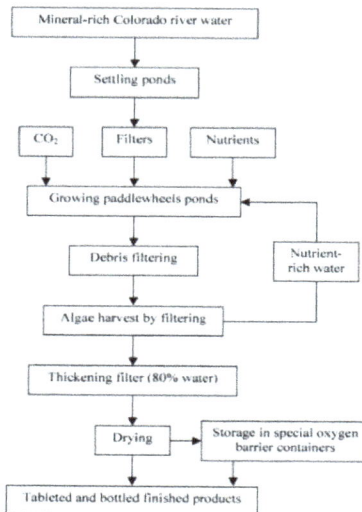

Figure 12. Earthrise Farms microalgal production process

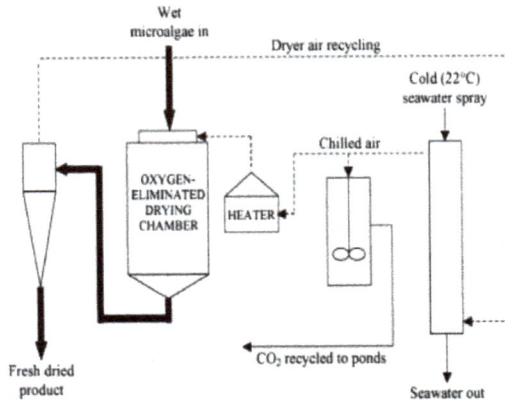

Figure 13. Cyanotech process for drying microalgae biomass

Figure 14. Glass tube photobioreactor (700 m3) producing Chlorella biomass (Klötze, Germany)

9. Conclusion

Meeting the increasing water demands with limited resources advocates Egypt to find innovative and sustainable approaches for management. It is essential to maximize the benefits of the available resources and to minimize the wastes and losses, not only in water resources but also in all economical and social resources, and in an integrated framework believing that everything is related to everything. So would it not be possible to kill several birds with one stone, using algae for absorbing CO_2 at the same time as providing nutrient recovery from food industrial effluents and domestic wastewater and producing renewable energy (fuels), as well as other pharmaceutical products, food, feed and fertilizer from the biomass? In recent years, microalgal culture technology is a business oriented line owing to their different practical applications. Innovative processes and products have been introduced in microalgal biotechnology to produce vitamins, proteins, cosmetics, health foods and animal feed. For most of these applications, the market is still developing and the biotechnological use of microalgae will extend into new areas.With the development of algal cultures and screening techniques, microalgal biotechnology can meet the challenging demands of food, feed, pharmaceutical industries, fuels and biofertilizers. The general needs of the human society are continuously increasing. We need every new compound which may be useful for the human society. More food, new drugs, and other goods are highly necessary for the benefit of humankind. The only question is the existence of sufficient natural and technical resources to fulfill these demands. Fortunately, in the area of the research of bioactive microbial products it seems that the ever expanding scientific and technical possibilities are increasing together with the continuously widening needs of the human.

Author details

Soha S.M. Mostafa

Microbiology Department, Soil, Water and Environment Research Institute, Agricultural Research Center, Giza, Egypt Acknowledgement

10. References

[1] Pratt DE (1992) Natural antioxidants from plant material. In: Huang MT, Ho CT and Lee CY (Editors), Phenolic Compounds in Food and their Effects on Health. II. American Chemical Society. ACS Symposium Series, 507.Washington. 54-71.

[2] Abou El alla FM and Shalaby EA (2009) Antioxidant Activity of Extract and Semi-Purified Fractions of Marine Red Macroalga. Gracilaria Verrucosa. Aust J. Bas. App. Sci. 3:3179-85.

[3] Li HB, Cheng KW, Wong CC, Fan KW, Chen F and Jiang Y (2007) Evaluation of antioxidant capacity and total phenolic content of different fractions of selected microalgae. Food Chemistry. 102:771–776.

[4] Richmond A. (2004) Handbook of microalgal culture: biotechnology and applied phycology. Blackwell Science Ltd.

[5] Brennan L and Owened PAuthor Vitae (2010) Biofuels from microalgae-Areview of technologies for production, processing and extractions of biofuels and co-products. Renewable and Sustainable Energy Reviews. 14(2):557-577.

[6] Dos Santos MD, Guaratini T, Lopes JLC, Colepicolo P and Lopes NP (2005) Plant cell and microalgae culture. In:Modern Biotechnology in Medicinal Chemistry and Industry. Kerala, India: Research Signpost.

[7] Burja AM, Banaigs B, Abou-Mansour E, Burgess JG and Wright PC (2001) Marine cyanobacteria - a prolific source of natural products. Tetrahedron. 57:9347-9377.

[8] Tyagi S, Singh G, Sharma A and Aggarwal G (2010) Phytochemicals as candidate therapeutics: An over view. International Journal of Pharmaceutical Sciences Review and Research. 3(1):53-55.

[9] Lipton AP (2003) Marine bioactive compounds and their application in mariculture. Marine Ecosystem, Univ. of Kerala, Kariavattom, 2(4): 695-581.

[10] Iwata K, Inayama T and Katoh T (1990) Effect of *Spirulina platensis* on plasma lipoprotein lipase activity in fructose induced hyperlipidemia in rats. J. Nutr. Sci. Vitaminol. 36:165-171.

[11] Barsanti L and Gualtieri P (2006) Algae and men. In: Algae: Anatomy, Biochemistry, and Biotechnology. Taylor and Francis Group LLC, CRC Press. 251–291.

[12] Valeem EE and Shameel M (2005) Fatty acid composition of blue-green algae of Sindh, Pakistan. Int. Journal of Phycology and Phycochemistry. 1:83-92.

[13] Funk CD (2001) Prostaglandins and leukotrienes: advances in eicosanoids biology. Science. 294:1871–1875.

[14] Wen Z-Y and Chen F (2003) Heterothrophic production of eicosapentaenoic acid by microalgae. Biotechnol. Adv. 21:273–294.

[15] Mozaffarian D, Ascherio A, Hu FB, Stampfer MJ, Willett WC, Siscovick DS and Rimm EB (2005) Interplay between different polyunsaturated fatty acids and risk of coronary heart disease in men. Circulation, 111(2):157-164.

[16] Eussen S, Klungel O, Garssen J, Verhagen H, van Kranen H, van Loveren H, Rompelberg C (2010) Support of drug therapy using functional foods and dietary supplements: focus on statin therapy. Br. J. Nutr. 103(9): 1260-1277.

[17] Petkov G and Garcia G (2007) Which are fatty acids of the green alga *Chlorella*? Bioch. Systemat. Ecol. 35:281-285.

[18] Abbadi A, Domergue F, Meyer A, Riedel K, Sperling P, Zank T and Heinz E (2001) Transgenic oilseeds as sustainable source of nutritionally relevant C20 and C22 polyunsaturated fatty acids? European Journal of Lipid Science and Technology. 103:106–113.

[19] Sayanova OV and Napier JA (2004) Eicosapentaenoic acid: biosynthetic routs and the potential for synthesis in transgenic plants. Phytochemistry. 65:147–158.

[20] Ponomarenko LP, Stonik IV, Aizdaicher NA, Orlova TY, Popvskaya GI, Pomazkina GV and Stonik VA (2004) Sterols of marine microalgae Pyramimonas cf. cordata (prasinophyta), Atteya ussurensis sp. nov. (Bacollariophyta) and a spring diatom bloom form Lake Baikal. Comp. Biochem. Physiol., B 138, pp.65–70.

[21] Froehner S, Martins RF and Errera MR (2008) Assessment of fecal sterols in Barigui River sediments in Curitiba, Brazil. Environment Monitoring Assessment. DOI:10.1007/s10661-008-0559-0.

[22] Volkman JK, Kearney P and Jeffrey SW (1990) A new source of 4-methyl sterols and 5a(H)-stanols in sediments: prymnesiophyte microalgae of the genus Pavlova. Organic Geochemistry. 15:489-497.

[23] Meyers SP, Latscha T (1997) Carotenoids. In: D'Abramo, L.R., Conklin, D.E., Akiyama, D.M. (Eds.), Crustacean Nutrition, Advances in World Aquaculture, vol. 6. World Aquaculture Society, Baton Rouge, LA, 164–193.

[24] Sommer TR, D'Souza FML and Morrissy NM (1992) Pigmentation of adult rainbow trout, Oncorhynchus mykiss, using the green alga Haematococcus pluvialis. Aquaculture. 106: 63–74.

[25] Chen BH, Chuang JR, Lin JH and Chiu CP (1993) Quantification of pro-vitamin A compounds in Chinese vegetables by high-performance Liquid Chromatography. J Food Prot. 56 (1): 51-54.

[26] Pulz O, Scheibenbogen K and Grob W (2001) Biotechnology with cyanobacteria and microalgae. In: Rehm, H. J., Reed, G. editors. A Multi-Volume Comprehensive Treatise Biotechnology Germany. Wiley-VCH Verlag GmbH. 10:107-136.

[27] Ben-Amotz A (1999) *Dunaliella* β-carotene: from science to commerce. In: Seckbach J (ed) Enigmatic microorganisms and life in extreme environments. Kluwer, Netherlands. 401–410.

[28] Higuera-Ciapara I, Félix-Valenzuela L, Goycoolea FM (2006) Astaxanthin: a review of its chemistry and applications. Crit. Rev. Food Sci. Nutr. 46:185–196.

[29] Del Campo JA, García-González M and Guerrero MG (2007) Outdoor cultivation of microalgae for carotenoid production: current state and perspectives, Appl Microbiol Biotechnol. 74:1163–1174.

[30] Takano S, Nakanishi A, Uemura D and Hirata Y (1979) Isolation and structure of a 334 nm UVabsorbing substance, porphyra-334 from the red alga Porphyra tenera Kjellman. Chem. Lett. 8:419-20.

[31] Nakamura H, Kobayashi J and Hirata Y (1982) Separation of mycosporine-like amino acids in marine organisms using reverse-phase high performance liquid chromatography. J. Chromatogr. 250:113-18.

[32] Klisch M and Häder D (2008) Mycosporine-Like Amino Acids and Marine Toxins - TheCommon and the Different. Marine Drugs. 6: 147-163.

[33] Büyükokurog˜lu ME, Gülçin I, Oktay M and Küfreviog˜lu ÖI (2001) In vitro antioxidant properties of dantrolene sodium. Pharmacol. Res. 44:491-495.

[34] Karawita R, Senevirathne M, Athukorala Y, Affan A, Lee YJ, Kim SK, Lee JB and Jeon YJ (2007) Protective effect of enzymatic extracts from microalgae against DNA damage induced by H_2O_2. Mar. Biotech. 9: 479-490.

[35] Lee SH, Karawita R, Affan A, Lee JB, Lee BJ and Jeon YJ (2008) Potential antioxidant activites of enzymatic digests from benthic diatoms *Achnanthes longipes, Amphora coffeaeformis*, and *Navicula* sp. (Bacillariophyceae). J. Food Sci. Nutr. 13: 166-175.

[36] Abd El-Baky H Hanaa, El Baz FK and El-Baroty GS (2004) Production of antioxidant by the green alga *Dunaliella salina*. Intl. J. Agric. Biol. 6(1): 49-57.

[37] Suzuki N and Mittler R (2006) Reactive oxygen species and temperature stresses: A delicate balance between signaling and destruction. Physiol. Plantarum, 126:45-51.

[38] Mittler R, S Vanderauwera, M Gollery and F Van Breusegem (2004) Reactive oxygen gene network of plants. Trends. Plant Sci. 9: 490-498.

[39] Tausz RM, Soledad M and Grille D (1998) Antioxidative defence and photoprotection in pine needles under field conditions. A multivariate approach to evaluate patterns of physiological responses at natural sites. Physiol. Planta. 104:760-768.

[40] Polle A and Rennenberg H (1994) Photooxidative stress in trees. In: Foyer, C.H. and F.M. Mullineaux (Eds.). Causes of photoxidative stress and amelioration of defense Systems in Plants. Boca Raton, FL: CRC Press.

[41] Foyer CH and Noctor G (2000) Oxygen processing in photosynthesis: Regulation and signaling. Rev. New Phytol. 146:359-388.

[42] Zhang J, Zhan B, Yao X, Gao Y, Shong J. Evaluation of 28 marine algae from the Qingdao coast for antioxidative capacity, determination of antioxidant efficiency, total phenolic content of fractions and subfractions derived from Symphyocladia latiuscula (Rhodomelaceae). J Appl Phycol 2007; 19: 97-108

[43] Shanab SMM, Mostafa SSM , Shalaby EA and Mahmoud GI (2012) Aqueous extracts of microalgae exhibit antioxidant and anticancer activities. Asian Pacific Journal of Tropical Biomedicine. 608-615.

[44] Glazer AN (1994) Phycobiliproteins: a family of valuable, widely used fluorophores. J. Appl. Phycol. 6:105-112.

[45] Rice-Evans CA, Miller NJ, Bolwell PG, Bramley PM, Pridham JB. The relative antioxidant activities of plant-derived polyphenolic flavonoids. Free Radic Res 1995; 22(4): 375-383.

[46] Shalaby EA, Shanab SMM and Singh V (2010) Salt stress enhancement of antioxidant and antiviral efficiency of Spirulina platensis; J. Med. Plants Res. 4(24):.2622-2632.

[47] Chinery R, Brockman JA, Peeler MO, Shyr Y, Beauchamp RD and Coffey R J (1997) Antioxidants enhance the cytotoxicity of chemotherapeutic agents in colorectal cancer-a p53-independent induction of p21(WAF1/CIP1) via c/ebp-beta. Nat. Med. 3:1233–1241.

[48] Mazhar D, Ang R and Waxman J (2006) COX inhibitors and breast cancer. Br. J. Cancer. 94: 346–350.

[49] Singh SC, Sinha RP and Häder DP (2002) Role of lipids and fatty acids in stress tolerance in cyanobacteria. Acta protozoologica. 41:297-308.

[50] Aboul-Enein AM, Shalaby EA, Abul-Ela F, Nasr-Allah AA, Mahmoud AM, El-Shemy HA, et al. (2011) Back to nature: Spotlight on cancer therapeutics. Vital Signs. 10: 8-9.

[51] Wang H, Liu Y, Gao X, Carter CL and Liu ZR (2007) The recombinant beta subunit of C-phycocyanin inhibits cell proliferation and induces apopotosis. Cancer Letters. 247(1):150-158.

[52] Mayer AMS and Hamann MT (2005) Marine pharmacology in 2001–2002: marine compounds with antihelmintic, antibacterial,anticoagulant, antidiabetic, antifungal, anti-inflammatory, antimalarial, antiplatelet, antiprotozoal, antituberculosis, and antiviral activities; affecting the cardiovascular, immune and nervous systems and other miscellaneous mechanisms of action. Comparative Biochemistry and Phycology, Part C.140: 265-286.

[53] Mendiola JA, Torres CF, Martín-Alvarez PJ, Santoyo S, Toré A, Arredondo BO, Señoráns FJ, Cifuentes A, Ibáñez E (2007) Use of supercritical CO_2 to obtain extracts with antimicrobial activity from Chaetoceros muelleri microalga. A correlation with their lipidic content. European Food Research and Technology. 224:505-510.

[54] Herrero M, Ibañez E, Cifuentes A, Reglero G, Santoyo S (2006) Dunaliella salina microalga pressurized liquid extracts as potential antimicrobials. J. Food Protection. 69:2471-2477.

[55] Tramper J, Battershill C, Brandenburg W, Burgess G, Hill R, Luiten E, Müller W, Osinga R, Rorrer G, Tredici M, Uriz M,Wright P and Wijffels R (2003) What to do in marine biotechnology? Biomolecular Engineering. 20:467-471.

[56] Guedes AC, Barbosa CR, Amaro HM, Pereira CI, Malcata FX (2011) Microalgal and cyanobacterial cell extracts for use as natural antibacterial additives against food pathogens. International J. Food Sci. and Technol. 46:862-870.

[57] Sánchez F, Fernández JM., Acien FG, Rueda A, Perez-Parra J, Molina E (2008) Influence of culture conditions on the productivity and lutein content of the new strain Scenedesmus almeriensis. Process Biochemistry. 43:398-405.

[58] Borowitzka MA (1995) Microalgae as sources of pharmaceuticals and other biologically active compounds. J. Appl. Phycol.7:65-68.

[59] Damonte EB, Pujol CA, Coto CE (2004) Prospects for the therapy and prevention of Dengue virus infections. Advances in Virus Research. 63:239-285.

[60] Huleihe M, Ishanu V, Tal J and Arad S. (2001) Antiviral effect of red microalgal polysaccharides on Herpes simplex and Varicella zoster viruses. Journal of Applied Phycology.13:127-134.

[61] Geresh S and Arad (Malis) S (1991) The extracellular polysaccharides of the red microalgae: chemistry and rheology. Bioresouce Technology. 38:195-201.

[62] Evans L, Callow M, Percival E, Fareed V (1974) Studies on the synthesis and composition of extracellular mucilage in the unicellular red alga *Rhodella*. J.Cell Science. 16:1-21.

[63] Ramus J and Robins DM (1975) The correlation of Golgi activity and polysaccharide secretion in Porphyridium. J. Phycol. 11:70-74.

[64] Tsekos I (1999) The sites of cellulose synthesis in algae: diversity and evolution of cellulose-synthesizing enzyme complexes. Journal of Phycology. 35:635-655.

[65] Keidan M, Friedlander M and Arad S (2009) Effect of brefeldin A on cell-wall polysaccharide production in the red microalga Porphyridium sp. (Rhodophyta) J. Appl. Phycol. 21:707-717.

[66] Ghannoum MA and Rice LB (1999) Antifungal agents: mode of action, mechanisms of resistance, and correlation of these mechanisms with bacterial resistance. Clinical Microbiology Reviews.12:501-517.

[67] Pratt R, Daniels TC, Eiler JB, Gunnison JB, Kumler WD et al (1944) Chlorellin, an antibacterial substance from *Chlorella*. Science. 99:351-352.

[68] Ghasemi Y, Yazdi MT, Shafiee A, Amini M, Shokravi S, Zarrini G. Parsiguine (2004) a novel antimicrobial substance from Fischerella ambigua. Pharmaceutical Biology. 42:318-322.

[69] Ördög V, Stirk WA, Lenobel R, Bancírová M, Strand M and van Standen J (2004) Screening microalgae for some potentially useful agricultural and pharmaceutical secondary metabolites. J. Appl. Phycol. 16:309-314.

[70] Ghasemi Y, Moradian A, Mohagheghzadeh A, Shokravi S, Morowvat MH (2007) Antifungal and antibacterial activity of the microalgae collected from paddy fields of Iran: characterization of antimicrobial activity of Chlorococcus dispersus. J. Biol. Sci.7: 904-910.

[71] Desbois AP, Mearns-Spragg A, Smith VJ (2009) A fatty acid from the diatom Phaeodactylum tricornutum is antibacterial against diverse bacteria including multi-resistant Staphylococcus aureus (MRSA) Marine Biotechnology. 11:45-52.

[72] Benkendorff K, Davis AR, Rogers CN and Bremner JB (2005) Free fatty acids and sterols in the benthic spawn of aquatic molluscs, and their associated antimicrobial properties. J. Experimental Marine Biology and Ecology. 316:29-44.

[73] Smith VJ, Desbois AP and Dyrynda EA (2010) Conventional and unconventional antimicrobials from fish, marine invertebrates and micro-algae. Marine Drugs. 8:1213-1262.

[74] Hewedy MA, Rahhal MMH and Ismail IA (2000) Pathological studies on soybean damping-off disease. Egypt. J. Appl. Sci. 15:88-102.

[75] Noaman NH, Khaleafa AF and Zaky SH (2004) Factors affecting antimicrobial activity of Synechococcus leopoliensis. Microbiol. Res. 159:395-402.

[76] Kiviranta J, Abdel-Hamid A, Sivonen K, Niemelä SI and Carlberg G (2006) Toxicity of cyanobacteria to mosquito larvae-screening of active compounds. Environ. Toxicol. Water Qual. 8: 63-71.

[77] Hussien Manal Y, Abd El-All Azza AM and Mostafa Soha S M (2009) Bioactivity of algal extracellular byproducts on cercospora Leaf spot disease, growth performance and quality of sugar beet. The 4th Conference on Recent Technologies in Agriculture: Challenges of Agriculture Modernization, Nov. 3rd –5th , Special Edition of Bull. Fac. Agric., Cairo Univ. 1:119-129.

[78] Oka Y, Koltai H, Bar-Eyal M, Mor M, Sharon E, Chet I and Spiegel Y (2000) New strategies for the control of plant parasitic nematodes. Pest Manage. Sci. 56:983-988.

[79] Dhanam M, Kumar AC and Sowajanya AS (1994) Microcoleu vaginatus (Oscillatoriaceae), a blue-green alga (Cyanobacterium) parasitizing plant and soil nematodes. Indian Journal of Nematology. 24:125-132.

[80] Shimizu Y (2003) Microalgal metabolites. Current Opinion in Microbiology. 6:236–242.

[81] Shawky, Samaa M, Mostafa, Soha SM and Abd El-All, Azza AM (2009) Efficacy of algae, azolla and compost extract in controlling root knot nematode and its reflection on cucumber growth. Bull. Fac. Agric., Cairo Univ. 60:443-459.

[82] Lardans, V and Dissous C (1998) Snail control strategies for reduction of schistosomiasis transmission. Parasitology Today 14:413-417.

[83] Ahmad, R, Chu WL, Ismail Z, Lee HL and Phang SM (2004) Effect of ten chlorophytes on larval survival, development and adult body size of the mosquito Aedes aegypti. Southeast Asian J. Trop. Med. Public health. 35(1):79-87.

[84] Mostafa Soha SM and Gawish Fathia A (2009) Towards to Control Biomphalaria alexandrina Snails and the Free Living Larval Stages of Schistosoma mansoni Using the Microalga Spirulina platensis. Aust. J. Basic and Appl. Sci. 3(4):4112-4119.

[85] Yamaguchi K (1997) Recent advances in microalgal bioscience in Japan, with special reference to utilization of biomass and metabolites: a review. J. Appl. Phycol. 8:487-502.

[86] Liang S, Xueming L, Chen F, and Chen Z (2004) Current microalgal health food R & D activities in China. Hydrobiologia. 512:45-48.

[87] Soletto D, Binaghi L, Lodi A, Carvalho JCM, and Converti A (2005) Batch and fedbatch cultivations of *Spirulina platensis* using ammonium sulphate and urea as nitrogen sources. Aquaculture. 243:217-224.

[88] Rangel-Yagui, CO, Godoy Danesi ED, Carvalho J CM and Sato S (2004) Chlorophyll production from *Spirulina platensis*: cultivation with urea addition by fed-batch process. Bioresour. Technol. 92:133-14.

[89] Metting FB (1996) Biodiversity and application of microalgae. J. Ind. Microbiol. 17:477-489.

[90] Pugh N and Pasco DS (2001) Characterization of human monocyte activation by a water soluble preparation of *Aphanizomenon flos-aquae*. Phytomedicine, 8:445-453.

[91] Delisle H, Alladsoumgué M, Bégin F, Nandjingar K and Lasorsa C (1991) Household food consumption and nutritional adequacy in wadi zones of Chad, Central Africa. Ecol. Food Nutrit. 25:229-248.

[92] Becker W (2004) Microalgae in human and animal nutrition. In: Richmond, A (ed.), microalgal culture. Handbook. Blackwell, Oxford. 312-351.

[93] Muller-Feuga A (2004) Microalgae for aquaculture: the current global situation and future trends. In: Richmond A (ed) Handbook of microalgal culture. Blackwell Science. Oxford.pp 352–364.

[94] Muller-Feuga A (2000) The role of microalgae in aquaculture: situation and trends. J. Appl. Phycol. 12: 527-534.

[95] Horrigan LRS, Lawrence and Walker P (2002) How Sustainable Agriculture Can Address the Environmental and Human Health Harms of Industrial Agriculture. Environ Health Perspect. 110:445-456.

[96] Mahdi SS, Hassan GI, Samoon SA, Rather HA, Dar SA and Zehra B (2010) Bio-frtilizers in organic agriculture. Journal of Phytology. 2(10): 42-54.

[97] Silva JA, Evensen CI, Bowen RL, Kirby R, Tsuji GY and Yost RS (2000) Managing Fertilizer Nutrients to Protect the Environment and Human Health. In: Plant Nutrient Management in Hawaii's Soils, Approaches for Tropical and Subtropical Agriculture. J. A. Silva and R. Uchida, eds. College of Tropical Agriculture and Human Resources, University of Hawaii at Manoa. Capter 1:7-22.

[98] Earanna N and Govindan R (2002) Role of biofertilizers in mulberry production-A review. Indian J. Seric. 41(2):92-99.

[99] kleiner KT and Harper KT (1977) Soil properties in relation to cryptogamic ground cover in Canyon-lands National park. J. Range Managem. 30: 202-205.

[100] Aly MHA, Abd El-All Azza AM and Mostafa Soha SM (2008) Enhancement of sugar beet seed germination, plant growth, performance and biochemical compounds as contributed by algal extracellular products. J. Agric. Sci., Mansoura Univ. 33(12):8429-8448.

[101] Wake H, Akasata A, Umetsu H, Ozeki Y, Shimomura and Matsunaga (1992) T. Promotion of plantlet formation from the somatic embryos of carrot with a high molecular weight extract of marine cyanobacterium. Plant cell Report.11:6265.

[102] Bapat VA,Iyer RK and Rao PS (1996) Effect of cyanobacterial extract on somatic embryogenesis in tissue culture of sandalwood (Santalum album) J. Medicinal and Aromatic Plant Science. 18(1):1014.

[103] Storni DC, Zaccaro M, Cristina M, Ileana G, María SA and Gloria ZDC (2003) Enhancing rice callus regeneration by extracellular products of Tolypothrixtenuis (Cyanobacteria) World J. Microbiology and Biotechnology. 19(1):2934.

[104] Zaccaro MC, Kato A, Zulpa G, Storni MM, Steyerthal N, Lobasso K and Stella AM (2006) Bioactivity of *Scytonema hofmanni* (Cyanobacteria) in *Lilium alexandrae* in vitro propagation. Electronic Journal of Biotechnology. 9(3):210214.

[105] Selvarani V (1983) Studies on the influence on nitrogen fixing and non-nitrogen fixing blue green algae on the soil, growth and yield of paddy (*Oryza sativa*-IR 50) M. Sc., Madurai Kamaraj University, Madurai.

[106] Mahmoud AA, Mostafa Soha SM, Abd El-All, Azza AM and Hegazi AZ (2007) Effect of cyanobacterial inoculation in presence of organic and inorganic amendments on carrot yield and sandy soil properties under drip irrigation regime. Egypt. J. of Appl. Sci. 22(12B): 716-733.

[107] Ali Laila KM and Mostafa Soha SM (2009) Evaluation of potassium humate and *Spirulina platensis* as a bio-organic fertilizer for sesame plants grown under salinity stress. The 7th International Conference of Organic Agriculture. 13-15 December, Egypt. J. Res. 87(1):369-388.

[108] Hegazi Amal Z, Mostafa Soha SM and Hamdino MIA (2010) Influence Of different cyanobacterial application methods on growth and seed production of common Bean under various levels of mineral nitrogen fertilization. Nature and Science. 8(11): 202-212.

[109] Packer M (2009) Algal capture of carbon dioxide; biomass generation as a tool for greenhouse gas mitigation with reference to New Zealand energy strategy and policy. Energy Policy. 37(9):3428-37.

[110] Brennan L and Owende P (2010) Biofuels from microalgae- A review of technologies for production, processing, and extractions of biofuels and co-products. 14:557-77.

[111] Li Y, Horsman M, Wu N, Lan CQ and Dubois-Calero N (2008) Biofuels from microalgae. Biotechnol Prog 24(4):815–820.

[112] Haiduc AG et al (2009) SunCHem: an integrated process for the hydrothermal production of methane from microalgae and CO_2 mitigation. J. Appl. Phycol. 21:529–541.

[113] Mostafa SSM, Shalaby EA and Mahmoud GI (2012) Cultivating Microalgae in Domestic Wastewater for Biodiesel Production. Not Sci Biol. 4(1):56-65.

[114] Aslan S and Kapdan IK (2006) Batch kinetics of nitrogen and phosphorus removalfrom synthetic wastewater by algae. Ecological Engineering. 28(1):64–70.

[115] Gonzales LE, Canizares RO and Baena S (1997) Efficiency of ammonia and phosphorusremoval from a Colombian agroindustrial wastewater by the microalgae *Chlorealla vulgaris* and *Scenedesmus dimorphus*. Bioresource Technology. 60:259–62.

[116] Lee K and Lee CG (2001) Effect of light/dark cycles on wastewater treatments bymicroalgae. Biotechnology and Bioprocess Engineering. 6:194–9.

[117] Martı́nez ME, Sańchez S, Jimeńez JM, El Yousfi F, Muńoz L (2000) Nitrogen andphosphorus removal from urban wastewater by the microalga Scenedesmusobliquus. Bioresource Technology. 73(3): 263–72.

[118] Olguı́n EJ, Galicia S, Mercado G and Perez T (2003) Annual productivity of *Spirulina* (Arthrospira) and nutrient removal in a pig wastewater recycle process under tropical conditions. J. Appl. Phycol. 15:249–57.

[119] Laliberte G, Lessard P, Noue J, Sylvestre S (1997) Effect of phosphorus addition onnutrient removal from wastewater with the cyanobacterium Phormidiumbohneri. Bioresource Technology. 59:227–33.

[120] Dumas A, Laliberte G, Lessard P, De la Noue J (1998) Biotreatment of fish farm effluents using the cyanobacterium *Phormidium bohneri*. Aquacultural Engineering. 17(1):57–68.

[121] Bilanovic D, Andargatchew A, Kroeger T and Shelef G (2009) Freshwater and marine microalgae sequestering of CO_2 at different C and N concentrations—response surface methodology analysis. Energy Conversionand Management. 50:262–7.

[122] Salih FM and Haase RA (2012) Potentials of microalgal biofuel production. Review, J. Petroleum Technol. and Alternative Fuels. 3(1):1-4.

[123] Chisti Y (2007) Biodiesel from microalgae. Biotechnol Adv. 25(3):294-306.

[124] Afify AMR, Shalaby EA and Shanab SMM (2010) Enhancement of biodiesel production from different species of algae. Grasas Y Aceites. 61(4):416-422.

[125] Andersen RA (2005) Algal Culturing Techniques. San Diego, Elsevier Inc, p. 678.

[126] Chen P Min M, Chen Y, Wang L, Li Y, Chen Q, Wang C, Wan Y, Wang X, Cheng Y, Deng S, Hennessy K, Lin X, Liu Y, Wang Y, Martinez B and Ruan R (2009) Review of the biological and engineering aspects of algae to fuels approach. Int. J. Agric. Biol. Eng. 2(4):1-30.

[127] Gadd GM and White C (1992) Removal of thorium from simulated acid process streams by fungal biomass-potential for thorium desorption and reuse of biomass and desorbent. J Chem Technol Biotechnol. 55:39–44.

[128] Figueira MM, Volesky B, Ciminelli VST and Roddick FA (2000) Biosorption of metals in brown seaweed biomass. Water Res. 34:196-204.

[129] Sheng, Ping Xin, Lai, Heng Tan, Paul, Chen J and Yen- Peng, Ting (2005) Biosorption performance of two brown marine algae for removal of chromium and cadmium. J. Dispersion Sci. Technol. 25(5):679-686.

[130] Lamaia C, Kruatrachuea M, Pokethitiyooka P, Upathamb ES and Soonthornsarathoola V (2005) Toxicity and Accumulation of Lead and Cadmium in the Filamentous Green Alga *Cladophora fracta* (O.F. Muller ex Vahl) Kutzing: A Laboratory Study, Science Asia. 31, 121-127.

[131] Shanab SMM and AM Essa (2007) Heavy metals tolerance, biosorption and bioaccumulation by some microalgae (Egyptian isolates) N. Egypt. J. Microbiol. 17:65-77.

[132] Atici TO, Obali A, Altindag S, Ahiska and Aydin D (2010) The accumulation of heavy metals (Cd, Pb, Hg, Cr) and their state in phytoplanktonic algae and zooplanktonic organisms in Beysehir Lake and Mogan Lake, Turkey. African J. Biotechnol. (9):475-487.

[133] Lefebvre DD, Kelly D and Budd K (2007) Biotransformation of HgII by cyanobacteria. Appl. Environ. Microbiol. 73:243-249.

[134] Kelly DJA, Budd K and Lefebvre DD (2006) Mercury analysis of acid- and alkaline-reduced biological samples; identification of meta-cinnabar as the major biotransformed compound in algae. Appl. Environ. Microbiol. 72: 361-367.

[135] Lengke MF, Ravel B, Fleet ME, Wanger G, Gordon RA and Southham G (2006) Mechanisms of gold bioaccumulation by filamentous cyanobacteria from gold (III)-chloride complex. Environ. Sci. Technol. 40:6304-6309.

[136] Essa AMM and Mostafa Soha SM (2011) Biomineralization of some heavy metals by cyanobacterial biogas. Egpt. J. Botany. In Press

[137] Desmorieux H and Decaen N (2005) Convective drying of *Spirulina* in thin layer. J. Food Eng. 66: 497–503.

[138] Jong-Yuh C and Mei-Fen S (2005) Potential hypoglycemic effects of *Chlorella* in streptozotocin-induced diabetic mice. Life Sci. 77:980–990.

Permissions

The contributors of this book come from diverse backgrounds, making this book a truly international effort. This book will bring forth new frontiers with its revolutionizing research information and detailed analysis of the nascent developments around the world.

We would like to thank Dr. Sudam Charan Sahu and Dr. Nabin Kumar Dhal, for lending their expertise to make the book truly unique. They have played a crucial role in the development of this book. Without their invaluable contribution this book wouldn't have been possible. They have made vital efforts to compile up to date information on the varied aspects of this subject to make this book a valuable addition to the collection of many professionals and students.

This book was conceptualized with the vision of imparting up-to-date information and advanced data in this field. To ensure the same, a matchless editorial board was set up. Every individual on the board went through rigorous rounds of assessment to prove their worth. After which they invested a large part of their time researching and compiling the most relevant data for our readers. Conferences and sessions were held from time to time between the editorial board and the contributing authors to present the data in the most comprehensible form. The editorial team has worked tirelessly to provide valuable and valid information to help people across the globe.

Every chapter published in this book has been scrutinized by our experts. Their significance has been extensively debated. The topics covered herein carry significant findings which will fuel the growth of the discipline. They may even be implemented as practical applications or may be referred to as a beginning point for another development. Chapters in this book were first published by InTech; hereby published with permission under the Creative Commons Attribution License or equivalent.

The editorial board has been involved in producing this book since its inception. They have spent rigorous hours researching and exploring the diverse topics which have resulted in the successful publishing of this book. They have passed on their knowledge of decades through this book. To expedite this challenging task, the publisher supported the team at every step. A small team of assistant editors was also appointed to further simplify the editing procedure and attain best results for the readers.

Our editorial team has been hand-picked from every corner of the world. Their multi-ethnicity adds dynamic inputs to the discussions which result in innovative outcomes. These outcomes are then further discussed with the researchers and contributors who give their valuable feedback and opinion regarding the same. The feedback is then collaborated with the researches and they are edited in a comprehensive manner to aid the understanding of the subject.

Apart from the editorial board, the designing team has also invested a significant amount of their time in understanding the subject and creating the most relevant covers. They scrutinized every image to scout for the most suitable representation of the subject and create an appropriate cover for the book.

The publishing team has been involved in this book since its early stages. They were actively engaged in every process, be it collecting the data, connecting with the contributors or procuring relevant information. The team has been an ardent support to the editorial, designing and production team. Their endless efforts to recruit the best for this project, has resulted in the accomplishment of this book. They are a veteran in the field of academics and their pool of knowledge is as vast as their experience in printing. Their expertise and guidance has proved useful at every step. Their uncompromising quality standards have made this book an exceptional effort. Their encouragement from time to time has been an inspiration for everyone.

The publisher and the editorial board hope that this book will prove to be a valuable piece of knowledge for researchers, students, practitioners and scholars across the globe.

List of Contributors

Mohammad Ali Malboobi, Ali Samaeian, and Tahmineh Lohrasebi
Department of Plant Biotechnology, National Institute of Genetic Engineering and Biotechnology, Tehran, I.R. Iran

Mohammad Sadegh Sabet
Department of Plant Breeding and Biotechnology, Faculty of Agriculture, Tarbiat Modares University, Tehran, I.R. Iran

Vakhrushev A.V., Fedotov A.Yu. and Vakhrushev A.A.
Institute of Mechanics, Ural Branch the Russian Academy of Sciences, Izhevsk, Russia

Golubchikov V.B and Golubchikov E.V.
Join Stocks Company Nord, Perm, Russia

Beata Koim-Puchowska, Monika Wieloch and Karolina Bombolewska
Nicolaus Copernicus University, Collegium Medium in Bydgoszcz, Department of Ecology and Environmental Protection, Bydgoszcz, Poland

Piotr Kamiński
University of Zielona Góra, Faculty of Biological Sciences, Institute of Biotechnology and Environment Protection, Department of Biotechnology, Zielona Góra, Poland
Nicolaus Copernicus University, Collegium Medium in Bydgoszcz, Department of Ecology and Environmental Protection, Bydgoszcz, Poland

Piotr Puchowski
Government Forestry in Toruń; Zamrzenica Forestry District, Bysław, Poland

Leszek Jerzak
University of Zielona Góra, Faculty of Biological Sciences, Institute of Biotechnology and Environment Protection, Department of Environmental Protection and Biodiversity, Zielona Góra, Poland

Victor Irogue Omorusi
Plant protect Division, Rubber Research Institute of Nigeria, PMB, Iyanomo, Benin City, Nigeria

José Renato Stangarlin, Odair José Kuhn, Lindomar Assi and Cristiane Cláudia Meinerz
Western Paraná State University – UNIOESTE, Marechal Cândido Rondon, Paraná, Brazil

Clair Aparecida Viecelli
Assis Gurgacz Foundation – FAG, Cascavel, Paraná, Brazil

Kátia Regina Freitas Schwan-Estrada
Maringá State University – UEM, Maringá, Paraná, Brazil

Roberto Luis Portz
Paraná Federal University – UFPR, Palotina, Paraná, Brazil

Heike Bücking, Elliot Liepold and Prashant Ambilwade
South Dakota State University, Biology and Microbiology Department, Brookings, USA

Camilo López and Álvaro L. Perez-Quintero
Universidad Nacional de Colombia, Bogotá, Departamento de Biología, Bogota D.C.,Colombia

Boris Szurek
UMR 186 IRD-UM2-Cirad, Résistance des Plantes aux Bioagresseurs (RPB), Institut de Recherche pour le Développement, Montpellier Cedex 5, France

Çimen Atak and Özge Çelik
Istanbul Kultur University, Faculty of Science and Letters, Department of Molecular Biology and Genetics, Ataköy, Istanbul, Turkey

Gregory P. Pogue
Kentucky BioProcessing, LLC, Owensboro, KY, USA
IC2 Institute, The University of Texas at Austin, Austin, Texas, USA

Steven Holzberg
Folsom Lake College, Folsom, CA, USA

Muhammad Shafiq Shahid and Masato Ikegami
NODAI Research Institute, Tokyo University of Agriculture, Tokyo, Japan

Pradeep Sharma
Division of Crop Improvement, Directorate of Wheat Research, Karnal, India

A. Bakrudeen Ali Ahmed, S. Mohajer, E.M. Elnaiem and R.M. Taha
Institute of Biological Sciences, Faculty of Science, University of Malaya, Kuala Lumpur, Malaysia

Soha S.M. Mostafa
Microbiology Department, Soil, Water and Environment Research Institute, Agricultural Research Center, Giza, Egypt